Science without God?

Rethinking the History of Scientific Naturalism

Edited by
PETER HARRISON
and
JON H. ROBERTS

OXFORD
UNIVERSITY PRESS

OXFORD
UNIVERSITY PRESS

Great Clarendon Street, Oxford, OX2 6DP,
United Kingdom

Oxford University Press is a department of the University of Oxford.
It furthers the University's objective of excellence in research, scholarship,
and education by publishing worldwide. Oxford is a registered trade mark of
Oxford University Press in the UK and in certain other countries

First Edition published in 2019

Published in the United States of America by Oxford University Press
198 Madison Avenue, New York, NY 10016, United States of America

British Library Cataloguing in Publication Data
Data available

Library of Congress Control Number: 2018951982

ISBN 978–0–19–883458–8

For Ronald L. Numbers

Acknowledgements

This volume had its origins in a conference held at Florida State University in February 2013. The meeting was to mark the retirement of the distinguished historian of science and medicine, Ronald L. Numbers, and to explore some of the themes of his seminal work on the historical relations between science and religion. One of those themes was scientific naturalism, and it became clear over the course of the meeting that there was room for a volume that dealt in detail with the historical origins of scientific naturalism, covering different historical eras and various scientific disciplines—hence, this present collection. A number of the contributors to this volume were present at that meeting. Other attendees offered helpful commentary and critique of those early drafts, and joined in what were extremely productive discussions of key issues. There is a long list of people to thank for those contributions: Terrie Aamodt, Keith Benson, Jon Butler, Ted Davis, Matt Day, Noah Efron, John Evans, Dana Freiburger, Fred Gregory, Florence Hsia, Judith Leavitt, Sue Lederer, David Livingstone, Jay Malone, Gregg Mitman, Blair Neilson, Efthymios Nicolaidis, Shawn Peters, Bob Richards, Todd Savitt, Rennie Schoepflin, Adam Shapiro, Hugh Slotten, Elliot Sober, John Stenhouse, Rod Stiling, William Trollinger, Steve Wald, John Harley Warner, and Stephen Weldon. Particular thanks are due to Jeffrey Jentzen, and to the Department of Medical History and Bioethics of the University of Wisconsin-Madison, for their financial support. Michael Ruse was a congenial and entertaining host at FSU, and also made a generous contribution, both materially and intellectually. Finally, we must express a special debt of gratitude to The Historical Society, Boston University, and Donald A. Yerxa, who were enthusiastic supporters of this project from the start, and whose generosity has made possible this collection.

Like that original meeting, this volume is dedicated to Ron Numbers, an outstanding scholar and dedicated teacher, a source of encouragement and inspiration to generations of historians, and, for many of us, a valued colleague and dear friend.

Contents

List of Figures		xi
Notes on Contributors		xiii
	Introduction *Peter Harrison*	1
1.	'All Things are Full of Gods': Naturalism in the Classical World *Daryn Lehoux*	19
2.	Naturalist Tendencies in Medieval Science *Michael H. Shank*	37
3.	Laws of God or Laws of Nature? Natural Order in the Early Modern Period *Peter Harrison*	58
4.	Between Isaac Newton and Enlightenment Newtonianism: The 'God Question' in the Eighteenth Century *J. B. Shank*	77
5.	God and the Uniformity of Nature: The Case of Nineteenth-Century Physics *Matthew Stanley*	97
6.	Chemistry with and without God *John Hedley Brooke*	111
7.	Removing God from Biology *Michael Ruse*	130
8.	Christian Materialism and the Prospect of Immortality *Michelle Pfeffer*	148
9.	The Science of the Soul: Naturalizing the Mind in Great Britain and North America *Jon H. Roberts*	162
10.	Down to Earth: Untangling the Secular from the Sacred in Late-Modern Geology *Nicolaas Rupke*	182
11.	Naturalizing the Bible: The Shifting Role of the Biblical Account of Nature *Scott Gerard Prinster*	197

12. Anthropology and Original Sin: Naturalizing Religion, Theorizing the Primitive 216
Constance Clark

13. The Theology of Victorian Scientific Naturalists 235
Bernard Lightman

Index 255

List of Figures

1.1. Sidney Harris on Method. © ScienceCartoonsPlus.com 36

4.1. Étienne-Louis Boullée, imagined cenotaph for Isaac Newton, interior view with orrery above tomb. Bibliothèque Nationale de France. 78

4.2. *God as Architect/Builder/Geometer/Craftsman, The Frontispiece of Bible Moralisée* (mid-thirteenth century), ÖNB Vienna, Codex Vindobonensis 2554, f.1 verso. Wiki-Commons. 82

4.3. *Christ Salvator Mundi*, artist unknown, lower Rhine, 1537–45. Getty Images. 83

4.4. William Blake, *Newton. Source*: Tate Images. 96

12.1. Jacket of William K. Gregory, *Our Face from Fish to Man* (New York: G. P. Putnam's Sons, 1929). 217

Notes on Contributors

John Hedley Brooke is Andreas Idreos Professor Emeritus of Science and Religion, University of Oxford. He has published extensively on the interface between science and religion, the history of natural theology, and the Darwinian revolution. His books include *Science and Religion around the World*, edited with Ronald L. Numbers (Oxford, 2011), *Heterodoxy in Early Modern Science and Religion*, edited with Ian Maclean (Oxford, 2005), and *Science and Religion: Some Historical Perspectives* (Cambridge, 1991).

Constance Clark is an Associate Professor of History in the Department of Humanities and Arts at Worcester Polytechnic Institute. She is the author of *God—or Gorilla: Images of Evolution in the Jazz Age* (Johns Hopkins, 2008). Her current research focuses on the history of popularizations of evolution, palaeontology and anthropology, especially in museums and zoos, and on the visual culture of popular science.

Peter Harrison is an Australian Laureate Fellow and Director of the Institute for Advanced Studies in the Humanities at the University of Queensland, Australia. His six books include *The Territories of Science and Religion* (Chicago, 2015)—based on his 2011 Gifford Lectures, *The Bible, Protestantism, and the Rise of Natural Science* (Cambridge, 1998) and *Wrestling with Nature: From Omens to Science* (Chicago, 2011), co-edited with Michael H. Shank and Ronald L. Numbers.

Daryn Lehoux is Professor of Classics at Queen's University, Kingston, Ontario. He is the author of *What Did the Romans Know?* (Chicago, 2012) and *Astronomy, Weather, and Calendars in the Ancient World* (Cambridge, 2007) as well as the co-editor, with A. D. Morrison and Alison Sharrock, of *Lucretius: Poetry, Philosophy, Science* (Oxford, 2013). He has published numerous papers and chapters on the ancient sciences and on the epistemology of nature.

Bernard Lightman is Professor of Humanities at York University, Toronto. Lightman's most recent publications include *Victorian Popularizers of Science* (Chicago, 2007), *Victorian Scientific Naturalism* (co-edited with Gowan Dawson; Chicago, 2014), *Evolution and Victorian Culture* (co-edited with Bennett Zon; Cambridge, 2014), and *The Age of Scientific Naturalism* (co-edited with Michael Reidy; Pickering & Chatto, 2014). He is currently working on a biography of John Tyndall and is one of the editors of the John Tyndall Correspondence Project, an international collaborative effort to obtain, digitalize, transcribe, and publish all surviving letters to and from Tyndall.

Michelle Pfeffer is a PhD candidate at the University of Queensland's Institute for Advanced Studies in the Humanities. Her research explores the multifaceted discussions about the human soul in the seventeenth and eighteenth centuries, which embraced scientific, philosophical, medical, theological, and historical discourses. Her particular interests lie in the contentious debates over the immortal and immaterial nature of the human soul. Other research interests include early modern medicine, the history of scholarship, and religiously motivated responses to contemporary 'materialist' science.

Scott Gerard Prinster is a PhD candidate at the University of Wisconsin-Madison, under the guidance of Ronald L. Numbers. His dissertation addresses the shifting relationships between biblical knowledge and scientific knowledge in the nineteenth-and early twentieth-century United States. Other research interests include American intellectual history, the popular spread of science-and-religion dialogue, and the construction of expertise and attitudes towards educated authority and intellectualism.

Jon H. Roberts is the Tomorrow Foundation Professor of History at Boston University. He has written a number of articles dealing primarily with the history of the relationship between science and religion, as well as the book *Darwinism and the Divine in America: Protestant Intellectuals and Organic Evolution, 1859–1900* (Notre Dame, 2001) which received the Frank S. and Elizabeth D. Brewer Prize from the American Society of Church History. He has also co-authored with James Turner *The Sacred and the Secular University* (Princeton, 2001). He is currently working on a book dealing with American Protestant thinkers' treatment of the mind between the outset of the settlement of North America and 1940.

Michael Ruse is the Lucyle T. Werkmeister Professor of Philosophy and the Director of the Program in History and Philosophy of Science at Florida State University. The author or editor of over fifty books, most on or around the history and philosophy of evolutionary theory, his next two books are *Atheism: What Everyone Needs to Know*, (Oxford), and (co-authored with Robert J. Richards) *Debating Darwin: Mechanist or Romantic?* (Chicago). He is now writing a book on evolution and literature.

Nicolaas Rupke recently retired from the Chair of the History of Science at Göttingen University and now holds the Johnson Professorship of History at Washington and Lee University. A geologist and historian of geology, his publications cover issues that pertain to the relationship of Bible and science. Among his books are *Richard Owen: Biology without Darwin* (Chicago, 2009), *Alexander von Humboldt: A Metabiography* (Chicago, 2008) and the edited volume *Eminent Lives in Twentieth-Century Science and Religion* (Peter Lang, 2009). Currently he is working on the non-Darwinian tradition in evolutionary biology.

J. B. Shank is Associate Professor in the Department of History at the University of Minnesota, and the Director of the University of Minnesota Center for Early Modern History and the Mellon Foundation funded Consortium for the Study of the Premodern World. His research focuses on the history of the mathematical and physical sciences in early modern Europe and their relation to the broader intellectual and cultural history of the period. He is the author of *The Newton Wars and the Beginning of the French Enlightenment* (Chicago, 2008).

Michael H. Shank is Professor of History of Science and Integrated Liberal Studies at the University of Wisconsin-Madison. He has broad interests in the physical sciences (and their analogues and contexts) before 1700. His primary research interests focus on late medieval natural philosophy and astronomy. He is the author of *'Unless You Believe, You Shall Not Understand': Logic, University, and Society in Late Medieval Vienna* (Princeton, 1988), and co-editor (with Peter Harrison and Ronald Numbers), of *Wrestling With Nature: From Omens to Science* (Chicago, 2011).

Matthew Stanley is an Associate Professor at New York University's Gallatin School of Individualized Study. He teaches and researches the history and philosophy of science. He is the author of *Practical Mystic: Religion, Science, and A. S. Eddington* (Chicago, 2007), which examines how scientists reconcile their religious beliefs and professional lives. He has held fellowships at the Institute for Advanced Study, Princeton, the British Academy, and the Max Planck Institute. He currently runs the New York City History of Science Working Group.

Introduction

Peter Harrison

In 1922, Canadian philosopher Roy Wood Sellars confidently declared that 'we are all naturalists now'.[1] While Sellars's announcement was perhaps a little premature, it is difficult to deny, almost one hundred years later, that commitment to some form of naturalism is the default position in virtually all departments of human knowledge. 'Naturalism', of course, can mean a number of different things. But what most forms of scientific naturalism have in common is a commitment to the methods of the natural sciences and to the reliability of the knowledge generated by those methods. As Sellars himself expressed it, naturalism is not so much a philosophical system as 'a recognition of the impressive implications of the physical and biological sciences'.[2] When we inquire further into what, specifically, is naturalistic about the sciences, the simplest answer is that their methods involve a rejection of supernatural or spiritual explanations and a focus on what is explicable in terms of natural causes, forces, and laws. Naturalism and science thus go hand in hand. This volume is about that partnership, and its long and intriguing history.

THE VARIETIES OF NATURALISM

While the subject of this book is the history of scientific naturalism, it is helpful to begin in the present with contemporary debates about naturalism and its relation to the natural sciences. Modern discussions of scientific naturalism

[1] Roy Wood Sellars, *Evolutionary Naturalism* (New York: Russell and Russell, 1922), p. i.

[2] Sellars, *Evolutionary Naturalism*, p. i. Sellars's son, Wilfred Sellars, put it even more starkly: 'science is the measure of all things': Wilfrid Sellars, *Science, Perception and Reality* (London: Routledge, 1963), p. 173. On the varieties of naturalism see Geert Keil, 'Naturalism', in *The Routledge Companion to Twentieth Century Philosophy*, ed. Dermot Moran (London: Routledge, 2008), pp. 254–307; David Papineau, 'Naturalism', in *Stanford Encyclopedia of Philosophy* (Spring 2014 edn), ed. Edward N. Zalta, http://plato.stanford.edu/entries/naturalism/.

usually distinguish between *methodological* naturalism and *metaphysical* (or ontological) naturalism.[3] The former, as the name suggests, refers to the methods employed by the scientist, which involve a bracketing out of any supernatural or non-physical explanations in the pursuit of an understanding of the world. Adoption of methodological naturalism is thus usually thought to be consistent with having religious commitments—it is just that these commitments are deemed irrelevant to the conduct of science. This tidy way of setting aside the religious views of scientific practitioners has been standard practice within the scientific community for some time, and theistic scientists have been among its most enthusiastic proponents. The expression 'methodological naturalism' was coined by theologian and philosopher Edgar S. Brightman in a 1936 address to the American Philosophical Association. While the term was not widely adopted at the time, since the 1990s it has become increasingly prominent in discussions about the nature of modern science, partly owing to its endorsement and popularization by Christian philosopher Paul de Vries.[4]

Adoption of methodological naturalism is now widely regarded as one of the key ways of demarcating science from non-science. In legal battles over the status of scientific creationism and intelligent design, successive US courts over a period of three decades have endorsed the principle of methodological naturalism as an essential feature of science. When rendering his 2005 verdict on the 'unscientific' status of intelligent design, Judge John E. Jones thus declared that 'expert testimony reveals that since the scientific revolution of the sixteenth and seventeenth centuries, science has been limited to the search for natural causes to explain natural phenomena'. The judge went on to say that this avoidance of reference to the supernatural was a 'self-imposed convention of science' and 'is referred to by philosophers as "methodological naturalism" '.[5] The expert testimony to which the judge referred included statements from theologian John Haught, philosopher Robert T. Pennock, and biologist Kenneth R. Miller.

If *methodological* naturalism is not necessarily inimical to theism, the same cannot be said for *metaphysical* naturalism. This is the position that goes

[3] For this distinction see, e.g., Robert Pennock, *Tower of Babel: The Evidence Against the New Creationism* (Cambridge, MA: MIT, 1999), p. 191; Maarten Boudry, Stefaan Blancke, and Johan Braeckman, 'How Not to Attack Intelligent Design Creationism: Philosophical Misconceptions about Methodological Naturalism', *Foundations of Science* 15 (2010): pp. 227–44; B. Forrest, 'Methodological Naturalism and Philosophical Naturalism: Clarifying the Connection', *Philo* 3 (2000): pp. 7–29.

[4] 'Such a universal naturalism—common to idealists and realists, to naturalists and theists alike—may be called scientific or methodological naturalism. But methodological naturalism is sharply to be distinguished from metaphysical naturalism.' Edgar Sheffield Brightman, 'An Empirical Approach to God', *Philosophical Review* 46 (1937): pp. 157–8; Paul de Vries, 'Naturalism in the Natural Sciences', *Christian Scholar's Review* 15 (1986): pp. 388–96. See also Ronald L. Numbers, 'Science without God: Natural Laws and Christian Beliefs', in *Science and Christianity in Pulpit and Pew* (New York: Oxford University Press, 2007), pp. 39–58.

[5] Tammy Kitzmiller et al. v. Dover Area School District et al., 400 F.Sup2d 707 (2005). No. 04cv2688.

beyond a mere methodological stance to assert that *in reality* there are no supernatural agents or forces, and that the natural sciences have the potential to explain everything. Naturalism in this sense, explains philosopher Kai Nielsen, 'denies that there are any spiritual or supernatural realities'.[6] Fellow philosopher David Papineau concurs with this characterization, adding that what follows from it is that all natural phenomena are physical and that all natural events are explicable, in principle, in terms of the laws of physics.[7] Metaphysical naturalists reject the existence of anything beyond the physical world and often represent themselves as being resolutely opposed to religion, superstition, and obscurantism. Papineau suggests that 'the great majority of contemporary philosophers' fall into this camp.[8]

While these definitions might seem clear enough in theory, in fact, understandings of scientific naturalism—its dual forms and their implications—have become the subject of considerable controversy.[9] As we have already seen, one of the chief sites of contestation in the United States concerns the scientific status of intelligent design and 'scientific creationism'. The distinction between methodological and metaphysical naturalism also invites the question of how they are related, and whether one might lead to the other. The advocacy of methodological naturalism by religiously committed philosophers and scientists would suggest that the principle is, at the very least, consistent with religious belief. This is because naturalistic methods are understood as a self-imposed limitation upon science, and one that restricts its competence to explanations of physical realities. The spiritual and supernatural, on this view, lie beyond the scope of scientific investigation. It follows that methodological naturalism insulates the realm of theology and the supernatural from scientific scrutiny. The US National Academy of Sciences puts it this way:

> Because science is limited to explaining the natural world by means of natural processes, it cannot use supernatural causation in its explanations. Similarly, science is precluded from making statements about supernatural forces because these are outside its provenance.[10]

This stance is consistent with a more general position on the relations between science and religion that sees them as operating in independent spheres or 'non-overlapping magisteria', to use the well-worn phrase of Stephen J. Gould.[11]

[6] Kai Nielsen, 'Naturalistic Explanations of Religious Belief', in *A Companion to Philosophy of Religion*, ed. Philip L. Quinn and Charles Taliaferro (Oxford: Blackwell, 1996), p. 402.

[7] David Papineau, *Philosophical Naturalism* (Oxford: Blackwell, 1993), p. 16.

[8] Papineau, 'Naturalism'.

[9] For difficulties of definition see Hans Halvorson, 'What is Methodological Naturalism?', in *The Blackwell Companion to Naturalism*, ed. Kelly James Clark (Oxford: Blackwell, 2006), pp. 136–49; Michael Rea, 'Naturalism and Material Objects', in *Naturalism: A Critical Analysis*, ed. William J. Craig and J. P. Moreland (London: Routledge, 2016), pp. 110–32.

[10] National Academy of Sciences, *Teaching About Evolution and the Nature of Science* (Washington, DC: National Academy Press, 1998), p. 124.

[11] Stephen J Gould, *Rock of Ages* (New York: Ballantine Books, 1999), p. 5.

This irenic arrangement has not found favour with all scientists and philosophers, however. Some critics have labelled this stance 'accommodationism', arguing that advocacy of methodological naturalism in the sense outlined above concedes far too much to religion, and is itself a form of theology.[12] It has also been suggested that methodological naturalism, far from insulating religion from the potentially corrosive influence of the natural sciences, ultimately points to the truth of thoroughgoing metaphysical naturalism. The reasoning goes like this: modern science assumes no supernatural causes; modern science has been remarkably successful; hence, its working assumption must be correct and metaphysical naturalism is true.[13] This argument evinces a significantly different understanding of methodological naturalism to the one set out above, regarding it as a provisional hypothesis that is subsequently justified in light of the successes of modern science. If this view is correct, it follows that the realm of the supernatural cannot be quarantined from scientific investigation in the way that many theistic scientists suggest. Rather, methodological naturalism is 'a provisory and empirically grounded attitude of scientists which is justified in virtue of the consistent success of naturalistic explanations and the lack of success of supernaturalistic explanations in the history of science.... Science does have a bearing on supernatural hypotheses, and its verdict is uniformly negative.'[14] The logic of this position shows why metaphysical naturalists typically reject the existence of the supernatural and at the same time argue for the omnicompetence of science. Clearly, then, there are two opposing understandings of the implications of methodological naturalism.

Another area of contemporary controversy has to do with whether methodological naturalism is an appropriate investigative strategy for committed theists. While religious scientists have been prominent advocates of the standard version of methodological naturalism, a few of their coreligionists differ. Most obviously, advocates of intelligent design have argued that there may be scientific grounds for thinking that such mechanisms as natural selection offer inadequate explanations of some features of living things. Naturalistic explanations of certain complex features of living things are argued to be

[12] See, e.g., Jerry A. Coyne, 'Science, Religion and Society: The Problem of Evolution in America', *Evolution* 66–8 (2012): pp. 2654–63.

[13] See, e.g., Forrest, 'Methodological Naturalism,' esp. p. 21; Pennock, *Tower of Babel*, p. 191; Michael Ruse, *Darwinism and its Discontents* (Cambridge: Cambridge University Press, 2006), p. 48; Coyne, 'Science, Religion and Society'.

[14] Boudry, Blancke, and Braeckman, 'How Not to Attack Intelligent Design Creationism', p. 227. A similar line of argument may be found in Paul Draper, 'God, Science, and Naturalism', in *The Oxford Handbook of Philosophy of Religion*, ed. William Wainwright (Oxford: Oxford University Press), pp. 272–303; Forrest, 'Methodological Naturalism'; Alexander Rosenberg, 'Disenchanted Naturalism', in *Contemporary Philosophical Naturalism and its Implications*, ed. Bana Bashour and Hans D. Muller (London: Routledge, 2014), pp. 17–36; Brian L. Keeley, 'Natural Mind', in *The Blackwell Companion to Naturalism*, ed. Kelly James Clark (Oxford: Blackwell, 2016), pp. 196–208. For a critique see Peter Harrison, 'Naturalism and the Success of Science', *Religious Studies*, forthcoming.

impossible in principle, and it follows that complete explanation must involve recourse to design.[15]

A related line of argument comes from the Christian philosopher, Alvin Plantinga, who suggests that we think about methodological naturalism as a constraint on the 'evidence base' of scientific enquiry. Plantinga proposes that the evidence base of Christians should include the existence of God and presumably of divine actions. This expanded evidence base would in principle give rise to a different kind of investigative activity that he has termed 'Augustinian Science'. Plantinga concedes that his perspective is 'unpopular and heretical', but nonetheless thinks it worth pursuing. This is partly because of his conviction that, in spite of its neutral pretensions, science as currently practised is in fact incipiently atheistic. To this extent he seems to agree with those who argue that the success of methodological naturalism points to the truth of metaphysical naturalism.[16]

In sum, contemporary arguments about naturalism go to the heart of the nature of modern science, and have a significant bearing on such varied issues as the legitimate bounds of scientific explanation, the plausibility of religious beliefs, and the content of school science curricula. Yet controversies about naturalism show little sign of abating, and there are deep-seated differences between the various parties to the debate.

What contribution might a history of the sciences and their relation to naturalism make to these discussions?

Apart from the intrinsic interest of the episodes considered in this volume, taken together they amount to an assessment of the historical claims made in the context of the various arguments outlined above. Thus the common claim that naturalism in some form has characterized science 'since the scientific revolution of the sixteenth and seventeenth centuries' is one that can be assessed, and a number of chapters deal directly with this issue. A related historical claim made by advocates of metaphysical naturalism concerns 'the consistent success of naturalistic explanations and the lack of success of supernaturalistic explanations in the history of science'. This suggests that a 'hard' naturalism that denies supernatural realities is not just an uncritical starting point or an unwarranted premise, but a stance for which supporting evidence can be provided. David Papineau maintains that 'familiarity with the relevant scientific history' will lead to the conclusion that metaphysical naturalism is more than just a

[15] For a concise account of the intelligent design movement, see Ronald L. Numbers, *The Creationists: From Scientific Creationism to Intelligent Design* (Cambridge, MA: Harvard University Press, 2006), ch. 17.

[16] Alvin Plantinga, 'Methodological Naturalism?', *Perspectives on Science and Christian Faith* 49 (1997): pp. 143–54. Plantinga has also advanced arguments suggesting an incompatibility between evolutionary theory and ontological naturalism. See James K. Beilby (ed.), *Naturalism Defeated: Essays on Plantinga's Evolutionary Argument against Naturalism* (Ithaca, NY: Cornell University Press, 2002); Alvin Plantinga and Michael Tooley, *Knowledge of God* (Oxford: Blackwell, 2008), pp. 31–51.

matter of 'nailing one's philosophical colors to the naturalist mast'.[17] Again, this volume provides such a 'relevant scientific history', although what that history demonstrates is not as clear-cut as Papineau supposes.

An even more fundamental task is to consider whether those who studied nature in the past thought in terms of an unqualified dichotomy between 'naturalistic explanation' and 'supernaturalistic explanation'. If not, the investigation of the putative long-term superiority of naturalistic over supernaturalistic explanations might turn out to be more complicated than at first thought. As we shall see, the distinction between natural and supernatural, which in modern discussions is regarded as largely unproblematic, has an important history that shows how interdependent and interactive these notions have often been. In the past, ideas about the relative self-sufficiency of the natural realm typically relied upon deeper metaphysical or theological assumptions that could not themselves be established by naturalistic methods. This issue will be briefly discussed below, and explored further in several of the chapters.

A final general question concerns another way in which a particular version of the history of science is taken to support some contemporary doctrine of naturalism. Most often this is a narrative that sees naturalism beginning with the ancient Greeks, going into decline in the Middle Ages, and being restored with the 'scientific revolution' of the seventeenth century. Essentially, this is a story about the connection between naturalism and human progress—one that not only attributes the success of the sciences to their naturalistic assumptions, but which also regards commitment to the supernatural as inimical to scientific progress. A number of the essays in this volume explore this narrative and offer challenges to it.

Rather than proceed at this point to a summary of the specific contents of each contribution, we will instead discuss them in relation to four prominent themes that emerge out of them: the natural–supernatural distinction; the idea of laws of nature; naturalist theories of the person; and the significance of naturalistic approaches in the historical and human sciences.

NATURAL AND SUPERNATURAL

The 'nature' of which contemporary naturalism is an 'ism' derives its primary sense from a contrast with the supernatural. Yet this now-familiar natural–supernatural distinction is by no means a self-evident one. Strictly speaking, it is the product of a set of reflections that took place in the Latin Middle Ages.

[17] Papineau, 'Naturalism'. For the contrary argument see Hilary Putnam, 'The Content and Appeal of "Naturalism"', in *Naturalism in Question*, ed. Mario De Caro and David Macarthur (Cambridge, MA: Harvard University Press, 2008), pp. 59–70.

Crucially, the appearance of this distinction does not coincide with what is usually regarded as the birth of scientific thought among the Greek philosophers of the fifth and sixth centuries BC. For these Presocratic thinkers, nature (*phusis*) seems to extend to everything there is: gods, human beings, animals, plants, and stones. As Daryn Lehoux demonstrates in his chapter on ancient science (Chapter 1), for the first Greek natural philosophers the gods were part of the furniture of the natural world, and hence our familiar natural–supernatural distinction was not then in play.[18] Subsequently, Plato (429–347 BC) was to suggest that there might be more to nature than his predecessors had supposed. As is well known, he suggested that the phenomena of the visible realm were dependent upon a more fundamental reality—the unseen and unchanging world of forms. Aristotle (384–322 BC), too, criticized his philosophical forebears for assuming that 'nature' was all-inclusive. In his threefold division of the sciences, nature became the subject matter of natural philosophy (analogous in many respects to what we now call 'natural science'), while the more elevated sciences of mathematics and theology dealt with unchangeable realities: mathematics with unchangeable things that were in some sense inseparable from matter; theology with that which was wholly independent and self-subsistent—God. While Aristotle largely focused his attentions on the material realm, in his scheme of things nature was still dependent on God.[19] As J. B. Shank points out in his contribution (Chapter 4), Greek science was both 'naturalistic and anchored in notions of the divine at the same time'.

The Greeks' way of dividing up intellectual territory already poses intriguing questions for metaphysical naturalists—how, for example, the truths of mathematics might be accounted for in terms of pure naturalism.[20] But more importantly, given that the Presocratic thinkers included the divine in their speculations about nature, this history generates some difficulties for the common narrative that traces the origins of scientific naturalism to ancient Greece. While Plato and Aristotle observed a distinction between the material world and what lay beyond it, they proposed that the more elevated sciences pertained to the latter realm. Furthermore, most versions of natural philosophy

[18] Neither is it clear that the distinction exists in other cultures. See, e.g., Lorraine Aragon, 'Missions and Omissions of the Supernatural: Indigenous Cosmologies and the Legitimisation of "Religion" in Indonesia', *Anthropological Forum* 13 (2003): pp. 131–40.

[19] Aristotle, *On the Heavens* 278b–279b; *Metaphysics* 1064a29–1064b13. There is still room for discussion of precisely what Aristotle meant by theology (*theologikê*), and what he imagined its relation to natural philosophy to be. See, e.g., Richard Bodéüs, *Aristotle and the Theology of the Living Immortals* (Albany, NY: State University of New York Press, 2000); Stephen Menn, 'Aristotle's Theology', in *The Oxford Handbook of Aristotle*, ed. Christopher Shields (Oxford: Oxford University Press, 2012), pp. 422–64.

[20] Philosophers such as W. V. Quine have suggested that the applications of mathematics provide an adequate empirical foundation, but this naturalistic solution is by no means universally accepted. See, e.g., James Robert Brown, *Platonism, Naturalism, and Mathematical Knowledge* (London: Routledge, 2001); Penelope Maddy, *Naturalism in Mathematics* (Oxford: Oxford University Press, 1997).

(excepting perhaps that of the Epicureans) invoked some principle beyond mere material nature in order to account for the intelligibility of the cosmos—a principle that was either immanent or transcendent. Finally, 'natural' was not then contrasted with 'supernatural'—a notion that had yet to be invented—but with what was artificial (or man-made), or 'violent', or to do with laws and human conventions.[21]

The Greek idea that there was something beyond nature proved congenial to later Christian thinkers, whose understanding of a transcendent, creating Deity posited a similar distinction between the mundane world of created things and the ultimate reality upon which that world depends. However, the explicit terminological distinction between natural and supernatural did not emerge until the twelfth century. The scholastic philosopher Peter Lombard (d.1164) sought to distinguish between two modes of causal activity in the world—one in which events unfold according to the order that God has implanted in things, and another in which God acts directly and without the mediation of created causes.[22] In characterizing this latter mode of divine activity he was to speak of a cause that was 'beyond nature' (*praeter naturam*) or 'preternatural'. Thomas Aquinas (1225–74) was to popularize the term 'supernatural' (*supernaturalis*) to label this mode of divine action.[23] The natural–supernatural distinction thus began to crystallize in the thirteenth century as a means of distinguishing two kinds of divine activity: one in which God works with the order he embedded into things; the other when he acts miraculously and independently of created causes.[24]

Two aspects of the social and intellectual context of this period are relevant to the emergence of this distinction. First was a concern to develop formal criteria for the miraculous, prompted by the procedures required for canonization. Earlier Christian thinkers such as Augustine of Hippo (354–430) maintained that there was no ultimate difference between miraculous and mundane events. Both were equally the direct work of God. So-called miracles were simply distinguished on the basis of their unusualness and the fact that they were beyond our present knowledge of nature. However, in the later Middle Ages a tightening up of canonization procedures—through which individuals were accorded

[21] See, e.g., Aristotle, *Physics* 192b9–193b23; *On the Heavens* 268b27–270a13; *Nicomachean Ethics* 1134b18–1135a6.

[22] Peter Lombard, *Sententiae in IV libris distinctae*, 2.18.5. On the emergence of this distinction see Robert Bartlett, *The Natural and Supernatural in the Middle Ages* (Cambridge: Cambridge University Press, 2008), pp. 1–33. For an anthropological perspective see Lucien Lévy Bruhl, *Le surnaturel et la nature dans la mentalité primitive*, new edn (Paris: PUF, 1963).

[23] Henri de Lubac, 'Remarques sur l'histoire du mot "Surnaturel"', *Nouvelle revue théologique* 61 (1934): pp. 350–70; Lubac, *The Mystery of the Supernatural* (New York: Crossroad, 1998).

[24] It also follows that Christian theology did not always require this formal distinction, and more recent theological understandings of God as 'the ground of being' suggest that it is dispensable. See, e.g., Niels Henrik Gregersen, 'Naturalism in the Mirror of Religion: Three Theological Options', *Philosophy, Theology and the Sciences* 1 (2014): pp. 99–129.

sainthood—brought with it a requirement for unambiguous evidence of the performance of a miracle.[25] This in turn called for a much more formal distinction between what could be accomplished through the powers vested in natural things by God (natural) and what was brought about solely by direct divine action (supernatural). Yet, even in instances of natural causation, God was typically imagined to be active both on account of his conservation of natural causes and his concurrence with their operation.[26]

A second relevant consideration was the great confluence of Christian and Aristotelian thought in the eleventh and twelfth centuries. As Michael Shank argues in Chapter 2, during this period the desire to appropriate ancient Greek philosophy led to the quest for a common intellectual ground, constituted by a tacit agreement to rely upon 'naturalistic' explanation alone. What this entailed, in essence, was a deliberate bracketing of appeals to divine revelation, and a quest for explanations that were in principle available to all irrespective of religious creed. Albert the Great (*c*.1200–1280), for example, put forward the idea of 'explaining the natural naturally' (*de naturalibus naturaliter*), by which he meant offering explanations of events without invoking miraculous divine activity. As Michael Shank suggests, this sounds very much like methodological naturalism as we now understand it. That said, this neutral and naturalistic territory, common to Greek, Islamic, and Christian thinkers, was still understood as entailing commitment to some version of theism. 'Natural' causes were themselves understood as reliant upon God for their efficacy. It was just that in the case of 'natural' occurrences God worked through the order that he had implanted in things, while in the case of the miraculous events he brought about effects directly. In short, naturalistic explanation was not opposed to theistic explanation per se, but merely to a particular kind of theistic explanation.

All of this suggests that recent philosophical discussions that stress the historical failure of 'supernatural explanations' when compared with 'naturalistic explanations' fail to take cognisance of the way in which this distinction functioned in the past. No significant medieval natural philosopher ever argued that supernatural explanations might offer an account of how nature usually operates. Indeed one reason for making the distinction was to make possible the identification of miraculous events, which become visible only against the background of the regularities of nature which were themselves attributable to divine providence.

The conceptual interdependence of Western conceptions of 'natural' and 'supernatural' from the Middle Ages onwards is a common theme of a number of

[25] For Augustine and Aquinas on miracles see Peter Harrison, 'Miracles, Early Modern Science, and Rational Religion', *Church History* 75 (2006): pp. 493–511. On the significance of canonization, see Laura Smoller, 'Defining the Boundaries of the Natural in the Fifteenth Century', *Viator* 28 (1997): pp. 333–59.

[26] Alfred J Freddoso, 'God's General Concurrence with Secondary Causes: Why Conservation is not Enough', *Philosophical Perspectives* 5 (1991): pp. 553–85.

the chapters. Only in the nineteenth century was there a concerted attempt to articulate a version of scientific naturalism that opposed itself to 'supernaturalism' and sought to eliminate it. As Bernard Lightman shows in Chapter 13, the first generation of self-styled scientific naturalists sought to recreate a history of naturalism, placing themselves in a tradition that harked back to the ancient Greeks and the seventeenth-century pioneers of modern science. In this they were largely successful, creating a familiar, if simplistic, narrative of the history of science that brought together science, naturalism (in their sense), and human progress. One of the goals of this volume is to challenge this distorted version of events.

LAWS OF NATURE

Related to the conceptual interdependence of natural and supernatural is the historical emergence of the idea of laws of nature. Today, naturalism is frequently defined as the view that the laws of physics are sufficient to explain everything that takes place within the world. Michael Ruse, in this book (Chapter 7) and elsewhere, has defined naturalism as the 'appeal to and reliance on law: blind, natural regularity'. David Papineau, as noted earlier, sees naturalism as entailing that all natural phenomena are physical and that all natural events are explicable, in principle, in terms of the laws of physics.[27] But there remains the question of whether the existence of the laws of physics is something that itself admits of a naturalistic explanation. This, of course, is a philosophical question, but it is relevant that for those who invented the notion of 'laws of nature' the answer was 'no'.

In the seventeenth century, as Peter Harrison shows in Chapter 3, René Descartes (1596–1650) proposed that the uniformity of nature was to be understood in terms of God's direct and unvarying influence on every 'natural' event. This was the birth of the modern conception of laws of nature, which came to replace the more relaxed Aristotelian notion of nature as that which happens 'always or for the most part'. Descartes derived the immutability of the laws of nature from the immutability of their divine source. In a sense, then, the invariable uniformity of nature came to be understood not as a consequence of God's withdrawal from the world but of his direct and incessant engagement with it. The regular operations of nature were thought of as a mode of divine activity. While English natural philosophers differed with Descartes on details of how laws of nature were to be discovered, they nonetheless agreed with his basic conception that laws of nature were simply God's regular willing of natural states of affairs. The idea of a rational natural order independent of God was a

[27] Michael Ruse, *But is it Science?* (New York: Prometheus, 1988), p. 21; Papineau, *Philosophical Naturalism*, p. 16.

vulgar idea, wrote the Newtonian philosopher Samuel Clarke, for the regular course of nature was nothing but 'the *Arbitrary Will and pleasure of God* exerting itself and acting upon Matter continually'.[28]

To be sure, Newtonian science could be appropriated for materialist and anti-religious purposes, as J. B. Shank notes in Chapter 4, but in England the idea that laws of nature had a necessary theological foundation persisted until well into the nineteenth century. As Matthew Stanley shows in Chapter 5, prominent men of science in the nineteenth century continued to attribute the regularities of nature to divine superintendence. Leading scientific theorists of the period, John Herschel (1792–1871) and William Whewell (1794–1866) thus insisted that the uniformity of nature, expressed in terms of immutable laws, was grounded in the constant and ubiquitous exercise of the omnipotent powers of God. John Brooke makes a similar point in Chapter 6, but in relation to the world of living things. His chapter highlights the 'non-naturalness' of naturalism, showing that even Charles Darwin spoke at times of a Creator who creates by means of laws in the organic realm.

Paradoxically, then, up to about the middle of the nineteenth century, we have a kind of naturalism that is explicitly grounded in theological assumptions about how God acts in the natural world. Thereafter, we see a growing tendency to regard natural laws themselves as an appropriate terminus for explanation, with those laws now regarded simply as brute features of the universe that simply need to be accepted. The historical derivation of the modern conception of laws of nature might lead us to wonder whether they offer a robust foundation for a philosophical naturalism.[29] Addressing this question, Ludwig Wittgenstein observed that 'at the basis of the whole modern view of the world lies the illusion that the so-called laws of nature are the explanations of natural phenomena'. He continued: 'people stop at natural laws as at something unassailable as did the ancients at God and Fate.' Wittgenstein concluded that ancients and moderns were equally mistaken, but that the ancients were more consistent since they reached an acknowledged terminus, while the moderns rested with a mere appearance of a complete explanation.[30]

Before moving on from laws of nature it is worth reflecting on how this theistically grounded conception of natural order differs from what came before. As we have seen, medieval scholastics tended to speak of an order implanted

[28] Samuel Clarke, *The Works of Samuel Clarke, D.D.*, 2 vols (London, 1738), vol. 2, p. 698.

[29] For contemporary philosophical doubts about the status of laws of nature see, e.g., Nancy Cartwright, *How the Laws of Physics Lie* (Oxford: Clarendon Press, 1983); Bas van Fraassen, *Laws and Symmetry* (Oxford: Clarendon Press, 1990); John Dupré, *The Disorder of Things: Metaphysical Foundations of the Disunity of Science* (Cambridge, MA: Harvard University Press, 1995).

[30] Ludwig Wittgenstein, *Tractatus Logico-Philosophicus* 6.371–2 (New York: Harcourt, Brace, 1922), p. 87. Wittgenstein, incidentally, was resolutely opposed to philosophical naturalism, remarking that adoption of the methods of the sciences 'leads the philosopher into complete darkness'. Wittgenstein, *The Blue and Brown Books*, 2nd edn (New York: Harper Torchbooks, 1960), p. 18.

into things by God. These internal principles were understood in terms of either the Stoic notion of seed-like inherent principles ('seminal principles'—Augustine, Peter Lombard), or the idea, derived from Aristotle, of internal properties that arise out the object's matter and form ('substantial forms'—Aquinas). Many of the moderns, beginning with Descartes, sought to expel these 'occult' properties from things, rendering natural things inherently inert. For a period thereafter, the animation of natural things was attributed directly to God who moved things directly and lawfully in accordance with his own will.

It is also instructive to reflect upon how the typical contrast cases for 'nature' (the violent, the artificial, and the conventional) changed in the modern period. As already noted, Aristotle and, more generally, the Greeks had observed two distinctions: natural *versus* artificial, and nature *versus* convention (or law). With the advent of the mechanical science of the seventeenth century both of these contrasts had been turned on their head by the idea of a God who made things. The distinction between nature and artefact was challenged, since God had created the machine of the world. At the same time, the nature/law distinction was dissolved, since God was understood to have promulgated the laws that directed the operations of natural bodies. These fundamental changes in the understanding of 'nature', evident in the new set of contrast cases, necessarily complicate any simple linear story about the history of naturalism.

One radical alternative to the early modern examples discussed in detail in the book warrants brief mention at this point. Baruch Spinoza (1632–77) rejected a sharp natural–supernatural dualism, denying the existence of a God who completely transcends nature. Spinoza is sometimes lauded as a kind of proto-modern naturalist who was ahead of his time. But rather than asserting that *nature* is all there is, Spinoza can be understood as asserting that *God* is all there is. For Spinoza, what we call nature must be part of God: 'whatever is, is in God... nothing can be conceived without God'.[31] Natural things are in some sense properties of God, to whom Spinoza still seems to maintain a religious attitude. Thus, 'the intellectual love of God' is the highest form of fulfilment to which humans can aspire.[32] Interestingly, Spinoza agreed with the Newtonians that God is the direct cause of all things. But laws of nature, for Spinoza, are not free divine choices imposed by a transcendent Deity onto the world that he has produced by the act of creation. Rather laws flow necessarily from the divine nature. They originate in God, but are immanent in nature rather than transcendent to it.[33] The historical record suggests that the more theologically

[31] Baruch Spinoza, *Ethics* pt. 1, prop. 15, *Complete Works*, ed. Michael L. Morgan, trans. Samuel Shirley (Indianapolis, IN: Hackett, 2002), p. 224. See also Alexander Douglas, 'Was Spinoza a Naturalist?', *Pacific Philosophical Quarterly* 96 (2015): pp. 77–99; Carlos Fraenkel, 'Spinoza's Philosophy of Religion', in *The Oxford Handbook of Spinoza*, ed. Michael Della Rocha (Oxford: Oxford University Press, 2013); Dominic Erdozain, 'A Heavenly Poise: Radical Religion and the Making of the Enlightenment', *Intellectual History Review* 27 (2017): pp. 71–96.

[32] Spinoza, *Ethics* pt. 5, prop. 33, p. 377.

[33] Spinoza, *Ethics* pt. 1, props. 16, 17, 18 (pp. 227–9). See also Jon Miller, 'Spinoza and the Concept of a Law of Nature', *History of Philosophy Quarterly* 20 (2003): pp. 257–76.

orthodox Newtonian understanding of laws of nature was the conception that was scientifically fruitful. But Spinoza's thinking about these matters is important nonetheless. His 'flattening' of the causal order into a single layer—a precondition for the emergence of modern naturalism—mirrors that of Descartes and Newton. More generally, he represents yet another way of reconceptualizing the relationship between natural and supernatural that will be reprised to some extent by modern neo-vitalists.

Again, this discussion of the origin of the idea of laws of nature shows that the idea of the 'natural' to which many contemporary naturalists presently defer is by no means a self-evident notion that is unproblematically given. The example of Spinoza, moreover, suggests that 'religious naturalist' need not be an oxymoron.[34]

NATURALISTIC THEORIES OF THE PERSON

The physical world is the domain most obviously explicable in terms of laws of nature, and hence it might seem to offer the conspicuous examples of historical naturalization. Yet the histories of medicine and psychology, with their focus upon the human subject, also seem to offer telling examples of a trend away from supernaturalistic explanation. Indeed, a strong case can be made for medicine as one of the most prominent sites of naturalization. As Ronald Numbers has expressed it: 'The most compelling instances of supernaturalism giving way to naturalism occurred not in physics or chemistry but in such areas as meteorology and medicine, in explanations of epidemics, eclipses, and earthquakes.'[35] The epidemics that ravaged North America in the eighteenth century, for example, had been routinely regarded as evidence of divine chastisement. However, with the success of inoculation—championed, as it turns out, by figures such as the puritan minister Cotton Mather (1663–1728)—the scourges of cholera, diphtheria, and yellow fever came to be regarded no longer as signs of divine displeasure. Rather, with the development of a variety of medical prophylactics, these came to be, simply, preventable diseases.[36] In this context, then, penitence and prayer could be displaced by the mundane methods of modern medicine.

During much the same period we also witness the beginnings of the medicalization of the soul, and of the apparently heterodox assertion that humans

[34] There are a number of varieties of contemporary religious naturalism. See, e.g., Wesley Wildman, 'Religious Naturalism: What It Can Be, and What It Need Not Be', *Philosophy, Theology, and the Sciences* 1 (2014): pp. 49–51; Fiona Ellis, *God, Value, and Nature* (Oxford: Oxford University Press, 2014).

[35] Numbers, *Science and Christianity in Pulpit and Pew*, p. 43.

[36] Numbers, *Science and Christianity*, p. 44.

are purely material beings.[37] These new, reductionist accounts of the human soul are often associated with the radical medical atheism of figures such as Julien Offray de La Mettrie (1709–51). La Mettrie's *Histoire naturelle de l'âme* (Natural History of the Soul, 1745) and *L'Homme Machine* (Man a Machine, 1748) scandalized even the most liberally minded of his contemporaries on account of their overt materialism and mechanical accounts of the operations of the soul. Yet materialism was not the sole preserve of radical French thinkers. Arguments for the mortality and materiality of the soul had been also championed by religious thinkers for whom the notion of an immaterial soul was an unbiblical and pagan conception. In Chapter 8, Michelle Pfeffer offers an account of a remarkable group of English thinkers who, in the seventeenth and eighteenth centuries, also insisted that the soul be conceived of in bodily and material terms. They did so not primarily from the standpoint of medicine and natural philosophy but by invoking scripture, theology, and history. The basic claim was that the genuinely Christian view of the person, and the position clearly set out in scripture, was of a purely material being. By contrast, the notion of an immaterial and immortal soul was said to be a pagan invention and its contemporary currency was simply evidence of the corruption of an older and more legitimate anthropology. This complicates a common association of materialism with religious scepticism.

In his contribution (Chapter 9), Jon Roberts takes up the later phases of the naturalization of the human soul, showing how dualist conceptions of the person were further subjected to serious challenges over the course of the nineteenth century. Advocates of the new 'science of mind' or the 'new psychology' sought to apply to the mind the powerful naturalistic methods that were proving successful in the physical sciences. Linking the physiology of the nervous system, a theory of organic evolution that minimized the difference between higher faculties of humans and animals, and laboratory practices of experiment and measurement, they sought to move the mind into the sphere of material nature. Scientific naturalism applied to the mind thus challenged the long-standing dichotomy between mind and matter. And if the mind were essentially reducible to the brain, there was no in-principle reason to deny that a purely naturalistic and materialistic account of its operations was possible. Not surprisingly these claims met with religious opposition. Roberts concludes that resistance to a purely materialistic account of mind is not restricted to those with religious commitments, and that the ontological status of the human mind remains an open question.

[37] See, e.g., Charles T. Wolfe, '*Tres medici, duo athei*? The Physician as Atheist and the Medicalization of the Soul', in *Early Modern Medicine and Natural Philosophy*, ed. Peter Distelzweig, Benjamin Goldberg, and Evan R. Ragland (Dordrecht: Springer, 2016), pp. 343–66; Ann Thomson, *Bodies of Thought: Science, Religion, and the Soul in the Early Enlightenment* (Oxford: Oxford University Press, 2008).

THE HISTORICAL AND HUMAN SCIENCES

While a number of the chapters in this volume focus upon areas of enquiry that we would now regard as part of the natural sciences, there is also good reason to consider naturalism in relation to history and the human sciences. This is partly because it was only in the nineteenth century, in Anglophone contexts at least, that 'science' came to refer more or less exclusively to the natural sciences.[38] Before this, it was used to label a variety of systematic bodies of knowledge including, for example, biblical criticism and natural theology.[39] Any comprehensive account of a history of scientific naturalism must take cognizance of the shifting meanings of 'science'. It is also significant that the areas most resistant to purely naturalistic explanations have been history, the historical sciences, and those areas that involved the study of human beings.[40] From the Middle Ages onwards the formal study of nature had called for a bracketing out of direct supernatural activity. However, denial of direct divine activity in the sphere of human actions was more problematic. This was because of the traditional Judeo-Christian belief that God was able to work immediately upon the human heart and hence indirectly exercise some influence on the course of history. Moreover, historical sciences such as geology and, from the nineteenth century onwards, evolution had the potential to clash with approaches associated with biblical accounts of the mutations of the earth and the origins of human beings. These latter approaches were non-naturalistic in the sense that they derived their authority from supernatural revelation or divine inspiration.

Perhaps the most controversial of all applications of naturalism was to the subject matter of religious history, and in particular the history recorded in the Hebrew Bible and New Testament. Here the question was whether the biblical accounts of the origins of life on earth, along with such geologically significant events as the universal deluge, could be read according to the canons of methodological naturalism, as if God was neither acting directly in the relevant events nor inspiring the biblical authors. In Chapter 10 Nicolaas Rupke gives an account of the naturalization of geological discourse, showing how references to God and the Bible gradually disappear from the geological literature over the course of the eighteenth century. As the Bible lost its privileged status as an impeccable record of past events, the history of the earth ceased to be a part of sacred history. The annals of nature came to be preferred to the annals of scripture as authorities for understanding the history of the earth.

[38] On the modern meanings of 'science' see Peter Harrison, *The Territories of Science and Religion* (Chicago: University of Chicago Press, 2015), ch. 6.

[39] See, e.g., William Whewell, *The Philosophy of the Inductive Sciences: Founded upon their History*, vol. 2, new edn (London: John Parker, 1847), p. 117; Nicholas Wiseman, *Lectures on the Connexion between Science and Revealed Religion*, 2 vols (London: Booker, 1836).

[40] Although as noted earlier, mathematics presents another area of difficulty.

Scott Prinster's contribution (Chapter 11) also deals with the topic of naturalism as it relates to the Bible. He offers an account of the influence of higher biblical criticism (then understood to be a science) as it developed in Germany in the early nineteenth century and found its way to Britain. The controversial collection *Essays and Reviews* (1860), written mostly by liberally minded Anglican clergymen, disseminated the principles of historical criticism to a somewhat scandalized audience in Britain. One of the central messages of the collection was the injunction to 'read Scripture like any other book'—which is to say, naturalistically.[41] We now tend to imagine that religious controversy in the 1860s was centred on Darwin's *Origin of Species* (1859) and its implications for our understanding of the status of human beings. Yet, as Prinster reminds us, *Essays and Reviews* sold more copies in two years—over twenty-two thousand—than Darwin's *Origin* did in two decades. Thus, while naturalistic approaches to the Bible were directly related to naturalistic readings of the book of nature, during the late nineteenth century it was the former rather than the latter that most exercised traditional religious believers.

It remains to say something about one of the other social scientific disciplines that deal with human beings—anthropology. The pattern of development of social sciences, in relation to metaphysical naturalism, was significantly different from that of the 'hard' sciences. Most of the canonical figures of the seventeenth-century scientific revolution had consciously articulated the theistic foundations of their enterprise, focusing on the quest for laws of nature but acknowledging the divine source of those laws. Theirs was a naturalism that was explicitly dependent on theistic considerations. By way of contrast, a number of the late eighteenth- and early nineteenth-century pioneers of the social sciences saw themselves as offering a naturalistic alternative to religious perspectives, both in their analysis of human affairs and in the normative prescriptions they offered for their improvement. They saw history as the sole product of human actions (rather than of divine providence) and regarded themselves as advocates of a project dedicated to the betterment of society in secular terms.[42] This latter project was to be grounded in 'scientific' rather than religious principles, and was intended to be naturalistic from the ground up. Arguably, then, while the natural sciences have tended to maintain a strict distinction between metaphysical and methodological naturalism, the social sciences have had an incipient commitment to a version of metaphysical naturalism from the start. Underscoring this commitment is the idea that the social sciences can trump religious worldviews by exposing their naturalistic

[41] Benjamin Jowett, 'On the Interpretation of Scripture', in *Essays and Reviews*, 2nd edn, ed. John William Parker (London: Parker and Son, 1860), p. 338. For other advocacies of naturalism in the volume see, e.g., pp. 111–12, 143, 155.

[42] See, e.g., a number of the contributions in Christian Smith (ed.), *The Secular Revolution: Power, Interests, and Conflict in the Secularization of American Public Life* (Berkeley, CA: University of California Press, 2003).

foundations: religions themselves are just products of nature. In short, the social sciences have sought to bring collective human action into the sphere of the natural, to offer a replacement for religion, and to explain the ubiquity of religion in naturalistic terms. In these respects the social sciences differ significantly from the physical sciences.

Anthropology, even in its earliest stages, thus sought to bring naturalistic explanations to bear on the phenomena of religion, as Constance Clark shows in Chapter 12. But anthropology was imagined to be a naturalistic and naturalizing enterprise in two further senses. First, it allotted to religion a specific role in 'primitive' societies and associated its gradual demise with a general theory of social development. The processes of social evolution were originally regarded as 'natural' in the sense that they conformed to natural laws conceptualized along the lines of the laws of biological development. Anthropology derived its scientific status from this focus on putatively universal laws of social progress. Second, anthropology was naturalistic in the sense that it was seen to provide a replacement for theological accounts of human origins. Thus, and in spite of the religious convictions of many of the pioneers of the discipline, anthropology was naturalistic insofar as it was imagined to offer an alternative version of events to the influential biblical narrative of Eden and a fall away from an original perfection. Anthropology thus enabled nineteenth-century thinkers to divest themselves of the theological motif of degeneration, allowing them to adopt the alternative notion of progressive organic evolution in which primitive societies would (or could) develop into civilized ones. Ironically, though, as Clark suggests, the influence of traditional Christian conceptions still informed the new naturalistic accounts of human origins and development. Arguably, for example, the spectre of the idea of original sin continues to haunt the discipline of anthropology.

Summing up: the essays in this volume demonstrate the great variety of ways in which naturalistic explanation has been characterized in the past and how these modes of explanation contributed to the scientific enterprise. While they are not intended to represent a comprehensive history of scientific naturalism, they nonetheless point towards three general conclusions. First, while ideas about what is natural have changed over time, throughout Western history 'natural' occurrences have most often been understood as requiring divine activity. It follows that in the past the supernatural–natural distinction did not map directly onto the exclusive disjunction: 'caused by God' or '*not* caused by God'. Second, and following on from this, it is not clear that history is characterized by an ongoing competition between 'supernaturalistic explanations' and 'naturalistic explanations', with only the latter proving successful in the long run.[43] It is certainly true that for most of Western history up until the eighteenth

[43] We are leaving aside here the question of whether 'successful' scientific explanations are truth tracking. One view, based on examples drawn from the history of science, would suggest

century, the possibility of miraculous events was almost universally accepted. But it was precisely because these events were imagined to be beyond the powers of nature that their 'supernaturalistic' explanation was held to be consistent with naturalistic explanations of the regularities of nature. Moreover, the possibility of coherent naturalistic explanation had itself typically been regarded as dependent upon some theological, or at the very least, metaphysical foundation. This is because both philosophical and religious thinkers believed that the orderliness of nature necessarily required an explanation that was itself somehow 'outside' or 'beyond' nature. Third, and finally, these essays represent a challenge to the history of naturalism that the nineteenth-century scientific naturalists invented for themselves, and which still remains current in certain popular versions of the history of science.[44] It is simply not the case that the ancient Greeks invented scientific naturalism; that this naturalism went into decline in the Middle Ages, was revived in the scientific revolution, and reached fulfillment in the nineteenth century. Rather, a version of naturalism flourished in the Middle Ages, to be replaced during the scientific revolution with a version of supernaturalism. In all of this what we see is a somewhat paradoxical pattern in which religious considerations laid the foundations for modern scientific naturalism. As the historian of science and medicine Ronald L. Numbers has expressed it: 'scientific naturalism was largely made in Christendom by pious Christians'.[45]

not. See Larry Laudan, 'A Confutation of Convergent Realism', *Philosophy of Science* 48 (1981): pp. 19–49; Peter Dear, *The Intelligibility of Nature: How Science Makes Sense of the World* (Chicago: University of Chicago Press, 2006); Peter Vickers, 'Historical Magic in Old Quantum Theory?', *European Journal for Philosophy of Science* 2 (2012): pp. 1–19.

[44] For the nineteenth-century versions, see especially John Tyndall's 'Belfast Address' (1874) in John Tyndall, *Fragments of Science*, vol. 2, 8th edn (London: Longmans, Green, and Co., 1892), pp. 145–6; Thomas Henry Huxley, *Science and the Christian Tradition* (New York: Appleton, 1894); Friedrich Lange, *Geschichte des Materialismus* (Iserlohn: J Baedeker, 1866). Also see Bernard Lightman's chapter (Chapter 13) in this volume.

[45] Numbers, 'Science without God', p. 58. This is not to deny, of course, that forms of naturalism flourished in other monotheistic traditions and beyond.

1

'All Things are Full of Gods'

Naturalism in the Classical World

Daryn Lehoux

All things are full of gods.
—Thales of Miletus

IN THE BEGINNING

Science is taken, on what we might call 'the conventional view', to be a rejection of the supernatural in favour of strictly naturalistic explanations of phenomena. When in Hesiod (Greece's great seventh or eighth century BC poet), lightning is said to have been given to Zeus by Brontés, Steropés, and Argés who were born of Earth (Gaia) after she had lain with Heaven (Uranus), we are to understand that—however lightning existed before this—it came into the province of our world by a supernatural agency.[1] By contrast, when the early philosopher Anaximander says, a century or so later, that lightning and thunder 'happen from wind', we are supposed to see an entirely naturalistic line of agency.[2] Modern commentators are quick to point out the importance of this change as the signal move in the birth of science:

> In a realm where mythology had provided divine beings as causes for the structure and events of the world,... Anaximander offers elemental bodies and natural events.... In modern terms, Anaximander provides a kind of paradigm

[1] Hesiod, *Theogony*, lines 138–9.

[2] H. Diels and W. Krantz (eds), *Die Fragmente der Vorsokratiker*, 6th edn (Berlin: Weidmannsche Verlagsbuchhandlung, 1951), fragment A23.

of explanation that sets the problems and the limits for a scientific understanding of the world.[3]

There is, however, a more profound reason to start natural philosophy with the Greeks rather than the older cultures, despite their many accomplishments. Although these older cultures had technical knowledge, keen observational skills, and vast resources of material and information, they failed to create natural philosophy because they did not separate the natural world from the supernatural world.[4]

[In Anaximander, the] basic explanatory factors are no longer more or less anthropomorphic gods. Instead, the genesis of the cosmos is explained in terms of recognizable elements of nature—in other words, the approach is *naturalistic*.[5]

So far, the conventional wisdom seems conventional enough. But in recent years cracks have begun to show in it, and it is worth teasing out the ways in which some of the key elements of this account play out.

In the first instance, although many histories of science and of philosophy try to downplay the fact, the gods never really go away in ancient science (nor does mythology, for that matter, but that question moves us beyond the bounds of this chapter).[6] As David Sedley has shown, a great number of ancient philosophers, including many of the ones most likely to find their way into histories of the sciences, saw purposive divine agency at work in the cosmos. Many were creationists of one stripe or another.[7] Even Anaximander, singled out in the above quotations as our first genuine naturalist, may well have thought there was a divine lawmaker behind the 'paying of reparations' that he claimed governed the regularities in nature, and we have two (closely related) sources that ascribe to him the idea that the stars are 'gods'.[8] One of these sources may even tie these stellar divinities back to Anaximander's supposedly naturalistic 'elemental body', the so-called 'unlimited': 'Anaximander said that the unlimited heavens are gods.'[9] Perhaps we should not read too much into the instance of '*apeiron*' (unlimited)

[3] Daniel W. Graham (ed. and trans.), *The Texts of Early Greek Philosophy: The Complete Fragments and Selected Testimonies of the Major Presocratics*, Part I (Cambridge: Cambridge University Press, 2010), pp. 45–6.

[4] Andrew Ede and Lesley B. Cormack, *A History of Science in Society*, 2nd edn (Toronto: University of Toronto Press, 2012), p. 3.

[5] Keimpe Algra, 'The Beginnings of Cosmology', in *The Cambridge Companion to Early Greek Philosophy*, ed. A. A. Long (Cambridge: Cambridge University Press, 1999), p. 48.

[6] On mythology, see, e.g., Luc Brisson, *How Philosophers Saved Myths: Allegorical Interpretation and Classical Mythology*, trans. Catherine Tihanyi (Chicago: University of Chicago Press, 2004) [French edn, *Introduction à la philosophie du mythe*, vol. 1: *Sauver les mythes* (Paris: J. Vrin, 1996)]; R. Buxton (ed.), *From Myth to Reason?: Studies in the Development of Greek Thought* (Oxford: Oxford University Press, 1999); Kathryn A. Morgan, *Myth and Philosophy from the Presocratics to Plato* (Cambridge: Cambridge University Press, 2000); Liba Taub, *Aetna and the Moon* (Eugene, OR: Oregon State University Press, 2000).

[7] David Sedley, *Creationism and its Critics in Antiquity* (Berkeley, CA: University of California Press, 2007).

[8] Sedley, *Creationism and its Critics*, p. 6.

[9] Stobaeus, *Anthology* 1.1.29b. All translations, unless otherwise noted, are my own.

here, but there is no denying the gods in this passage.[10] However we may read it, it certainly complicates any account of Anaximander as the first pure naturalist, and it is perhaps unsurprising that this passage almost always goes unquoted and unacknowledged in modern commentary. Indeed, it is a core part of my argument in this chapter that the imposition of a naturalism–supernaturalism divide in ancient science frequently relies on such selective blinkering.

Furthermore, the question of what we might even mean by using the loaded term 'supernatural' raises its head. After all, if we define science as the domain of the natural and label one set of causal agents as by definition *beyond* that domain—*super* it, in the Latin—we are certainly setting up a clear demarcation, but at the non-trivial risk of begging the question. Or perhaps the gods are just part of the natural domain in the first place. If we try to refortify the natural–supernatural distinction by now arguing that the gods as causal agents do not, in point of fact, exist (or more cautiously, have never been proven to exist), we simultaneously close off a considerable portion of what we might otherwise want to accept as historical science, pre-modern as well as modern, since so very many historically posited causal entities turn out to be just as non-existent: N-rays, phlogiston, psychic pneuma—the list is endless. A closely related question revolves around who gets to define the category of the natural in the first place: us or them? After all, as Geoffrey Lloyd has repeatedly pointed out, the idea of 'the natural' itself has a history.[11]

Many modern accounts tell us that the Presocratic philosophers (the loose grouping of early philosophers with whom Anaximander is usually categorized) were reacting against the mytho-poetic genealogies that populated the world with anthropomorphic deities. But when we try to chase this claim back to the ancient evidence itself, we find that the idea of a reaction-against is very much a product of modern scholarship: the Presocratics themselves don't talk this way at all. There are a few potential exceptions, however, insofar as some of the Presocratics did single out Homer and Hesiod for criticism. Leaving the atomists aside for the moment, we find only three of these early philosophers explicitly critical of the poets. First, the sometimes-curmudgeonly philosopher Heraclitus (early fifth century BC) says in several places how one or the other of them was foolish (he adds that Homer should be thrashed), but he rarely gives us his explicit reasons, and so we cannot presume that naturalism was even part of what was at issue. (And in the one instance where he fleshes out the objection, he tells us that Hesiod did not recognize that 'the road up and the road down were one', for what that is worth.)

[10] Not least because the word 'unlimited' is replaced by the word 'stars' in the second of the 'heavens are gods' sources. Pseudo-Plutarch, *Opinions of the Philosophers* 1.7.12.

[11] G. E. R. Lloyd, *Magic, Reason, and Experience* (Indianapolis, IN: Hackett, 1979); Lloyd, 'The Invention of Nature', in *Methods and Problems in Greek Science*, ed. G. E. R. Lloyd (Cambridge: Cambridge University Press, 1991), pp. 417–34.

The second explicit Presocratic reaction against Homer and Hesiod comes in Heraclitus's older contemporary Xenophanes, who mocks the traditional anthropomorphization of the gods by supposing that if horses had hands and could draw, then their gods would inevitably look like horses, and the gods of cows would look like cows. His motivation for this seems to have been that he thought Homer and Hesiod to have been impious for supposing that the gods had human foibles such as lust, jealousy, and a propensity for deception. It is significant that he nowhere objects to the idea of supernatural causation itself, but instead to a trivialization of the nobility and power of the gods. Indeed, Xenophanes's theology posits a supreme deity which, although it lacks human physical and emotional attributes, is explicitly said to have causal efficacy in the world: 'Withdrawn from toil, he moves all things by the will of his mind.'[12] Elsewhere, we are also told that Xenophanes thought that God could 'do everything he wishes', and that God 'sees and hears'.[13] Thus, for all his bluster against the depiction of the gods in the great poets, Xenophanes's objection has nothing to do with the primacy of naturalistic causation. Quite the opposite, in fact: it seems that he is arguing for an even more powerful and universal line of divine causation than Homer and Hesiod would allow.

Beyond Xenophanes and Heraclitus, the only other criticism of the poets we find in the (non-atomist) Presocratics is a cavil attributed to Thales about the date of the rising of the Pleiades.[14] Thus far, the disagreement on naturalism between the Presocratics and the poets is clearly a later superposition on the evidence as we have it.

IMPERIUM

That the emphasis on naturalism at the birth of science may be barking up the wrong tree can further be shown by looking at how the sciences developed over the course of antiquity, and so I would like to jump ahead by a few hundred years, to the height of the Roman empire and the fully developed and very well attested natural philosophies of the early centuries AD, to see how divinity interacts with nature in the accounts that eventually grew out of those first

[12] Simplicius, *Commentary on Aristotle's Physics* 23.20. The verb for 'move' here *(kradainei)* means to swing or brandish, as one might a sword, or to vibrate, as a bell; the poetic phrase 'withdrawn from toil' *(apaneuthe ponoio)* could either be meant to imply that the god himself needs make no effort to move things, or that he is far removed from the toils of this world in doing so. I tried to capture the ambiguity in the translation.

[13] Pseudo-Aristotle, *On Melissus, Xenophanes, and Gorgias* 977a35; Pseudo-Plutarch, *Miscellanies*, 4.

[14] Pliny, *Natural History* 18.213.

philosophies. I hope to show that not only was divinity immanent and active in nature in most accounts at this later date, but that it had been so all along.

If we look at the various philosophies on offer in imperial Rome, we find considerable overlap between a number of different schools on some of the most important questions.[15] This tendency has often been referred to (sometimes with disapproval) as the period's *eclecticism*. In the past couple of decades, however, scholars have increasingly begun to see this eclecticism less as the dabbling of dilettantes and more as a sophisticated shared intellectual background among the educated classes in Rome.[16] There are, of course, exceptions: people who disagreed rather sharply on what the fundamental makeup of the world was (the atomist Epicureans), and those who were radically sceptical of the possibility of obtaining knowledge about the world at all (Pyrrhonians). We will return to these presently.

What we see with the remaining schools—Stoics, Aristotelians, and many Platonists—is a good deal of overlap on the broad strokes of how the world is composed, and this consensus is so widespread that often it is impossible to tell with which (if any) school a particular scientific author has an affiliation. Many Roman-era authors in fact betray no particular school affiliation, but instead a general agreement on the standard philosophical and theological tropes of the age, coupled with their own individual elaboration of finer points when relevant. So the first-century-AD encyclopaedist Pliny the Elder begins his *Natural History* with a paean to a cosmic divinity that might have come from the pen of almost any educated Roman:

> The world, *this* (according to whatever other name you want to call the heavens by which everything is embraced round), is rightly believed to be a god, eternal, immeasurable, never born nor ever perishing.... It is sacred, eternal, immeasurable, everything in everything. Indeed it is itself the everything, finite but as though infinite, certain in all things but as though uncertain, the whole within and without encompassed in itself, both the product of the nature of things, and the nature of things itself.[17]

This is not to impute every detail of this account to every Roman, and indeed even within what I have elsewhere called the 'concentric' schools of Stoicism, Aristotelianism, and middle Platonism there is considerable room for fine-tuning,

[15] I have argued this point at length in Daryn Lehoux, *What Did the Romans Know?: An Inquiry into Science and Worldmaking* (Chicago: University of Chicago Press, 2012), esp. chap. 8.

[16] J. M. Dillon and A. A. Long (eds), *The Question of Eclecticism* (Berkeley, CA: University of California Press, 1988); David Sedley 'Philosophical Allegiance in the Greco-Roman World', in *Philosophia Togata*, ed. Miriam T. Griffin and Jonathan Barnes (Oxford: Oxford University Press, 1989), pp. 97–119; A. A. Long, 'Roman Philosophy', in *The Cambridge Companion to Greek and Roman Philosophy*, ed. David Sedley (Cambridge: Cambridge University Press, 2003), pp. 184–210; Christopher Gill, 'The School in the Roman Imperial Period', in *The Cambridge Companion to the Stoics*, ed. Brad Inwood (Cambridge: Cambridge University Press, 2003), pp. 33–58; Lehoux, *What Did the Romans Know?*

[17] Pliny, *Natural History* 2.1.1–2.

but any educated Roman would immediately recognize the assertions and motivations at play in this passage, and (again, with the exceptions of Epicureans and Pyrrhonists), would have found much to agree on. What is particularly striking for our present purposes is the emphasis on the rationality and the divinity of the cosmos as a whole.

We find this point emphasized again and again in Roman sources: Cicero, an Academic sceptic, finds something very like it 'most compelling' at the conclusion of his *On the Nature of the Gods*; the greatest physician and polymath of his own day, Galen, sees a purposive divine agency behind the flawless design of human and animal bodies; Seneca, the author of an extended treatise on physics that has been the subject of considerable scholarly attention in recent years,[18] again sees divine rationality as central to how the world works and is held together; Manilius, the great Augustan astrological author, sees knowledge of the heavens passed down to the first astronomers by God himself in what appears to be an act of divine revelation; Ptolemy fits the science of harmonics into a causal hierarchy with God as a cause of being at the top, and his fellow (and possibly contemporary) harmonic theorist Aristides Quintilianus goes so far as to say that the harmonies we perceive in nature were devised and created by a rational and unified divinity. At the outset to his *Almagest,* Ptolemy says that 'the first cause of the first motion of the universe... can be thought of as an invisible and motionless deity', gesturing back, one suspects, to Aristotle's account of the prime mover.

What we also find pervasively sown throughout Roman science is a conception of nature as law-like, which conception frequently finds itself rooted in the idea of God as divine lawmaker and ruler of the cosmos. So when Vergil says that 'nature has always imposed laws and edicts' to create natural regularity,[19] we *could* try to make the case that the active verb 'imposing' is merely metaphorical language, but the problem is that such talk is ubiquitous in accounts of the law-likeness of nature in antiquity, and its ubiquity should make us wonder whether there isn't more to it than mere metaphor. More importantly, in the (not infrequent) instances where we find the details fleshed out explicitly, we see the deliberate invocation of active divinity and we are forced to recognize that divine governance is in fact the underlying explanation, in a wide range of Roman philosophers, for nature's regularity.

[18] See e.g., Francesca Romana Berno, *Lo specchio, il vizio, e la virtù* (Bologna: Pàtron Editore, 2003); Brad Inwood, *Reading Seneca* (Oxford: Oxford University Press, 2005); Harry M. Hine, 'Rome, the Cosmos, and the Emperor in Seneca's *Natural Questions*', *Journal of Roman Studies* 96 (2006): pp. 42–72; Hine, 'Seneca's *Naturales Quaestiones* 1960–2005 (Part 1)', *Lustrum* 51 (2009): pp. 253–329; Hine, 'Seneca's *Naturales Quaestiones* 1960–2005 (Part 2)', *Lustrum* 52 (2010): pp. 7–160; Hine, *Lucius Annaeus Seneca: Natural Questions* (Chicago: University of Chicago Press, 2010); Lehoux, *What Did the Romans Know?*; Gareth D. Williams, *The Cosmic Viewpoint: A Study of Seneca's Natural Questions* (Oxford: Oxford University Press, 2012).

[19] Vergil, *Georgics* 1.60–1.

Divinities are frequently said to 'govern' or 'rule' over the world actively. Thus Cicero:

> So I perceive that it has been the opinion of the wisest that law has not been invented by the minds of men nor is it some kind of decree made by peoples, but something eternal, which rules the whole cosmos by the wisdom of its commands and prohibitions. Thus they say that this first and final law is the mind of God, compelling or forbidding everything by means of reason.[20]

Or Seneca, where we see both creation and maintenance:

> Nor did [the ancients] believe that Jupiter throws lightning-bolts with his hand, like the one we worship on the Capitol and in other temples. They recognize the same Jupiter as we do, the ruler and guardian of the universe, the mind and breath of the cosmos, the master and the craftsman of this creation, for whom every name will be appropriate .... You wish to call him nature? You will not be wrong: he it is from whom everything is born, by whose breath we live. You wish to call him the cosmos? You are not mistaken: for he himself is all that you see, contained in his own parts, sustaining both himself and his creation.[21]

This latter passage is doubly interesting, as Seneca begins it with his assertion (repeated several times in the *Natural Questions*) that the ancients didn't really believe the silly stories about the gods that had been bandied about by the poets (to which we might add the rather surprising observation that we don't, in fact, have much evidence that he was wrong).[22] We see Jupiter called 'nature' and 'cosmos', and are told that his role is as sustainer, container, genitor, master, mind, and breath of the cosmos, its ruler and guardian.

The gods are frequently characterized as running the cosmos 'for the best' and this for the simple reason that they care about us and about the world as a whole. Thus in Cicero's *On the Nature of the Gods*, the Stoic character Balbus sees divine providence behind the clever construction of the cosmos, from the stars on high down to the lowliest plants and animals on earth. Galen, in his great paean to the divine and beneficent goddess Nature, gives this utterly charming proof:

> Let me tell you what I felt the first time I saw an elephant .... In the place where other animals have a nose, the elephant has a narrow, free-swinging part, so long that it touches the ground. When I first saw this, I thought it superfluous and useless, but when I saw the animal using it like a hand, it no longer seemed

[20] Cicero, *Laws* 2.8.

[21] Seneca, *Natural Questions* 2.45. Translation modified slightly from Hine, *Natural Questions*.

[22] For my argument that poetic genealogical accounts may well have been seen by contemporaries as fiction rather than being taken to be literal explanations of the creation of the world, see Daryn Lehoux, 'Creation Myths and Epistemic Boundaries', *Spontaneous Generations* 3 (2009): pp. 28–34; Lehoux, *Ancient Science* (New York: Wiley, forthcoming). Compare also Paul Veyne, *Did the Greeks Believe in Their Myths?*, trans. Paula Wissing (Chicago: University of Chicago Press, 1988) [French orig. *Les Grecs, ont-ils cru à leurs mythes?* (Paris: Editions du Seuil, 1983)].

> so . . . . The elephant handles everything with the end of this part, folding it around what it receives, even the smallest coins, which it gives to its riders by stretching up to them its proboscis—for that is what they call the part of which we are speaking . . . . Now, since the animal performs the most useful actions with it, the part itself is shown to be useful, and Nature to be skillful . . . . And when the elephant died and I dissected the channels leading from the apertures up to the root of the part . . . I admired the skill of Nature more than ever. When I also learned that in crossing a river or lake so deep that its entire body is submerged, the animal raises its proboscis high and breathes through it, I perceived that Nature is provident not only because she constructed excellently all parts of its body but also because she taught the animal to use them.[23]

One could go on and on in this vein, but the point is abundantly clear: if the conventional wisdom of the birth of naturalism is right, then it has a lot of explaining to do for why and how all this divinity crept so widely and pervasively back into accounts of nature just a few hundred years later.

DEMARCATION

Creationist accounts—and here I follow David Sedley in using the term 'creationist' in its broadest possible sense to refer to any account where the cosmos was made or shaped by some kind of superhuman agency[24]—obviously present a straightforward challenge to the conventional wisdom, but one that is also perhaps all-too-easily circumvented, as the defender of naturalism can simply fall back on the definition of science as non-supernaturally causal. If it is creationist, the argument would run, then it is not science. Indeed, a similar tack will be straightforwardly applicable to worlds that are seen as 'governed by' divinity, and ultimately to virtually any theological interaction with natural philosophy, and so we should perhaps outline our reasons for not trusting such a move in the first instance.

We have already seen an impressive list of ancient authors who invoke divinities in nature in one way or another: Ptolemy, Galen, Aristotle, Plato, Pliny, Cicero, Seneca, as well as a number of Presocratics. (I made the case for something going on in Anaximander above, but we could also even more straightforwardly add Thales, Anaxagoras, Anaximenes, Heraclitus, Empedocles, Diogenes of Apollonia, all the Pythagoreans, Parmenides—himself possibly even a priest of Apollo—and others).[25] Even Democritus did not deny the gods, and in fact

[23] Galen, *On the Usefulness of the Parts of the Body* 17.1, trans. M. T. May.

[24] Sedley, *Creationism and its Critics.*

[25] We have a single inscription from Parmenides's home town of Elea that mentions him by name. It reads: 'Parmenides, son of Pyres. Priest of Apollo the Healer, natural philosopher'. For commentary, see Lehoux, *Ancient Science*; publication of the inscription is in P. Ebner, 'An Ancient

the only Presocratics I can find who may have done so are the otherwise unknown 'Hippo' mentioned as *atheos*, 'atheist', in Simplicius, and also possibly Prodicus of Cos.[26]

Among these, we find some, like Anaxagoras, Empedocles, Plato, and possibly others, who account for the existence of the world as we see it by supposing that it was created or shaped from pre-existing matter by the action of a divinity. If the matter from which the cosmos is formed is sometimes said, as in Plato, to have been pre-existing, that does not entirely 'naturalize' the account, at least insofar as divine agency is still responsible for the shape and characteristics of the world. A supernatural entity of one sort or another is clearly interacting with the system, and 'the natural order' itself is seen to be non-self-starting. The chain of natural physical causation, that is, is seen as insufficient to explain its own beginning.

If we try now to use the natural–supernatural boundary distinction as the demarcation criterion for what will count as science, then clearly Plato, Empedocles, and Anaxagoras are going to fall on the non-science side. Some scholars may be happy with this exclusion, but then we have to find some way of telling the story of the history of the early sciences that excludes or marginalizes both the author of the profoundly influential *Timaeus* and the inventor of the famous four-element theory of physics. Earth, air, water, and fire, after all, occur nowhere together as a closed group before Empedocles. In biology, Empedocles also offers us the earliest version of something like a theory of evolution through natural selection, which, because he did not think the increasing organization of matter over time to have been self-starting, was ultimately guided by a divine force. Attempts to scrub the divinity from this account have been made in the past, but the increasingly evident prominent strain of mysticism and religious imagery that is emerging in Empedocles is no longer possible to ignore.

If we were to disbar these influential thinkers from science on these grounds, what then do we do with those who do not offer creationist accounts as such but who instead see divinity as immanent in the cosmos in some way? Is the role of a governing or guiding nature any less of a supernatural intervention than that of a creator? Indeed, the lines between initial creation and ongoing governance are often very blurry, and it is difficult to tell with some authors whether some form of creationism may not be lurking in the background. So in Galen, Nature is said to be skilful and provident, actively designing anatomical structures, but we have no idea whether this is meant as a full-blown creationist account or whether it aspires to something more subtle. In those who advocate divine governance but deny creationism explicitly, we find their reasons are not

Medical Centre Identified at Velia', *Illustrated London News*, 31 August 1963, pp. 306–7. Note that in Ebner, illustrations 2 and 4 are inadvertently switched. The photograph of the Parmenides inscription actually appears as figure 4, but its text and caption under figure 2.

[26] Simplicius, *Commentary on Aristotle's Physics* 23.24. On Prodicus, see Graham, *Early Greek Philosophy*, p. 861.

because there is something teleologically suspicious about divine creators but because their physics simply demands an eternally existent—and therefore uncreated—cosmos. So the Stoics argue that the universe periodically burns up as the moisture in it is consumed by the stars (a purely physical reaction) only to be reborn as order reimposes itself and the causal chain of cosmos-formation begins again. In their strictly deterministic physics, where identical causes have identical effects, where the cosmos unfolds in a rigorously predictable way, it would seem that divinity is not needed to kick-start the system. Nevertheless, they see God everywhere and the cosmos itself becomes a huge, divine, rational, and providential divinity.

Looking beyond creationism and divine governance we find a third model, a little different from either, where some divine force gets called upon when the explanatory chain of physical causation seems to need an originary push of some sort, but neither creationism nor divine governance is explicitly used to describe it. So in the *Physics* Aristotle posits an entirely naturalistic chain of causation for everything in the heavens and on earth, but is unable to find a way to have the stars eternally self-moving. Since they are neither alive nor divided into parts, and since he cannot accept that they move according to their own wills (for then they could stop of their own accord too), he needs some external cause to account for stellar motion. In fact, he believes that he needs an actor *external to the whole system* to make it work.[27] This is no mere side-problem affecting only one part of the cosmos, however. It is, after all, the eternal motion of the stars that keeps the four elements down here on earth from settling out into a stagnant heap with earth at the centre surrounded by concentric and unmoving spheres of water, air, and fire. Left to their own devices, where earth and water move to the centre of the cosmos, air and fire away from it, that is what the natural motions of the four elements would give us. It is only the constant stirring of the four sublunar elements by means of the whirling of the heavens that prevents this from happening, and the constant whirling of the heavens is what needs a boost of some sort from outside to keep it going. The entire dynamic of the cosmos, from the earth on up, then, needs some kind of external cause.

Here is where things start to get a little more esoteric. Because the first mover cannot itself be moved, Aristotle starts to employ increasingly abstract and philosophically technical considerations in order to describe its role. What we end up with is a first mover that is eternal, completely unified, and without parts, without magnitude, and external to the cosmos as a whole. Finally, when

[27] Aristotle, *Physics* 8 254b32–3; see also *On the Heavens* 300b22; *On the Movement of Animals* 699a12; *Metaphysics* Λ.

he fleshes out the characteristics of the first mover in the *Metaphysics*, we find Aristotle at long last calling it what one has suspected all along: it is *divine*.[28]

We have already seen Ptolemy adopting something like this line in the *Harmonics* and the *Almagest*. This is all well and good, but if we found the exclusion of the creationists on the grounds of supernaturalism even a little worrying, we should be very concerned at this point, for we now have some of the biggest names in early science unabashedly touting divine causation in the cosmos, not just coincidentally or metaphorically, but deeply and centrally. What, we might ask, would the history of ancient, medieval, and even early modern science look like without Aristotle, Galen, and Ptolemy?

Clearly, the line from naturalism to science is not nearly as simple and straightforward as the conventional-wisdom account would have it.

CLEARER INSTANCES OF NATURALISM?

There are, however, two places in ancient science where naturalism in our sense seems to be more clearly and explicitly on the table and it will be worth looking at these before too-hastily concluding that naturalism is a complete non-starter as an issue in ancient science and philosophy. These are in Epicurean atomism, where the gods are explicitly said to be disinterested in (or possibly even incapable of) interacting with the cosmos, and the Hippocratic text *SD* (*The Sacred Disease*), where the author rails against quacks and incantation-healers in favour of what has often been championed as a purely naturalistic aetiology for the (now merely 'so-called') 'sacred' disease.

A fairly typical reading of *SD* can be found in Jouanna's (otherwise very good) *Hippocrates*. In it, Jouanna reads *SD* together with the opening of another Hippocratic text, *DG* (*Diseases of Girls*), as offering an entirely naturalistic account of an ailment that probably (though possibly not exclusively) maps onto the disease we now know as epilepsy. The author of *DG* (we have no idea if he was the same author as that of *SD*) says that those who make offerings to Artemis for relief of the seizures that sometimes attack girls at the onset of puberty are 'deceived'. Jouanna comments:

> The opposition between physicians and soothsayers hinted at here is radical. The seers believe in the divine origin of the disease and attribute it to the virginal Artemis. Furthermore, once the crisis is passed, they recommend that offerings be made to the goddess to thank her and to appease her, thereby preventing a recurrence of the attack. As against this advice, the Hippocratic author makes his own

[28] Aristotle, *Metaph*. 1074b1–2.

> recommendation: *unencumbered by taboos of morality or religion*, he urges the girl to marry as soon as possible in order that the obstacle preventing the blood from draining out be removed. The attack of the physician on the soothsayers, *whom he accuses of misleading the patient and her family*, is brief but vehement, and gives us a sense of how bitter the rivalry between physicians and soothsayers at the patient's bedside must have been.[29]

There are two aspects of this reading I am trying to draw out in the italics. First is just how strange the phrase 'unencumbered by taboos of morality' is in this context. After all, what the physician is offering us as a 'naturalistic' explanation is that the girl's menstrual blood needs to be prevented from blocking itself up in the diaphragm. The cure for this is marriage and pregnancy as quickly as possible after the onset of her first menses (the same recommendation is offered elsewhere in the Hippocratic corpus as a means of preventing the womb from wandering around the body and suffocating the girl: the frequent application of the husband's semen keeps the womb from drying out and then going in search of moisture elsewhere in her body).[30] Far from being morally unencumbered, this is in fact a *spectacular* instance of the 'naturalization' of cultural norms, where (male) Greeks were quite happy to find justification *in nature* for their common—and we would now say quite objectionable—cultural practice of marrying their female children off as young as twelve or fourteen years old, usually to much older men.

The second aspect of this passage is that Jouanna's claim that the author of *DG* accuses the soothsayers of (apparently deliberately) misleading the women reads more into the passage than may be warranted,[31] and a more careful reading takes some of the wind out of the sails of Jouanna's larger argument about the bitterness of the rivalry at the bedside. Here is what the text of *DG* actually says:

> When the girl regains her reason it is to Artemis that the women dedicate many offerings ... on the recommendation of the soothsayers, but they are deceived.[32]

Notice what is not said here: the author does not say that the soothsayers mislead, but only that (some) people in the story are misled. The participle for 'being misled' is indeed feminine,[33] and so would seem to apply primarily to the

[29] Jacques Jouanna, *Hippocrates*, trans. M. B. DeBevoise (Baltimore, MD: Johns Hopkins University Press, 1999), p. 183. Emphasis mine.

[30] *Diseases of Women* 1.

[31] Perhaps he is following Lloyd, *Magic, Reason, and Experience*, p. 29, although Lloyd's position developed into something more nuanced in later works.

[32] *Diseases of Girls* 1, translation modified from DeBevoise, who translates *exapateōmenai* as '*completely* deceived'—certainly this is a possible reading, but the extra force implied by 'completely' is not necessarily inherent in the Greek, and for the sake of the present discussion it is perhaps best left as an open question.

[33] There are variations in the spelling across MSS, but they do all seem to agree on the feminine ending.

women, but it is a not-uncommon feature of Greek adjectival expressions that they agree in number and gender with the most important actors in a sentence even when they apply to all. Everyone in this sentence, the soothsayers *and* the women, may be the deceived parties. If we decline to read a deliberate conspiracy to mislead on the part of the soothsayers into this passage, then what we have is not an accusation of villainy but a much more passive assertion, that the correct explanation is the one our author offers. That is to say, without bringing considerably more evidence to bear on this passage it is impossible to read some great physician–soothsayer bedside rivalry into it.

When we do cast around for outside evidence of the relationships between physicians and Greek religious practice, many scholars have been finding quite the opposite of rivalry. The cults of Asclepius and other healing deities, for example, see their rise and spread in concert with that of Hippocratic medicine, and there is increasing evidence that the two were seen throughout Greek antiquity to work together fruitfully. We now know that physicians had ritual responsibilities within Athenian cult practice, for example, and we have numerous dedications within Asclepian shrines that come from both doctors and their grateful (or hopeful) patients. So Nutton:

> All this suggests that notions of hostility between human and divine healers have been much exaggerated. The author of *Sacred Disease*, far from denying the gods any role in healing, professes his own high notion of piety towards them. He approves of the types of purification offered by most shrines, whereby the divinity washes away the stains, faults, and impieties of everyday life. He even appears to concede the propriety of making a dedication to a god if one's condition were the result of some divine punishment... although, of course, his investigation into epilepsy and mania has convinced him that these conditions have a purely natural cause.[34]

This is a radically different reading of the argument in *SD* than we find in many authors, but a very sensible one when we take off the blinkers of our own prejudice, where science and rationality must be purely naturalistic. If we allow the author of *SD* his own terms and categories then he is in a position not unlike that of a modern physician faced with an anti-vaccinationist: there is an assertion that one group of people ('quacks and magicians,' but never in *SD* 'religious people') misunderstand the evidence, but nowhere an outright denial by the physician of the overarching premise (religious efficacy in the Hippocratic instance, comparable perhaps to our modern physician's recognition that some drugs on the market are inefficacious or overly risky, just not the commoner vaccines). Indeed, I would, in light of *SD*'s conclusion, even go so far as to soften Nutton's apparent conclusion that the sacred disease is said to have a 'purely' natural cause (though to be fair, I suspect that given his argument

[34] Vivian Nutton, *Ancient Medicine* (New York: Routledge, 2004), pp. 111–13.

throughout his chapter Nutton did not mean it in our current sense). Here is how the author of *SD* words it:

> The disease that they call 'sacred' comes about from the same causes as the other diseases: from ingestions and evacuations, and from cold, the sun, and the changing winds that never rest. These things are divine, so that it is not necessary to think this disease should be set apart as more divine than the rest. Instead all are divine, and all are human.[35]

All are divine, and all human. That there is a natural cause to the sacred disease is not to deny divinity or divine agency in medicine, just to argue that we know what physical conditions in the body cause or bring on this particular disease, and that we can then try to counteract them physically to cure the patient. Nowhere—nowhere—is there a denial of the gods in medicine, as Nutton so forcefully shows.[36]

Turning now to the other candidates for 'pure' naturalism, the Epicurean atomists, we find that, finally, a good case can be made. (A caveat: what follows cannot be taken to apply straightforwardly to earlier forms of atomism—Democritus and Leucippus—where our evidence is more murky.[37] For present purposes I will stick to where the evidence is clear, which is to say, with Epicurus and Lucretius.) For the Epicureans, the cosmos was both infinite and eternal. They argued that all observable matter, all observable processes including life itself, could be accounted for by the interactions of invisible little particles called, from their indivisibility, *atoms*. These atoms whizzed around at inconceivable speed in an unlimited void, and their interactions were entirely due to collision and rebound ('entirely' needs some qualification, but for now it makes the point).

There were several aspects to Epicureanism that made it a radical outlier to the other Hellenistic and Roman schools which, as was remarked earlier, found much to agree on. One is the acceptance of an infinite cosmos. Most (if not all) other schools seem to have settled on a finite spherical cosmos, ending at the sphere of the fixed stars. Another radical aspect of Epicureanism was its claim that there was continuous void in the cosmos. No other school outside of some medical theorists and a (related) handful of mechanists and post-Aristotelians seems to have thought void could possibly exist within the cosmos, and even those believed only in tiny 'microvoids' (analogous, perhaps, to the little holes in sponges) that accounted for compressibility of some materials as well as

[35] *Sacred Disease*, 12.

[36] Nutton, *Ancient Medicine*. One could further point to the roles of prayer in other Hippocratic texts such as *Regimen IV*, to the praise of the creator in one of our most impressive Hippocratic anatomical texts, *The Heart*, or to the assertion that all of nature is 'set in order' by god in *Regimen I*, among many other passages.

[37] For what it is worth, Democritus does espouse praying for 'lucky' images in one fragment. Sextus Empiricus, *Adversus mathematicos* 9.19–24.

air and steam pressure within mechanical devices.[38] The other widespread objection to Epicureanism, however, was considerably more serious. It was theological, and it plays directly into the current question about naturalism.

The Epicureans argued that it would be beneath the dignity of the gods to concern themselves with the petty affairs of mortals, and so they set up a theology that saw the gods living in perfect bliss in the interstices between the many worlds scattered throughout the infinite void of the universe. These gods paid no attention whatsoever to our lives or to our prayers and entreaties for help. Although they always insisted that the gods existed, Epicureans simultaneously denied that the gods cared in any way about us, or that they ever interacted in the running of the world. Indeed, if Lucretius is to be believed, it is precisely the fear of divine retribution and meddling that Epicureanism was meant to cure. So far, the Epicurean cosmos appears to cleave to the ideal of those who would seek a 'pure' naturalism in antiquity.

Why, we might then ask, do the gods need to exist at all, if they never interact with the world, never answer our prayers, never show themselves in any way? After all, for the vast majority of ancient thinkers, the proof of the existence of the gods was to be found in the natural order and beneficence of the cosmos.[39] If the order of the cosmos had nothing to do with divinities (and indeed in Epicureanism was not seen as beneficent at all), how could we prove that the gods existed in the first place? The Epicurean answer was more of an assertion than a proof, but it seems to have been sincere for all that: the gods stood as moral exemplars for us to follow. Their perfect happiness and untroubled state was something we could emulate down here on earth as *the* ethical ideal.

Now, I said earlier that there was a qualification to be made to the word 'entirely' in my assertion that the interactions of Epicurean atoms were 'entirely' due to collision and rebound. This is not quite true. In Lucretius, we find a very interesting argument that seems to stem from the (not always clear) Epicurean argument for falling bodies. In most other ancient physics, where the universe is finite and spherical and the earth sits at the centre, falling bodies are not a problem at all: they simply move naturally toward the centre of the cosmos. But in the infinite and therefore centreless Epicurean cosmos this will not work. The solution seems to have been to assert the primacy of a single direction called 'down', toward which all atoms in the universe move naturally unless they are otherwise impeded or pushed. This presents a problem for Lucretius, insofar as there is no reason for atoms to interact if they are all simply falling at the same speed and in the same direction. All we have is an infinite shower in parallel, and therefore non-interacting trajectories. Lucretius's solution—and it

[38] See Sylvia Berryman, *The Mechanical Hypothesis in Ancient Greek Natural Philosophy* (Cambridge: Cambridge University Press, 2009); Daryn Lehoux, 'All Voids Large and Small', *Apeiron* 32 (1999): pp. 1–36.

[39] An excellent recent study of this is Clifford Ando, *The Matter of the Gods* (Berkeley, CA: University of California Press, 2008).

likely goes back to Epicurus—is to posit an (occasional? one-time?) random *swerve* to the trajectory of an atom, which causes it to incline in a different direction *for no reason whatsoever* and thus begin a chain-reaction of atomic collisions in atoms that would otherwise have just kept falling.[40] That the swerve is said to be 'uncaused' we have from Cicero's *De fato*, and the implication seems to be that, for Epicureans as for so many others, physical causation is not enough to explain all that we see in the universe. But where other schools brought in some form of divine agency as a solution, the Epicureans opted instead for randomness.

That this is a pretty serious compromise can be seen from how radically it breaks from the physicality of the rest of Epicurean physics, as well as from the widespread contemporary belief that physics should employ no uncaused motions. (Indeed in the most widespread physics of the Lucretius's day, Stoicism, the universe was just the opposite of random: it was instead completely deterministic, all actions strictly fated.)

Nevertheless, the Epicureans seem to have put a priority on natural explanations, and the swerve is certainly not what *we* would call supernatural, but it could well be argued that, as an uncaused motion, contemporaries would have seen it as at least *non*-natural, which may be an interesting complication in light of the question of this volume: is non-natural causation any more naturalistic than supernatural causation?

INSCRUTABILITY

Part of what drives that last question is my worry about a word that gets bandied about, almost invariably, in the common-wisdom accounts of naturalism that I have been challenging in this chapter. That word is *anthropomorphic*. Commentator after commentator who wants to see naturalism at the heart of the birth of philosophy and of science says that what is being denied in ancient philosophy are the causal roles of *anthropomorphic* deities. We have already seen Algra deploy this qualification in the opening section of this chapter, but it is not difficult to find other instances:

> These 'theologians' were presumably, like Hesiod, prepared to invoke divine, probably anthropomorphic, agents in their accounts. Such agents are notably absent from early Ionian cosmology.[41]

[40] On the import of the phrase *sponte sua* in describing the swerve's motion, see Monte Ransome Johnson, 'Nature, Spontaneity, and Voluntary Action in Lucretius', in *Lucretius: Poetry, Philosophy, Science*, ed. Daryn Lehoux, A. D. Morrison, and Alison Sharrock (Oxford: Oxford University Press), pp. 99–130.

[41] James Warren, *Presocratics* (Berkeley, CA: University of California Press, 2007), p. 24.

> There are two characteristic features of the Homeric-Hesiodic world-view that are of leading significance for a study of the 'origins of science'. These two features are connected, though one of them is very obvious and one rather less so. The obvious feature, overwhelmingly so to a modern reader, is the *centrality of anthropomorphic gods* [emphasis in original] (especially the Olympian gods) in the world. The less obvious feature is the *finitude* and the *vagueness about the limits*... of the world.[42]

> Prior to science,... when they attempted to explain their world, it was in terms of myths and anthropomorphic gods.[43]

Why should anthropomorphism be an issue so consistently? No one really elaborates, but I suspect that the problem has much to do with an attempt on the part of modern commentators to distinguish many of the theological commitments we have now seen on the part of ancient authors from the many other accounts of divinities circulating in antiquity (most obviously in myth and folklore). What I mean to say is that for all that I have tried to problematize and qualify naturalism, there is still an important point to be made that the gods of the philosophers are—universally, it would seem—not the same as the gods we think of when we think of Greek mythology. What I think anthropomorphism is meant to flag is just this distinction, but in a qualified way. No one today would have much riding on what kinds of pictures any particular Greek may or may not have painted of his or her gods, which is to say that the issue is not one of idolatry or some such, but is instead more akin to what Xenophanes was worried about: the philosophers did not, it seems, believe the gods acted capriciously. This is surely correct, so far as it goes, and is, I suspect, the intuition at the heart of the universal modern emphasis on the rejection of anthropomorphism.

This point becomes interesting when we try to open it out and ask what it is about capriciousness that is so objectionable. After all, when the agent is seen as a real one (as in dogs and humans), otherwise capricious action is not inherently or necessarily seen as unnatural. Indeed, much of twentieth-century biology, psychology, and behavioural science has been dedicated to disenchanting and naturalizing apparent capriciousness by reducing it to what we believe to be understandable and qualifiable (if normally hidden) structural, evolutionary, or instinctual motivations. Of course, the wills of the gods are not subject to such reduction (nor to double-blinded empirical study) and so their capriciousness remains a kind of 'pure' capriciousness, which is to say that their wills remain inscrutable, unpredictable, and from where we stand, apparently random. If such capricious gods are allowed to interfere in the cosmos, then we end up in the situation of the old Sidney Harris cartoon where a miraculous occurrence shows up as a crucial step in an otherwise sound mathematical proof (see Figure 1.1).

[42] Edward Hussey, 'The Beginnings of Science and Philosophy in Archaic Greece', in *A Companion to Ancient Philosophy*, ed. Mary Louise Gill and Pierre Pellegrin (Oxford: Blackwell, 2006), p. 3.

[43] Andrew Gregory, *Eureka! The Birth of Science* (Duxford: Icon Books, 2001), p. 1.

Figure 1.1. Sidney Harris on Method. © ScienceCartoonsPlus.com

The universe becomes unpredictably arbitrary in ways no naturalist should be willing to bear.

If I am right in unpacking the anthropomorphism qualification in this way, then something interesting has happened to our story. Where the ancient philosophers who invoke divinities do so, nearly universally, to account for nature's regularity, complexity, and beauty, the one school we have found to be pure naturalists, the Epicureans, are the one school who allow for just this kind of capriciousness—the random, uncaused swerve—in the cosmos. It may not be a personalized kind of capriciousness (just the opposite), but for all that it is exactly the kind of explanation that 'pure naturalism'—if such a thing even exists—was designed to avoid.

2

Naturalist Tendencies in Medieval Science

Michael H. Shank

The region between the Atlantic and the Indus Valley during the medieval period may seem a spectacularly barren domain for an examination of naturalism.[1] Many readers will stereotypically associate these space-time coordinates not merely with religion, but with the overweening dominance of religion, be it Christianity, Islam, or Judaism. After all, is this not the era in which priests, monks, mullahs, qādīs, and rabbis ruled their communities with their revealed scriptures in hand? And, as a famous quip misleadingly puts it, was not theology the 'queen of the sciences'?[2] A deep probe into the blogosphere is unnecessary to find claims that the religions of the book all accept miracles and therefore deny naturalism (i.e. are supernaturalist). The case is closed, and the inquiry is finished.

Or is it? Some of the historical evidence I present in this chapter challenges the standard expectation. Instead of surveying the entire medieval era, I illustrate a few late-medieval approaches to naturalism, primarily in Latin Europe roughly from the twelfth century to the fifteenth, with occasional examples pertinent to Judaism and Islam in that period.[3] After a brief introduction, my approach is initially institutional, then intellectual.

[1] For China, see Joseph Needham, *Science and Civilization in China*, vol. 2: *History of Scientific Thought* (Cambridge: Cambridge University Press, 1956). Needham famously associated Chinese naturalism with Taoism, Buddhism, and neo-Confucianism. See also Janghee Lee, *Xunzi and Early Chinese Naturalism* (Albany: State University of New York Press, 2005).

[2] On the curious roots of the expression, see Luca Bianchi, *Il vescovo e i filosofi: La condanna parigina del 1277e l'evoluzione dell'aristotelismo scolastico* (Bergamo: Pierluigi Lubrina, 1990), p. 111; Michael H. Shank, 'Natural Knowledge in the Latin Middle Ages', in *Wrestling with Nature: From Omens to Science*, ed. Peter Harrison, Ronald L. Numbers, and Michael H. Shank (Chicago: University of Chicago Press, 2011), pp. 83–115, esp. pp. 104–5.

[3] A valuable discussion of miracles and the natural order appears in Stephen McCluskey, 'Natural Knowledge in the Early Middle Ages', in *The Cambridge History of Science*, vol. 2: *Medieval Science*, ed. David C. Lindberg and Michael H. Shank (New York: Cambridge University Press, 2013), pp. 286–301, esp. p. 289.

First, I show how the new universities, thanks to their structural division into faculties, extended and reinforced approaches akin to what we now call methodological naturalism. This development did not occur on the sly or in the minds of a few individuals. Naturalistic explanation was embedded by statute in the very organization of the new universities. It thus became deeply entrenched and solidly institutionalized, not only in centres of learning but also in the wider culture of literate Europe.

Quantitatively and qualitatively, the thirteenth-century faculties of arts (arts and medicine in Italy) became the primary sites and main drivers of medieval naturalism, which diffused to the other faculties. Already since the twelfth century, natural philosophers took for granted that explanations of natural phenomena appealed to reason and sense perception within an 'ordinary course of nature'. As a matter of practice, they ruled out appeals to the supernatural as illegitimate. The programme of naturalist explanation in the faculty of arts thus benefited from the extraordinary success of the university and diffused with it. The faculties of arts (and medicine) of dozens of universities founded roughly between 1200 and 1500 exposed hundreds of thousands of students to such an outlook.[4] Those numbers help to explain how something akin to methodological naturalism became a leading component of the late-medieval intellectual landscape.

The masters of arts' vision of their methodological autonomy was not restricted to their own faculty. Strikingly, this trend was endemic among theology students themselves, most of whom had earned a master of arts degree and continued to teach natural philosophy in the arts faculty while earning their theology degrees. What is more, they carried their methodological naturalism into their theological work and writing. Encapsulating this attitude is the expression *de naturalibus naturaliter*, '[to treat] the natural naturalistically', first attributed to the master of arts and theologian Albertus Magnus.

Before turning to the main argument, a brief disclaimer. I do not intend to suggest a necessary linkage between science and naturalism. However much they may overlap, the connection between them is contingent, as one of the most impressive scientific achievements of antiquity illustrates. The Babylonian creators of the first predictive mathematical astronomy were engaged in celestial divination. Their goal was to decode signs from the gods, an objective that motivated their systematic empirical and theoretical inquiries into the heavens. However counter-intuitive to us, these motivations underlay their mathematical methods of predicting the changing velocity of the sun in the zodiac, the

[4] 'In Central Europe alone, from the middle of the fourteenth century to the end of the Middle Ages, almost a quarter of a million individuals acquired the status of student. A conservative calculation suggests approximately that three-quarters of a million students attended university in the whole of Europe over the same period.' Rainer Christoph Schwinges, 'Admission', in *A History of the University in Europe*, vol. 1: *Universities in the Middle Ages*, ed. Hilde de Ridder-Symoens (Cambridge: Cambridge University Press, 1992), pp. 171–94, esp. p. 181.

appearance of the first lunar visibility after the new moon, the recurrence of the stationary points in the retrogradations of the planets, and so on. The reason why naturalism had nothing to do with their achievements is simple: they seem not to have had a concept of (what we call) nature as an autonomous realm, distinguishable from the divine.[5] Something like that distinction seems to take off in several venues during the sixth to fifth centuries BC. It appears in the sparse opening lines of Genesis (the Priestly source), among a handful of Presocratic thinkers that Aristotle will call *physikoi* (literally 'naturalists', thinkers about nature, natural philosophers), and in some works of Hippocratic medicine.[6]

Thus, I focus on naturalist themes in late-medieval Europe in particular not to foist an ingredient of modern science onto the Middle Ages, but to highlight historical evidence that is contrary to stereotype: naturalist attitudes were already endemic and widespread and, for the most part, uncontroversial in late-medieval learned culture in particular (whether Christian, Muslim, or Judaic).

THE TWELFTH-CENTURY BACKGROUND TO MEDIEVAL LATIN NATURALISM

A crucial ingredient in this attitude was genetic: the scientific tradition had come from elsewhere. Thus, among medieval intellectuals, understandings of the natural world often went hand in hand with an attitude that was implicitly comparative and relativist. Whatever their religion, they were keenly aware that they were not the first to have thought systematically about the natural world or to invent tools to do so. They knew well that their predecessors—including assorted pagans, 'non-believers', or infidels—had impressive achievements in this area. This is a non-trivial point. Most medieval thinkers did not assume that shared theological beliefs were a prerequisite for understanding the natural world or for great acumen in explaining it. Overwhelmingly, disagreements about creed were irrelevant to the cogency of arguments about the constitution and functioning of the natural world. In twelfth-century Muslim Spain, the qādī and natural philosopher Ibn Rushd (Averroes) considered the pagan Aristotle to be the supreme human intellect. So did his disciple Ibn Tufayl. In twelfth-century Cairo, the physician, natural philosopher, and rabbi Maimonides adopted a similar attitude, often offering Aristotelian solutions to

[5] Francesca Rochberg, 'Natural Knowledge in Ancient Mesopotamia', in Harrison, Numbers, and Shank, *Wrestling with Nature*, pp. 9–36, esp. pp. 9–24; Giovanni Manetti, *Theories of the Sign in Antiquity*, trans. Christine Richardson (Bloomington, IN: Indiana University Press, 1993), chap. 1.

[6] G. E. R. Lloyd, 'The Invention of Nature', in *Methods and Problems in Greek Science: Selected Papers*, ed. G. E. R. Lloyd (Cambridge: Cambridge University Press, 1991), pp. 417–34, esp. pp. 418–22.

the perplexed for whom he wrote his *Guide*. In early twelfth-century France and England, the natural philosopher Adelard of Bath referred freely to his 'Arab masters'. By the time Averroes had been translated into Latin, he had come to stand for the autonomy of natural philosophy.[7]

For the three religions in the Abrahamic tradition, the key factor associated with the vicissitudes of naturalism was their exposure to Greek natural philosophy, mathematics, and logic. In what follows, I discuss primarily natural philosophy. One should not, however, underestimate the importance of mathematics and logic in forming naturalistic views. Being formal, both mathematics and logic in the Greek tradition rely on rules of thinking, reasoning, and proof that are general and usually independent of content. To the extent that various types of inquiry into the natural world either were designated explicitly as mathematical (typically astronomy, optics, and music) or drew systematically on logic (e.g. natural philosophy's reliance on the ideals of demonstration or inference), they participated in the neutrality of logic and mathematics. These formal approaches implicitly introduced into the disciplines in which they were used the idea of rules (*regulae*) and rule-following, that is to say, notions of regularity (or, rule-likeness) and law-likeness.[8]

Regardless of their religion, leading medieval thinkers treated this type of rationality (in the generic sense of 'using reason') as characteristic of humanity in general, and therefore as non-credal. Accordingly, Avicenna and Averroes could not, and did not, dismiss the impressive views of thinkers from Plato and Aristotle to Ptolemy merely for not being co-religionists, nor did their Latin successors dismiss the foregoing for being pagan or Muslim.

This relative autonomy of inquiry into natural philosophy, if not nature, was not new. Although he was no natural philosopher, Augustine (fifth century) had bluntly warned Christians not to make fools of themselves by arguing about the natural world, scripture in hand, against 'non-Christians' whose views were supported by 'most certain reason and experience'.[9] The latter, not

[7] *Ibn Tufayl's Hayy ibn Yaqzān: A Philosophical Tale*, trans. Lenn Evan Goodman (New York: Twayne, 1972); Moses Maimonides, *The Guide of the Perplexed*, trans. Shlomo Pines (Chicago: University of Chicago Press, 1963), e.g., p. 326 for sublunar physics; *Adelard of Bath: Conversations with His Nephew: On the Same and the Different, Questions on Natural Science and on Birds*, ed. and trans. Charles Burnett (Cambridge: Cambridge University Press, 1999), pp. 102–3; Stephen Gaukroger, *The Emergence of a Scientific Culture: Science and the Shaping of Modernity, 1210–1685* (Oxford: Oxford University Press, 2006), p. 101.

[8] Daryn Lehoux, *What Did the Romans Know? An Inquiry into Science and Worldmaking* (Chicago: University of Chicago Press, 2012), chap. 3.

[9] See his *Literal Meaning of Genesis* (bk. 1, chap. 19, sec. 39): 'It often happens that even a non-Christian may know something about the earth, the heavens, the elements of this world, the motion and revolution, and even the size and spacing of the stars... so that he holds [this] by the most certain reason and experience. It is extremely shocking and dangerous and to be avoided at all cost that he [the non-Christian] hear a Christian talking deliriously (*delirare*) about these matters as if he were expounding Christian scriptures, presumably giving the meaning of Holy Scripture, talking nonsense on these topics.... [The non-Christian] will barely be able to keep from laughing.' Augustine, *De Genesi ad litteram libri duodecim*, in *Oeuvres de Saint Augustin*, vol. 48

scripture, justified explanations of nature, a view roughly consistent with his picture of an ordinary course of nature and—in the grand scheme of things—his decided lack of interest in miracles.[10]

Given enough time and ingenuity, it is possible in principle to link theological conclusions with anything. But deducing all claims about the natural world from revealed texts was a tall order. For starters, the Torah, the Bible, and the Qur'an were silent on most detailed questions about the natural order. Did concentric spheres or epicycles best explain the motions of the planets? Was vision caused by rays that enter the eye or that are emitted from it? Are matter and form adequate tools for analysing such phenomena as human generation or the metamorphosis of the caterpillar into a butterfly? Why do some shore birds sleep while standing on only one leg? As many medieval scholars knew, revealed texts, related credal statements, and theological conclusions were in practice irrelevant to answering specific questions in natural philosophy or the mathematical sciences.

As William of Conches noted in the twelfth century, Genesis 1 asserts the existence of various beings (the facts) without explaining how they came into being (the causes). William therefore insisted that, by supplying the missing causal explanations, the natural philosopher (*physicus*, literally 'physicist') could not possibly be doing anything contrary to Genesis.[11] When nothing is said, no contradiction is possible. Implicit in this point was an attitude akin to methodological naturalism. Although William surely did not doubt that God was the ultimate cause of the universe or that Genesis mentions the creative word of God, he claimed that Genesis did not discuss causes. Implication: the invocation of God is not a satisfactory causal explanation for specific details of natural philosophy. Accordingly, William of Conches praised Plato's *Timaeus* for offering better explanations than Genesis and for raising many more questions about the natural world.

(Paris: Desclée de Brouwer, 1972), pp. 136–7 (my translation). See also David C. Lindberg, 'Science as Handmaiden: Roger Bacon and the Patristic Tradition', *Isis* 78 (1987): p. 523.

[10] In several places, Augustine defined the miraculous subjectively, as what *we* take to be unusual. Augustine, *De Genesi ad litteram*, bk. 9, 17.32; *Contra Faustum*, bk. 26, 3. See David C. Lindberg, 'Science and the Early Christian Church', *Isis* 74 (1983): pp. 509–30, esp. p. 527; Catherine Wilson, 'From Limits to Laws: The Construction of the Nomological Image of Nature in Early Modern Philosophy', in *Natural Law and Laws of Nature in Early Modern Europe: Jurisprudence, Theology, Moral and Natural Philosophy*, ed. Lorraine J. Daston and Michael Stolleis (Farnham: Ashgate, 2008), pp. 13–28, esp. pp. 20–1; Colin Brown, 'Issues in the History of the Debates on Miracles', in *The Cambridge Companion to Miracles*, ed. Graham Twelftree (Cambridge: Cambridge University Press, 2011), pp. 273–90, esp. p. 275; Rowan Greer, *The Fear of Freedom: A Study of Miracles in the Roman Imperial Church* (College Station, PA: Pennsylvania State University Press, 1989), pp. 170–2.

[11] Alexander Fidora and Andreas Neiderberger, 'Philosophie und Physik zwischen notwendigem und hypothetischem Wissen: Zur wissenschaftstheoretischen Bestimmung der Physik in der *Philosophia* des Wilhelm von Conches', *Early Science and Medicine* 6 (2001): pp. 25–6, 31–2.

Significantly, William occasionally tweaked some 'factual' claims as well. He interpreted as allegorical the biblical claim about waters above and below the firmament, which he considered to be 'against reason'.[12] In a characteristically naturalistic move, William allegorized an apparently literal biblical claim to make it consistent with reason. For William, what gave way in this case was the literal biblical interpretation, not the reason-based claim he endorsed.[13]

William of Conches was in the vanguard of a new trend within natural philosophy, in which wonders took a 'hit' as well. Whereas most intellectual traditions between the twelfth and the fifteenth centuries 'set the passion of wonder in a highly positive light', the natural philosophical tradition, from Adelard of Bath onward, was exceptional. In one pithy characterization, the natural philosopher's task was 'to make wonders cease',[14] specifically by discovering their natural causes.

THE MEDIEVAL IMPORTANCE OF ARISTOTELIAN NATURALISM

To these formal considerations must also be added matters of content. Here, the importance of Aristotle for medieval naturalism is difficult to overestimate. At issue are not the fine details of Aristotle's self-understanding but rather how he was read, particularly by medieval intellectuals committed to one or the other religions of the book. One fundamental aspect of Aristotle is closely tied to the previous point about the linkage between formal reasoning and notions of regularity. Aristotle could not, of course, have any empirical evidence for the eternity of the world. It was a genuine principle, one that presumed the universe to be so well regulated that natural phenomena will proceed in the future precisely as they proceed now, which is precisely as they have been proceeding in the past. The system is in such complete equilibrium and regularity (*regula*, i.e. rule, that is to say, law-likeness) that it cannot 'go out of whack', as the vernacular puts it. The presence of Aristotle's Unmoved Mover at the end of the causal chain takes nothing away from this point, for its existence is reached inductively from considerations about causality and motion. The important thing to notice is that, even without making concessions to Aristotle's metaphysics or the eternity of the world, anyone who practised natural philosophy along the lines of the Philosopher was committed to a programme of explanation akin to what we now call methodological naturalism, fully naturalist in its domain of natural philosophy.

[12] See the literature cited in Shank, 'Natural Knowledge in the Latin Middle Ages'.

[13] It is a serious mistake to read the hardened literalisms of the Reformation and Counter-Reformation (to say nothing of American Fundamentalism) back into the late-medieval period, where they would have been considered bizarre.

[14] Lorraine J. Daston and Katherine Park, *Wonders and the Order of Nature, 1150–1750* (Cambridge, MA: Zone Books, 1998), p. 109, p. 126.

Some readers may find this point counter-intuitive since the development of early modern science is widely believed to have required the overthrow of Aristotle.[15] To be sure, many specific Aristotelian explanations were rejected, some of them already by the later Middle Ages (impetus theory, the possibility of a rotating earth, etc.). The fact remains, however, that mastering and imitating Aristotle's approach to explanation in general had a profound effect on medieval understandings of nature, and especially on the project of 'explaining the natural naturally' (*de naturalibus naturaliter*). The initial tensions between the Aristotelian system, on the one hand, and the religions of the book and Platonism, on the other, were non-trivial. I argue that the accommodations that eventually occurred took place precisely on the terrain of naturalistic explanation, which survived in a stronger position.

Many prominent thinkers committed to one of the three religions of the book would eventually adopt the naturalist explanations that Aristotle's writings embodied, especially in the ninth to fourteenth centuries in Islamic civilization, and the twelfth to fifteenth centuries in the Latin world and among Jewish intellectuals scattered throughout the Mediterranean world (see below). Before examining them, however, I turn to the institutional side of my argument, with a focus on the new Latin university.

The archetypical domain in which naturalist explanation thrived was the medieval university, which separated the faculty of arts from the faculty of theology. The flowering of this naturalism in the universities after the thirteenth century was not unprecedented. It had begun to emerge already in the twelfth century in towns that hosted some of the livelier schools that had preceded the emergence of the universities as corporations—notably Paris with its cathedral school and its many schools of 'arts', and Bologna with its schools of law.[16]

After the emergence of the universities, natural philosophers who worked in the faculty of arts took for granted that matters of faith got bracketed, and they concentrated on chains of proximate causes. It does not matter whether the authors who exemplified this type of naturalism in their natural philosophical work propounded elsewhere full-blown metaphysical or theological systems that left no part of nature untouched. Rather, the point is, how did these authors handle limited-scope explanations? And what did they consider to be adequate ones?

[15] Pierre Duhem, who put the history of medieval science on the intellectual map, argued flamboyantly that the Parisian Condemnations of 1277 marked the beginning of modern science. They pushed masters of arts to examine conclusions that contradicted those of Aristotle, such as the possibility of vacua, the plurality of worlds, etc. See Luca Bianchi, '1277: A Turning Point in Medieval Philosophy?', in *Was ist Philosophie im Mittelalter?*, ed. Jan A. Aertsen and Andreas Speer (Berlin: de Gruyter, 1998), pp. 90–110, esp. pp. 105–10.

[16] See Edward Grant, *A History of Natural Philosophy from the Ancient World to the Nineteenth Century* (Cambridge: Cambridge University Press, 2007), p. xi, chaps 9–10. This book refutes the bizarre thesis that medieval natural philosophy, indeed all of natural philosophy until around 1800, was de facto religious because, whether or not it mentioned God, it was implicitly tied to a picture of the natural world as the work of the Creator.

THE EMERGENCE OF THE UNIVERSITY AS AN AUTONOMOUS INSTITUTION

Before the emergence of the universities in the late twelfth century, ecclesiastical authorities in Europe ran almost all the schools.[17] Abbots and bishops controlled, respectively, monastic and cathedral schools.[18] By the mid-thirteenth century, however, universities had emerged as largely autonomous institutions. This was a momentous change, even if it had been several generations in the making. The self-governing corporations of masters now controlled their membership, their curricula, and the criteria for the degrees they awarded.

During the twelfth century, the masters who taught in the schools of law and arts in particular (in Bologna, the students as well) began to organize themselves into corporations (*universitates*). Paris illustrates this trend most pointedly. The many masters in the schools of that city were primarily teachers of the 'liberal arts'[19] who placed much weight on the fundamental discipline of logic—'the art of arts, and the science of sciences' in the immortal words of Peter of Spain. They fought for legal privileges that gave their guilds the rights of associating, determining the standards of competence, and selecting their members by examination.

Unlike most of its peers, the university in Paris emerged in a town with a bishop and it also taught theology. This coincidence was rare: most universities before the late fourteenth century did not have a faculty of theology, and most before 1500 would avoid episcopal sees. The thirteenth-century Parisian conflicts between the masters of arts and the bishop were therefore particularly acrimonious. Although many were unaffiliated with the cathedral school, the masters lived in its shadow and with the bishop's assertions of control. The inroads of naturalist thinking in such an environment are therefore particularly telling.

An early twelfth-century school incident foreshadowed the thirteenth-century university trend toward autonomy and significantly boosted the independence of masters in Paris. Peter Abelard, a brilliant master and outstanding logician, broke away from the cathedral school of Notre Dame, on the Île de la Cité in the Seine. Abelard moved to the left bank of the river, where students flocked to

[17] This section builds on arguments that Edward Grant pioneered in his 'Science and the Medieval University', in *Rebirth, Reform and Resilience: Universities in Transition, 1300–1700*, ed. James M. Kittelson and Pamela J. Transue (Columbus, OH: Ohio State University Press, 1984), pp. 68–102.

[18] Among the early exceptions were some free schools of arts in Paris, the schools of law in Bologna, and of medicine in Salerno.

[19] This venerable expression, which we still use, traditionally denoted the seven disciplines of the classical trivium (grammar, rhetoric, logic) and quadrivium (arithmetic, geometry, music, and astronomy). In the thirteenth-century universities, 'the arts' had already escaped from this taxonomy. The faculty of 'arts' not only covered vastly more than the old liberal arts (which omitted the expanding domain of natural philosophy), but also weighted the older disciplines unequally (logic predominated).

hear him teach. On the Mont Sainte Geneviève, just outside the city walls, he was presumptively beyond the reach of the chancellor of Notre Dame, the bishop's appointee, who sought to control all Parisian schools. In short, throughout the second half of the twelfth century, students and masters came to Paris in ever-increasing numbers. 'Free schools' independent of the cathedral school and the monastic schools proliferated.[20] This large cohort of masters, many of whom were not affiliated with the cathedral school, began to organize around their common interests and grievances, and to function as a guild. The overwhelming majority of masters in Paris were masters of arts, who fought the bishop for control of their teaching. They eventually secured royal privileges (*privi-legia*, that is, laws specific to the guild treated as a person/body). These privileges literally constituted the autonomy ('a law unto themselves') of the masters' guild or corporation (the original, generic meaning of *universitas*). In brief, the legal autonomy of the 'university' (corporation) of masters gave them control of their organization, from membership and standards to teaching and curriculum.

Whether independently constituted or consciously emulated, specimens of the *universitas* of masters cropped up throughout Europe. With them came also the naturalistic explanations characteristic of the natural philosophy at the core of the curricula in the faculties of arts. As Joan Cadden has summarized the matter:

> Before the twelfth century, then, natural philosophy had no stable constitution, no particular social or institutional support system, and no significant corpus of authoritative and authorizing texts; and by the middle of the thirteenth century natural philosophy was a clearly defined domain of learning supported by a well-articulated set of institutions and an established group of recognized texts.[21]

At Paris, the initial period of tensions with the bishop encompassed the 1210s to the 1230s, when attempts to teach the newly translated Aristotelian natural philosophy and commentaries on it by Muslim scholars sporadically incurred prohibitions that tested the resolve of the masters. By 1255, however, the full curricular effects of the masters' struggles for autonomy were stunningly evident. The official arts faculty curriculum now *required* the recently translated Greco-Arabic logical, philosophical, and mathematical writings, including the natural philosophy of Aristotle and his Muslim commentator Averroes (d.1198). This material was obviously independent of specifically Christian theological claims. On some questions, however, such as the doctrine of the soul, Aristotle's *Metaphysics* (i.e. his Theology), and the *De caelo* (creation), the subject matter seemed to overlap. But it is important to realize that this shared turf worried

[20] Stephen Ferruolo, *The Origins of the University: The Schools of Paris and their Critics, 1100–1215* (Palo Alto, CA: Stanford University Press, 1985), pp. 23–4.

[21] Joan Cadden, 'Science and Rhetoric in the Middle Ages: The Natural Philosophy of William of Conches', *Journal of the History of Ideas* 56 (1995): p. 2.

the natural philosophers less than the theologians, who had to confront the interface of their natural philosophical training and conclusions with their new studies.

The masters of arts' control of their curriculum, examinations, and requirements for degrees had important institutional consequences. A thorough grounding in the faculty of arts (usually a master of arts degree) became *the* prerequisite for study in the higher faculties of law, medicine, and theology.

THE FACULTY OF ARTS, INSTITUTIONAL SEAT OF NATURALISTIC EXPLANATION

The organization of the universities into distinct faculties set up a division of labour, which led to a policing of boundaries which, in turn, had important consequences for naturalistic explanation. Masters of arts and their students were to keep to their subject matter—logic, natural philosophy, and the mathematical sciences—and were not to tread on theological terrain. The views and attitudes of the philosophers thus flowed mostly unidirectionally into the higher faculties with little reflux back into the faculty of arts.[22]

To grasp this point, one must first confront the widespread misunderstanding that the medieval university primarily taught theology. In fact, the overwhelming majority of matriculants never studied the subject. Most universities had no faculty of theology before the later fourteenth century. Even then, however prestigious their faculty, theologians were vastly outnumbered by students of the arts (arts and medicine in Italy) and often of law.[23]

By choice and design, the largest faculty taught exclusively 'secular' topics. The masters' engagement with ideas and texts written mostly by pagans and infidels was not a sideshow. It became the core of their duties. In the 'arts' disputations, arguments were to draw on the publicly accessible tools of logic, reason, and experience. It is for this reason that writers who, on confessional grounds, might be classed as pagan or infidel in fact enjoyed high standing and deep respect as natural philosophers. The mandate, self-understanding, and curriculum of the faculty of arts thus enshrined as its modus operandi an approach that was effectively functioning as methodological naturalism.

The faculty of arts—rightly called much later the 'philosophical faculty'—fostered this outlook not only by its procedures and curriculum but also by diffusing its attitudes and values both in the higher faculties and outside the

[22] Monica Asztalos, 'The Faculty of Theology', in *Universities in the Middle Ages*, pp. 409–41, esp. pp. 430–2.

[23] Antonio García y García, 'The Faculties of Law', in *Universities in the Middle Ages*, pp. 388–408, esp. pp. 400–1.

university. Each of the higher faculties typically required a master of arts degree for admission: even the road to theology ordinarily passed through natural philosophy.[24] All the budding theologians who were not in a religious order had imbibed naturalist conventions and attitudes while becoming masters of arts.

These new theology students were thus thoroughly versed in natural philosophy and committed to the approaches of the faculty of arts. Indeed, in most cases, they arguably deepened these links. Like today's graduate students who instruct undergraduates while earning a Ph.D., medieval theology students often financed their lengthy education by teaching in the arts faculty, to which they remained bound by oath. Some of these eternal students were among the longest serving and most experienced masters in the faculty of arts. By dint of teaching this material, they continued to improve their proficiency in these subjects. At the University of Vienna, for example, John of Gmunden never finished the theological studies he had nearly completed. Instead, he chose to teach astronomy and mathematics in the arts faculty for the rest of his life.[25] Among theologians who did finish, this dual role fostered a de facto methodological separation between theology and the 'arts' when teaching in the latter faculty. When they were doing theology, they could, and did, relax this separation. Already in the thirteenth century, but especially in the fourteenth, the consequence was a powerful influx of natural philosophical content and high-powered logic into theological disputations, leading traditionalist critics to object and to enact statutes limiting these inroads.[26]

For the reasons given above, theologians offered some of the most telling articulations of this naturalistic outlook. Before becoming theologians, the Dominicans Albertus Magnus and Thomas Aquinas (d.1273) were masters of arts. They loved natural philosophy, to which they devoted much thought and energy both before and during their theological careers. They not only advocated treating natural philosophy separately from theology, but did so in practice. When dealing with Aristotle's natural philosophy, whether in the *De caelo* or the *De generatione et corruptione*, Albertus Magnus famously said: 'Here [we are concerned about] the impossible and necessary, that is, [about] the world with regard to its essential and proximate causes' and also 'I am not concerned about the miracles of God, since I will discuss natural things'.[27]

[24] See Rémi Brague, 'Sens et valeur de la philosophie dans les trois cultures médiévales', in *Was ist Philosophie*, pp. 229–44, esp. pp. 242–3.

[25] Claudia Kren, 'Astronomical Teaching at the Late Medieval University of Vienna', *History of Universities* 3 (1983): pp. 19ff.

[26] John E. Murdoch, 'From Social into Intellectual Factors: The Unitary Character of Medieval Learning', in *The Cultural Context of Medieval Learning*, ed. John E. Murdoch and Edith D. Sylla (Dordrecht: D. Reidel, 1975), pp. 271–348.

[27] 'Dico quod nihil ad me de Dei miraculis, cum ego de naturalibus disseram.' Albertus Magnus, *De generatione et corruptione*, 1.1.22, in *B. Alberti Magni... Opera omnia*, ed. Auguste Borgnet (Paris: Louis Vivès, 1890–95), vol. 4, p. 363. See also Edward Grant, *The Nature of Natural*

Although they disagreed with Albertus Magnus and Aquinas on many points, the masters of arts Siger of Brabant and Boethius of Dacia agreed with them methodologically. Siger even rephrased Albertus Magnus's dictum in nearly identical terms while criticizing him: 'But the miracles of God do not concern us now, since we will discuss natural things naturalistically (*de naturalibus naturaliter*).'[28] Pietro d'Abano (d. *c*.1316), a famous Paduan professor of medicine who had trained at Paris, also cited Albertus's quip on disregarding miracles.[29]

ILLUSTRATIVE CONTROVERSIES: THE SO-CALLED 'DOUBLE TRUTH' AND THE ETERNITY OF THE WORLD

The vast number of propositions defended by masters of arts drew no censure outside of their own ranks.[30] Controversies about the few that did can help us to sharpen the picture. Indeed, the charge—now discredited—that a 'doctrine of double truth' thrived in thirteenth-century Paris derives precisely from the pervasiveness of this methodology. In 1270, Aquinas accused Siger of Brabant of holding one thing rationally and the opposite according to faith (Siger's known writings do not substantiate such a belief).[31] Although atypical, the positions of Siger and his colleagues are nevertheless very useful historically: they show clearly what some masters of arts not merely considered thinkable, but also thought. The controversial character of the propositions highlighted by the opponents of Siger (and of his colleague Boethius of Dacia) comes from their focus on issues deemed much more theological than most.[32] In the following,

Philosophy in the Late Middle Ages (Washington, DC: Catholic University of America Press, 2010), pp. 163–4, pp. 251–2.

[28] 'Sed nihil ad nos nunc de Dei miraculis, cum de naturalibus naturaliter disseramus' (my translation; also 'about physical things in a physical manner'). Siger of Brabant, *Quaestiones de anima intellectiva*, in Pierre Mandonnet (ed.), *Siger de Brabant et l'averroïsme latin au XIIIeme siècle*, part 2, *Textes inédits*, 2nd edn (Louvain: Institut Supérieur de Philosophie de l'Université, 1908), p. 154.

[29] Matthew Klemm, 'A Medical Perspective on the Soul as Substantial Form of the Body: Peter of Abano on the Reconciliation of Aristotle and Galen', in *Psychology and Other Disciplines: A Case of Cross-Disciplinary Interaction, 1250–1750*, ed. Paul J. J. M. Bakker, Sander De Boer, and Cees Leijenhorst (Leiden: Brill, 2012), pp. 275–95, esp. p. 277.

[30] Richard C. Dales, *Medieval Discussions of the Eternity of the World* (Leiden: Brill, 1990), chap. 7–9; Richard C. Dales and Omar Argerami (eds), *Medieval Latin Texts on the Eternity of the World* (Leiden: Brill, 1991).

[31] Aquinas, *On the Unity of the Intellect*; The Jesuit Inchofer would later accuse Galileo in similar terms, see Luca Bianchi, *Pour une histoire de la double vérité* (Paris: Vrin, 2008), pp. 15–16, 152–3.

[32] Indeed, there may have been some theological taunting here. As Richard Dales has noted, 'the Fourth Lateran Council in 1215 had made the temporal beginning of the world to be an article of faith', and he has shown that early thirteenth-century treatments of the eternity of the world originated in the faculty of theology, in response to patristic reports of arguments for the eternity of the world. Dales, *Eternity of the World*, pp. 50–1, pp. 86–7 *passim*, quotation on p. 50.

I survey first the controversies surrounding the eternity of the world at Paris; I then turn to the famous Condemnation of 1277 as an illustration of the nexus between institutional and intellectual tensions.

A pointed illustration of this interplay emerges in debates about one of the hottest topics at Paris in the 1250s–1270s: the eternity of the world. Questions about this topic were stimulated by Aristotle's *De caelo,* a staple text in the curriculum of the faculty of arts. The Aristotelian position was sometimes sharpened in responses to the question as to whether there was a first man. Remarkably, the negative answer found both expositors and defenders in the faculty of arts.[33]

After the bishop of Paris prohibited this thesis (and twelve others) in 1270, four treatments of the subject surfaced. As Richard Dales has noted, 'whatever the theoretical basis of the bishop's condemnation may have been, it seems to have been universally ignored'.[34] Nudged by a commission appointed by a papal legate, the Parisian faculty of arts in 1272 moderated the tensions between the faculties of arts and theology by enacting the following statute:

> No bachelor or master of our faculty should presume to determine or even to dispute any purely theological question, as concerning the Trinity or the Incarnation and similar matters, since this would be transgressing the limits assigned to him, for the Philosopher says that it is utterly improper for a non-geometer to dispute with a geometer.[35]

The statute pointedly says nothing about the hotly debated eternity of the world, which it evidently did not consider a 'purely theological question'. Indeed, discussion would continue for more than a generation. Instead, the statute sets some boundaries in language that suggests restraint. At stake are presumptions, assignments, and propriety. Significantly, the masters of arts justified this restraint by appealing to an Aristotelian principle, not to the authority of theology.

The faculty of arts continued its self-policing by acknowledging possible zones of conflict. In case someone did dispute 'any question which *seems* to touch both faith and philosophy' and determined it in a manner 'contrary to the faith', he would be 'deprived of our society as a heretic'.[36] Despite its heavy-handed rhetoric, this intriguing phrase in effect threatened only social ostracism. The faculty of arts might denounce a member, but it could not make him a heretic.[37]

[33] Dales, *Eternity of the World*, pp. 14–42.

[34] Dales, *Eternity of the World*, p. 129.

[35] Edward Grant (ed.), *A Source Book in Medieval Science* (Cambridge, MA: Harvard University Press, 1974), pp. 44–6.

[36] Grant, *Medieval Science*, p. 45 (my emphasis).

[37] When in his *Questions on the Physics*, John Buridan discussed the vacuum in the mid-fourteenth century, he used great caution: appealing to his oath to uphold the above statute, he judged the question, 'whether it is possible that a vacuum exist', to touch both faith and philosophy. It is not clear whether he was responding to sabre-rattling among the theologians or worrying about the vacuum and divine omnipresence.

Although this statute did not touch theologians disputing in their own faculty, many of the latter were masters of arts who continued methodologically to separate naturalistic from theological explanations even when they were not bound by a prohibition. By common consent, one practised natural philosophy by appealing to reason and experience, not to matters of faith.[38] Indeed, in the overwhelming number of cases, the latter were completely irrelevant to the discussions at hand.

Inevitably, it is not the statistically significant normalcy but rather the exceptional instances of conflict that draw all the attention. One of the most famous such instances in the middle of the thirteenth century concerned debates about the eternity of the world, an episode that helps us understand the spectrum of attitudes at the time. This fundamental tenet of Aristotle's worldview elicited different approaches to natural philosophy in the setting of the university. Three leading options illustrate the spectrum of thirteenth-century thought in this debate: the views of Bonaventure (d.1273) at the theological end of the spectrum and of Siger of Brabant at the Aristotelian end, with Thomas Aquinas treading the line between them. All three were active in Paris in the 1260s.

Bonaventure, a Parisian master of arts who later joined the Franciscan order and taught in the faculty of theology, endorsed the theological position. In his view, the eternity of the world was to be rejected for flatly contradicting the biblical doctrine of creation.[39] The stance of the master of arts Siger of Brabant is more complicated. Like many of his colleagues, Siger touted his identity as a philosopher. In his *On the Eternity of the World* he defended, and even amplified, a straightforwardly Aristotelian position, if only for pedagogical purposes in an environment that used the disputation heavily. Finally, Aquinas's larger vision was that of a systematizer who strove to resolve contradictions between various areas of knowledge. He showed how the eternity of the world could in fact be consistent with a notion of creation that distinguished causal priority from temporal precedence. That said, he too endorsed the biblical notion of creation in time.[40]

[38] See Edward Grant, *God and Reason in the Middle Ages* (Cambridge: Cambridge University Press, 2001), chap. 5, esp. pp. 205–6.

[39] 'According to Bonaventure, only two theories regarding the origin of the cosmos are really tenable: first, the theory of the pagan philosophers according to which the world is eternal and the matter of the universe is without ultimate causal origin; second, the Christian doctrine of creation according to which the universe depends entirely for its being on God, is produced "from nothing (*ex nihilo*)", and is temporally finite in the past.' Tim Noone, 'Bonaventure', *Stanford Encyclopedia of Philosophy* (Spring 2013 edn), ed. Edward N. Zalta, http://plato.stanford.edu/entries/bonaventure/.

[40] Calvin G. Normore, 'Who Was Condemned in 1277?', *Modern Schoolman* 72 (1995): pp. 273–81. In the ninth century, John Scot Eriugena had already promoted the notion that the world could be both eternal and made. Richard C. Dales, 'Robert Grosseteste's Place in Medieval Discussions of the Eternity of the World', *Speculum* 61 (1986): pp. 544–63, esp. p. 545.

THE CONDEMNATIONS OF 1277 AS A TURF BATTLE

It was against the background of these (and other) debates that Étienne Tempier, the bishop of Paris, famously prohibited 219 theological and philosophical propositions, twenty-seven of which concerned the eternity of the world.[41] Although the Condemnations were a slapdash job of some local importance, they have garnered much more attention and explanatory power than they deserve.[42] Their doctrinal aspects have disproportionately overshadowed the political and turf-related aspects of the conflict, which interest us here.

Long read as a straightforward case of theological repression of philosophy, the Condemnations of 1277 are in fact an important symptom of 'push-back' against the very naturalism that I am documenting here.[43] They indeed represent the reaction of a commission of some theologians and of the bishop against masters of arts and other theologians, whose utterances in various disputations (and elsewhere) appeared to tread on the turf of the theologians or otherwise upset them. Again, to avoid misunderstanding, they are not formal declarations of heresy; indeed, much to the chagrin of historians, the propositions are anonymous.

The theologians worried about theses associated with the faculty of arts at Paris, that is, views that were discussed and in some cases defended. The bishop's prologue to the Condemnations explicitly claimed that arts students at Paris were 'exceeding the boundaries of their own faculty'. It remains unclear where the commission learned about these theses, whether in writing, in disputation, or in taverns.[44]

The most simplistic view of the Condemnations of 1277 assumes that they represent the final word of 'the Church' and constitute yet another case study of

[41] Edward Grant, *The Foundation of Modern Science in the Middle Ages* (Cambridge: Cambridge University Press, 1996), p. 74.

[42] Pierre Duhem argued that the Condemnations of 1277 mark the beginning of modern science because they freed medieval natural philosophy from dogmatic Aristotelianism. Pierre Duhem, *Études sur Léonard de Vinci: Ceux qu'il a lu et ceux qui l'ont lu*, 3 vols (Paris: Hermann, 1906–13), vol. 2, p. 412; vol. 3, p .vii. Gaukroger recently made the astonishing claim that 1277 'shaped the intellectual landscape of Europe for the next 350 years'. Gaukroger, *Emergence of a Scientific Culture*, p. 48. A more moderate view appears in Grant's writings. Most apt, however, is the debunking work of John E. Murdoch, '1277 and Late Medieval Natural Philosophy', in *Was ist Philosophie*, pp. 111–21.

[43] For an overview and corrections of the historiography of the Condemnations, see J. M. M. H. Thijssen, *Censure and Heresy at the University of Paris, 1200–1400* (Philadelphia, PA: University of Pennsylvania Press, 1998).

[44] Some members of the faculty of arts were targets, as was evidently the case for the theologian Thomas Aquinas. Roland Hissette has found some of the theses in the faculty of arts, but the origins of most remain far from clear. In any event, in 1325, soon after Aquinas was canonized (1323), a later bishop of Paris revoked any articles from the 1277 Condemnation 'insofar as they touch or are asserted to touch the teaching of Saint Thomas'. Intriguingly, he did not specify which they were; see Thijssen, *Censure and Heresy*, pp. 55–6; Hans Thijssen, 'Condemnation of 1277', *Stanford Encyclopedia of Philosophy* (Winter 2012 edn), ed. Edward N. Zalta, http://plato.stanford.edu/entries/condemnation/.

ecclesiastical repression. This view fails to understand the contexts of either the Condemnations of 1277 or their aftermath. With respect to their immediate context, what the Condemnations show is precisely the depth of the naturalism that some masters of arts and their students in Paris defended when left to their own devices. Since the relation of the Condemnations to subsequent developments in natural philosophy is a contentious topic, both analytically and historiographically, I will not discuss it here. Suffice it to say that one of the most careful analyses of the Condemnations of 1277 notes that there was nothing final about them, and that the likes of Ockham (d.1347) believed that the Parisian Condemnations had been imprudent and 'damned the truth'. Indeed the questions continued to be discussed into the fourteenth century by the likes of Ockham and his colleagues in Paris and Oxford.[45]

One other clarification is in order on a point that has confused many commentators. A goodly number of the propositions prohibited in 1277 included statements about what God could not do (e.g. move the universe rectilinearly, create a vacuum). It has therefore been tempting for historians to link these Condemnations with the references to the 'absolute power of God' in fourteenth-century natural philosophy, and in some instances to misinterpret these references as the heavy intrusion of theology into natural philosophy. In later-medieval natural philosophy, an invocation of the absolute power of God functioned as a logical and analytical tool, equivalent to the principle of non-contradiction (since God was presumed able to do anything that did not violate the principle of non-contradiction). The absolute power of God was thus put to use to explore hypothetical situations, counterfactuals, and possible worlds. Indeed, such appeals constituted only a fraction of arguments *secundum imaginationem*.[46] To see a theological point in this usage is to misunderstand its goal, namely that of exploring possible scenarios by testing mutually consistent hypothetical situations with logical thought experiments.

Another telling symptom of naturalism in the Latin Middle Ages concerns answers to the intriguing question: 'whether theology is a science?' *Utrum theologia sit scientia* was frequently raised in late-thirteenth- and fourteenth-century commentaries on Peter Lombard's *Sentences*, the traditional apex of the degree in theology. The question implied that the answer was not obvious. Indeed, the standard for a *scientia* was set by Aristotle's criteria in his *Posterior Analytics*.[47] Whereas Aquinas answered the question with a partial 'yes', Ockham denied

[45] Murdoch argues for a limited effect on the fourteenth century in Murdoch, '1277', pp. 111–21, esp. 115 n. See also Dales, *Eternity of the World*, pp. 222–7.

[46] Murdoch, '1277', pp. 116–17; Grant, *God and Reason*, p. 190.

[47] Edith D. Sylla, 'Autonomous and Handmaiden Science: St. Thomas Aquinas and William of Ockham on the Physics of the Eucharist', in *Cultural Context of Medieval Learning*, pp. 348–96. Note that theology itself had changed significantly in the later medieval period. Formerly identified with biblical studies, it had become a highly rational enterprise that made heavy use of the most sophisticated logical techniques available, both Aristotelian and post-Aristotelian.

that theology was a science. The premises of a science had to be self-evident or otherwise incontrovertible, whereas the articles of faith that formed the premises of theology were anything but self-evident to human minds.

TWO SHORT CASE STUDIES: LEVI BEN GERSON AND NICOLE ORESME

To complement the institutional case presented above, my argument concludes with two fourteenth-century episodes, the value of which lies precisely in their specificity. A thorough argument would include thinkers from the three religions of the book to illustrate that the trend of naturalization transcended the limits of any one tradition in medieval Europe. For reasons of space I limit myself to two.

The astronomical work of the polymath rabbi Levi ben Gerson (d.1344) is a prime example of mitigated methodological naturalism, all the more interesting because he was a rabbi and biblical commentator. His naturalism transpires in his interpretation of 2 Kings 20 (see also Isaiah 38) in which King Hezekiah asks for a divine sign to certify the prediction that he will recover from the illness that left him near death. God offers Hezekiah a choice between two signs associated with a gnomon that his father had built on the palace steps. Does Hezekiah wish the shadow to go down ten degrees (or steps), or to go up by the same amount? Hezekiah rejects the first option on the grounds that 'it is easy for the shadow to lengthen ten steps' (2 Kings 20:10).

When discussing this sign, which King Hezekiah chose because it went against his expectation, Levi ben Gerson reveals the depth of his naturalist understanding of the celestial motions. Comparing the sign given to Hezekiah with the miracles of Moses, he notes the (apparent) sharp contrast between the two. Whereas the miracles of Moses all involved the sublunar world, 2 Kings 20 seems to suggest a miracle in the heavens. Levi, however, deems it inconceivable that the cause of the shadow's aberrant behavior could lie in an aberrant solar motion. He thus restricts the cause of the phenomenon to the sublunar realm, indeed to the shadow itself. Here too he points to a likely natural explanation for the occurrence (namely, refraction owing to clouds or vapours). There is no change in the ordinary course of nature in either realm. A sublunar interpretation also appears in the biblical commentary of Paul of Burgos (d.1435), a Jewish convert to Christianity who became bishop of Burgos. Like Levi, he excludes the possibility of a change in the Sun, preferring to call on a miraculous sublunar light affecting the shadow (on a sundial rather than steps).[48]

[48] Bernard Goldstein, 'Galileo's Account of Astronomical Miracles: A Confusion of Sources', *Nuncius* 5 (1990): esp. pp. 3–16, esp. pp. 7–14. Hezekiah's sign continued to stir debate and to stimulate

Levi's naturalism is the more pronounced of the two, since it eschews any suspension of the course of nature when another explanation suffices. Both men concur, however, in taking the regularity of the celestial motions to be unshakable and in restricting the divine sign to highly local circumstances.

Keeping the explanations of wonders terrestrial, if not local, was a concern that also drew the attention of some Latin natural philosophers. One of the most systematic efforts to 'make wonders cease' is *On the Causes of Marvels* (*c.*1370) by Nicole Oresme (d.1382), who reduces to natural causes many 'marvels' involving vision, hearing, taste and touch, and 'operations of the soul and body'.[49] Oresme was a master of arts who eventually studied theology, translated Aristotelian works into French, and became bishop of Lisieux. *On the Causes of Marvels* probably originated in a 'quodlibetal' university disputation, one in which a master had to answer questions raised by anyone 'on any topic' (*de quodlibet*). Oresme's comments illustrate nicely the vigorous promotion of naturalism in the faculty of arts, especially telling in a future theologian and bishop. Intriguingly, Levi ben Gerson and Oresme had both been associated with the circle around the music theorist (after 1351, Bishop) Philippe de Vitry (d.1361).[50]

Oresme opens his treatise by proposing:

> to show the causes of some effects that seem to be marvels and to show that the effects occur naturally, as do the others at which we ordinarily do not marvel. There is no reason to have recourse to the heavens, the last refuge of the wretched (*ultimum et miserorum refugium*), or to demons, or to our glorious God as if he would produce these effects directly, more so than those effects whose causes we believe are well known to us.[51]

Particularly striking is Oresme's concern with explanation *in general*. Several features of his statement deserve special notice. Oresme's first sentence relativizes the marvellous. His concern is '*some* effects which *seem* to be marvels'. Although he is not dealing with all marvels, the end of the sentence highlights his reductionist goal. He will show that these effects 'occur naturally', that is, that they have natural causes and therefore ought to be removed from the category of the marvellous. Oresme's second sentence puts Ockham's Razor to work.[52] 'There is no reason' (that is, no justification) for invoking three different kinds of cause: the heavens, demons, or God. The three dismissed causes all belong to

invention into the sixteenth century. In a naturalizing move, Christoph Schissler (*c.*1531–1608), an instrument maker from Augsburg, built a bowl sundial that, when filled with water, mimicked the description of the miracle (now in Philadelphia, American Philosophical Society).

[49] Daston and Park, *Wonders*, p. 126, pp. 130–3.

[50] Edward Grant (trans. and annotator), 'Part I of Nicole Oresme's *Algorismus proportionum*', *Isis* 56 (1965): pp. 327–41; Marshall Clagett, *Nicole Oresme and the Medieval Geometry of Qualities and Motions* (Madison, WI: University of Wisconsin Press, 1968), pp. 447, 471.

[51] Bert Hansen, *Nicole Oresme and the Marvels of Nature: De causis mirabilium* (Toronto: Pontifical Institute of Medieval Studies, 1985), pp. 136–7, modified slightly, including the rendering of *miserorum* as 'wretched'.

[52] Ockham's classic formulation reads: 'It is vain to do with more, what can be done with fewer.'

one class in the sense that Oresme disapproves of them in these circumstances, but the reason for his disapproval is not completely obvious. Not all medieval thinkers would have considered 'the heavens' or demons supernatural. Each is, however, remote and invisible and, as Oresme implies, constitutes an illegitimate form of explanation—'the last refuge of the wretched'.

Not unlike Augustine, Oresme treats the marvellous as a mental attitude resulting from the ignorance of the observer rather than as a property of the world. Underlying his argument is a reductionist analogy. We see plenty of effects to which we attribute natural causes. When we believe the latter to be well known, we do not marvel at them. Conversely, when onlookers deem a particular effect marvellous, it is because they do not know or understand its cause. How is the effect to be explained? Oresme characterizes the standard response among his contemporaries as locating the cause of such effects in celestial influence, or demons, or God. The treatment of all three causes as illegitimate constituted an attack not only on 'the wretched' generically but also on such notorious non-wretched predecessors as Albert Magnus, who freely appealed to celestial causes to explain the diversity of effects in the lower world.

A few of the many issues that Oresme considers are whether Socrates has seen his dead father (no—only the 'species', as in a dream), whether maniacs are demoniacs (wrong), or 'whether your imagining would move me, when I am unwilling, or would move a stone' (no compelling reason for positing this, and contrary to Aristotle).[53]

Finally, the opening paragraph of Oresme's conclusion to the treatise also deserves to be quoted in full:

> The above chapters are sufficient to demonstrate to a person of understanding that it is not necessary to have recourse, because of the diversity and marvellousness of effects, to the heavens and *unknown influence*, or to demons, or to our glorious God as the cause more than for any other things whatsoever, since it has been sufficiently demonstrated in the above chapters that effects just as marvellous (or nearly so) are found here below. And for finding the causes of these, people do not have recourse to the aforesaid [i.e., the heavens, demons, and God] as causes, but are well satisfied with natural causes.[54]

The preceding pages have emphasized first the significance of the faculties of arts in the medieval universities as a locus of attitudes that resemble methodological naturalism and the leading centres for its cultural diffusion. The primary example was Paris, the university with the most tension between the faculties of arts and theology and the bishop. My point is quantitative as well as qualitative. Arts was by far the most populous faculty in the ninety-odd universities founded between 1200 and 1500. (Some sixty remained in existence in 1500.) These institutions exposed large numbers of students to naturalistic explanation,

[53] Hansen, *Nicole Oresme*, pp. 155, 263, 315.

[54] Hansen, *Nicole Oresme*, pp. 360–1 (my emphasis; I modified Hansen's translation slightly).

including many from the small minority of masters of arts who went on to study theology. To conclude, I illustrated naturalistic trends in several leading intellectuals from different religions of the book, who show an interest in explaining naturalistically occurrences that their predecessors and contemporaries treated as marvellous or miraculous. When confronted by a phenomenon they did not understand, their first reaction was not to call it a miracle, but to assume that it could be explained naturally.

Against this background, we can understand better the predicament of Galileo. In the seventeenth century, he was still working with, and within, the accepted model of the university and its boundaries. As a mathematician and natural philosopher, he felt at liberty to present arguments on any topic pertinent to 'the arts', including the constitution of the cosmos and the possible motions of the Earth. He rightly expected arguments against his views to remain on the same terrain, that is, in the domains of logic, natural philosophy, and mathematics. As Galileo understood the traditional rules of engagement, if people wished to refute him, they were duty-bound to present rational and empirical arguments against his views. For this reason, Galileo complained bitterly when the arguments against him alleged contradictions between scripture, which was not relevant in natural philosophy, and heliocentrism, which had nothing to do with the faith.[55] Bringing scriptural or theological arguments into the discussion violated the standard rules for doing natural philosophy.

Newton begged to differ but in a very modest way. When scholars cite his statement, 'to treat of God from phenomena is certainly a part of natural philosophy', they often forget several important pieces of contextual information.[56] First, when the very religious Newton first wrote and published his *Mathematical Principles of Natural Philosophy* he said nothing of the sort. He made one passing reference to God in the body of book 3, but evidently regretted it. He suppressed it in his own interleaved copy of the first edition, and deleted it from the main text of the second and third editions (1713, 1725). It was to book 3 in these last two that he appended the 'General Scholium' with its expansive comments on an omnipresent God, the prelude to his rejection of hypotheses about the cause of gravity and his speculations on an all-pervasive subtle spirit. As I. B. Cohen noted, Newton's remarks on not feigning hypotheses and on the propriety of discussing God in experimental philosophy appeared together and for related reasons.[57]

As Amos Funkenstein has noted, however, the seventeenth century is characterized by a curious intellectual development—the emergence of a 'secular

[55] The best English translation of Galileo's *Letter to Castelli* appears in *The Essential Galileo*, ed. and trans. Maurice Finocchiaro (Indianapolis, IN: Hackett, 2008), pp. 103–9, esp. pp. 105–7.

[56] Newton, *Isaac Newton: The Principia*, trans. I. Bernard Cohen and Anne Whitman (Berkeley, CA: University of California Press, 1999), p. 943; I. Bernard Cohen, *Introduction to Newton's Principia* (Cambridge: Cambridge University Press, 1971), pp. 240–5.

[57] Cohen, *Introduction to Newton's* Principia, pp. 155–6, 240–5; also Grant, *God and Reason*, pp. 203–4.

theology of sorts' developed by such laymen as Galileo, Descartes, Leibniz, Newton, Hobbes, and Vico: 'never before or after were science, philosophy, and theology seen as almost one and the same occupation'.[58] This phenomenon has recently been seriously misinterpreted as a holdover from the Middle Ages, which allegedly embedded theology in natural philosophy. As the foregoing pages suggest, it was not medieval at all. Funkenstein was right to see this attitude as a peculiar aberration of seventeenth- and eighteenth-century Europe.

Indeed, one could go a step further. The dabbling in theology by seventeenth-century natural philosophers without formal training in the subject is arguably another symptom of growing naturalization. In an era in which theology had long been professionalized, natural philosophers were growing ever more uppity. Like their late-medieval predecessors, but with greater confidence, they took their natural philosophical conclusions, reached by sense perception and rational argument, to be so reliable, if not unassailable, as to constitute the basis for reinterpreting scripture. Who better to do so than they?

[58] Amos Funkenstein, *Theology and the Scientific Imagination, from the Middle Ages to the Seventeenth Century* (Princeton, NJ: Princeton University Press, 1986), pp. 3–4.

3

Laws of God or Laws of Nature?

Natural Order in the Early Modern Period

Peter Harrison

In one of the most celebrated passages of the third edition of the *Principia mathematica* (*Mathematical Principles of Natural Philosophy*, 1726), Isaac Newton announced that 'to treat of God from the phenomena is certainly a part of natural philosophy'.[1] This unambiguous declaration, bearing the authority of the leading natural philosopher of the period, might seem to settle once and for all the question of whether 'natural philosophy'—the systematic or, as we would now say, 'scientific' study of the natural world—can be pursued without reference to God. Indeed, one influential twentieth-century commentator, taking his cue from passages such as this, suggested that a theological orientation is a distinctive feature of the enterprise. Natural philosophy is characterized as being fundamentally concerned with 'God's achievements, God's intentions, God's purposes, God's messages to man'.[2] This is said to be a key distinguishing feature of *natural philosophy*, and one that clearly sets it apart from modern *science*, which from about the middle of the nineteenth century is increasingly prosecuted without reference to God or the supernatural.

Yet Newton's insistence that treatment of God is certainly part of the scientific study of nature leaves a number of questions unanswered. For a start, the fact that Newton found it necessary to make this claim at all might suggest that some of his contemporaries had expressed alternative views.

Secondly, there is a difference between making reference to God and invoking elements of revealed theology of the kind found in scripture. As Michael Shank has demonstrated in the previous chapter, natural philosophy, as understood by scholastic thinkers, was distinct from revealed theology and the boundaries

[1] Isaac Newton, 'General Scholium', in *Isaac Newton: The Principia*, trans. I. Bernard Cohen and Anne Whitman (Berkeley, CA: University of California Press, 1999), p. 943.

[2] Andrew Cunningham, 'Getting the Game Right: Some Plain Words on the Identity and Invention of Science', *Studies in History and Philosophy of Science* 19 (1988): p. 384.

between them had been carefully policed. References to the divine could certainly occur in the philosophical discussions that took place in the medieval Arts faculties, not least because such references were relatively commonplace in the pagan sources that made up the curriculum (as Daryn Lehoux makes clear in Chapter 1). But mere reference to God is not the same as reference to God's specific purposes and intentions, which is more properly the subject matter of revealed theology.

Thirdly, and following on from this, Newton gives only hints about precisely what role God is supposed to play in natural philosophy. There is an obvious clue in his references to 'design and dominion' and 'the counsel of intelligent Agent', both of which suggest that natural philosophy can provide evidence of divine design in the natural world.[3] But the systematic quest for the design in the natural world was recognized as an activity related to natural philosophy, yet distinct from it—namely, natural theology. While natural philosophy might well provide data for natural theology, it could presumably fulfil that purpose while remaining separate from it, and indeed might fulfil it better the more it was perceived to be theologically neutral.

Finally, it is worth noting that Newton's remark appears in the brief 'General Scholium', or commentary, that was appended to the second and third editions of the *Principia*. The first edition of this classic work makes only a single reference to God. A similar pattern is encountered in Newton's other major scientific work, the *Opticks* (1704), in which theological reflections make a belated appearance in the 'Queries'. If treatment of God were truly integral to natural philosophy, one might wonder why theological references were not peppered throughout Newton's natural philosophical writings, rather than being relegated to brief appendices as if they were a pious afterthought.[4]

In sum, then, Newton's assertions about God and natural philosophy, rather than settling the issue, raise a number of further questions. Was there a consensus in the early modern period about this issue, and was it encapsulated in Newton's claim? If there was a consensus, did it represent a significant change from what came before? And, if God were implicated in natural philosophy, as Newton clearly implies, what precise role did he play?

The last of these questions is important for understanding the others. When we look closely at how early modern philosophers make reference to God, we encounter two main approaches: God makes an appearance at the beginning of the exercise, as a premise or presupposition that makes natural philosophy possible; or, more commonly, God appears at the end, as a kind of obvious conclusion to be drawn from natural philosophy (which, nonetheless, might be

[3] Newton, *Principia*, p. 940; Newton, *Opticks*, 4th edn (1730; repr. New York: Dover, 1979), p. 402.

[4] For a discussion of the significance of these apparently sparse references to God in Newton's scientific writings see Stephen Snobelen, '"The Light of Nature": God and Natural Philosophy in Isaac Newton's Opticks', *Estudios de Filosofía* 35 (2007): pp. 15–53.

conducted largely without overt theological assumptions). In the seventeenth century these two options are represented, respectively, by René Descartes and Isaac Newton. Descartes and Newton represent not only two models of God's involvement in natural philosophy but they advocated different methods, and their competing models of the cosmos dominated seventeenth-century natural philosophical discussions. In addition to these two main options, which will frame the discussion in this chapter, there are a few other ways in which God makes an appearance in natural philosophical discourse. God is frequently credited with motivating individual natural philosophers, who come to regard this activity as a divinely sanctioned vocation. This may make no difference to the content or methods of natural philosophy, but accounts for why individuals might pursue it, or regard it as a legitimate activity. Thinking more of natural philosophy as a human activity rather than as simply a theoretical discipline, God might be seen as helping prepare the human mind to be the recipient of knowledge about the natural world, or, conversely, natural philosophy might be regarded as a formative enterprise that prepares the individual mind to receive the more elevated truths of theology. After considering the contrasting approaches of Descartes and Newton, this chapter will discuss the question of God's direct activity in the world, before returning in conclusion to a discussion of God and the question of the religious and intellectual formation of the natural philosopher.

GOD AND THE FOUNDATIONS OF NATURAL PHILOSOPHY

René Descartes (1596–1650) is the most prominent early modern thinker for whom God acts as a kind of presupposition for natural philosophy. In his best-known work, the *Meditations* (1641), Descartes announces that knowledge of natural things must begin with contemplation of the nature of God, in whom, as Descartes put it 'all the treasures of wisdom and the sciences lie hidden'.[5] This seems to be much more than gratuitous dissimulation, moreover, since the whole argumentative structure of the *Meditations* seems to require it. Descartes must first establish the existence of God. He does this with an ingenious logical argument that suggests the existence of God is as self-evident as the three-sidedness of a triangle. To have a clear and distinct idea of God is to have the

[5] Descartes, 'Meditations', in *The Philosophical Writings of Descartes*, ed. and trans. John Cottingham, Robert Stoothoff, Dugald Murdoch, and (for vol. 3) Anthony Kenny, 3 vols (Cambridge: Cambridge University Press, 1984–91), vol. 2, p. 37; Descartes, 'Objections and Replies', in *Philosophical Writings*, vol. 2, p. 290. Descartes's readers would have recognized this as a reference to Col. 2:3 which speaks of God 'in whom are hid all the treasures of wisdom and knowledge' (KJV) (*in quo sunt omnes thesauri sapientiae et scientiae absconditi*, Vulgate).

idea of his necessary existence, just as to have a clear idea of a triangle is to have the idea of its three-sidedness. Much ink has been spilt on this 'proof', but for our purpose its significance is that Descartes is then able to proceed to further discoveries about his own mind and the workings of the world.[6] For Descartes, God acts as a kind of guarantor of the reliability of all of our knowledge.

Descartes's more explicitly scientific writings also make God foundational. In his treatise *The World*, which he completed in 1633 but withdrew from publication on learning of Galileo's condemnation by the Holy Office in that same year, Descartes asks his readers to imagine a hypothetical world that consists solely of mechanical operations—matter in motion. God brings this world into existence by creating matter and then imposing upon it certain laws that will perpetually govern its motions. The constancy of these laws along with the fact that they hold universally is derived from the unchangeableness of their divine source: 'God is immutable and... acting always in the same way, he always produces the same effect.'[7] Descartes's three laws of motion, or 'laws of nature' as he calls them, thus derive their content and character from God. While *The World* was not published during Descartes's lifetime, many of its basic ideas were rehearsed in later writings that were. In the *Discourse on the Method* (1637), Descartes again spoke of laws of nature that had their source in God and which could be deduced from the divine nature alone: 'I showed what the laws of nature were, and without basing my arguments on any principle other than the infinite perfections of God, I tried to demonstrate all those laws.'[8] Again in the *Principles of Philosophy* (1644), which he had intended as a natural philosophy textbook, Descartes explains that 'from God's immutability we can also know certain rules or laws of nature, which are the secondary and particular causes of the various motions we see in particular bodies.'[9] Here, again, Descartes describes three basic laws of nature.

It is difficult to overstate the novelty of Descartes's arguments in these works. He resurrected a logical argument for God's existence that had been unfashionable since Thomas Aquinas's critique of it in the thirteenth century. This argument controversially relied purely on logic, rather than beginning with observations about the natural world. More importantly for our present purposes, Descartes essentially invented the modern idea of 'laws of nature'. Although there would be much discussion still to come about the nature and status of these laws, the concept came to play a central role in scientific explanation.[10] As Daryn Lehoux

[6] Somewhat confusingly, there are two 'ontological' or a priori arguments for God's existence, one in Meditation 3 and one in Meditation 5. These comments pertain mostly to the second argument. Descartes, *Philosophical Writings*, vol. 2, pp. 44–9.

[7] Descartes, 'The World', in *Philosophical Writings*, vol. 2, p. 96.

[8] Descartes, 'Discourse on the Method', in *Philosophical Writings*, vol. 1, p. 132.

[9] Descartes, 'Principles of Philosophy', in *Philosophical Writings*, vol. 1, p. 240.

[10] See, e.g., John Henry, 'Metaphysics and the Origins of Modern Science: Descartes and the Importance of Laws of Nature', *Early Science and Medicine* 9 (2004): pp. 73–114; Friedrich Steinle, 'The Amalgamation of a Concept—Laws of Nature in the New Sciences', in *Laws of Nature: Essays*

has suggested in the opening chapter, the vocabulary of 'laws of nature' is encountered as early as the classical period. However, nature had usually been understood as governed by the internal properties or essences of natural things, even if these internal tendencies had been implanted in things by God. In Descartes's mechanical and atomistic world, however, natural things are essentially inert, and are moved according to external laws imposed directly by God.

Descartes's attribution to God of an immediate role in the operations of nature may seem about as far from a purely naturalistic approach as can be imagined. Yet this appeal to the divine nature has some rather paradoxical implications. Firstly, while God's activity in nature might be constant and ongoing, because God is immutable, his action never changes. This is important for guaranteeing the intelligibility of the natural order and the consistency of its operations, but it also seems to rule out any special activity on God's behalf. As Descartes himself put it, 'God will never perform any miracle in the new world, and... intelligences or rational souls... will not disrupt in any way the ordinary course of nature'.[11] While Descartes means this assumption to apply to a hypothetical world, his whole purpose is to suggest that this imaginary world is, in fact, essentially the same as ours, for his rules are meant to hold for all possible worlds. It is very unlikely that Descartes was sceptical about the possibility of miracles, and his proscription of miracles is introduced as a hypothetical simplifying condition. But his system could certainly be interpreted as ruling them out. Related to this, because God's action never changes (since he is immutable), true explanations for changes in the natural world must be sought in the laws of nature, rather than in God himself. As Descartes explains:

> Note, in the first place, that by 'nature' here I do not mean some goddess or any other sort of imaginary power. Rather, I am using this word to signify matter itself, in so far as I am considering it taken together with all the qualities I have attributed to it, and under the condition that God continues to preserve it in the same way that he created it. For it follows of necessity, from the mere fact that he continues thus to preserve it, that there must be many changes in its parts which cannot, it seems to me, properly be attributed to the action of God (because that action never changes), and which therefore I attribute to nature. The rules by which these changes take place I call the 'laws of nature'.[12]

on the Philosophical, Scientific and Historical Dimensions, ed. Friedel Weinert (Berlin: de Gruyter, 1995), pp. 316–68; Peter Harrison, 'Laws of Nature in Seventeenth-Century England: From Cambridge Platonism to Newtonianism', in *God, Man, and the Order of Nature: Historical Perspectives*, ed. Eric Watkins (Oxford: Oxford University Press, 2013), pp. 127–48. But cf. Daryn Lehoux, 'Laws of Nature and Natural Laws', *Studies in History and Philosophy of Science* 37 (2006): pp. 527–49.

[11] Descartes, 'The World', in *Philosophical Writings*, vol. 1, p. 97.

[12] Descartes, 'The World', pp. 92–3. On the significance of this passage see Theo Verbeek, 'The Invention of Nature', in *Descartes' Natural Philosophy*, ed. Stephen Gaukroger, John Schuster, and John Sutton (London: Routledge, 2000), pp. 149–67.

In sum, Descartes's invocation of God as the ultimate efficient cause of the natural world seems to allow for an almost completely 'naturalized' explanatory account of the world. Nature could be explained purely in terms of laws of nature, and provided one does not inquire further into the metaphysical basis of those laws, these explanations might be regarded as self-sufficient. Descartes thus relies upon God to establish the intelligibility and regularity of nature, and moves to a new form of explanation in terms of divinely imposed laws of nature. All of this is consistent with Descartes's own lifelong commitment to Catholicism.[13] But arguably his position could be conducive to a kind of theologically premised naturalism. For subsequent thinkers, many of whom were habituated into thinking of laws of nature as an unproblematic feature of the universe that were simply a 'given', Descartes left open the possibility of terminating explanation of the natural world with those laws themselves. Indeed, naturalism today is often characterized in precisely that way, as involving appeals to 'blind' laws of nature and nothing else.

A number of Descartes's contemporaries were attuned to the fact that elements of his philosophy might be interpreted in naturalistic ways. Blaise Pascal (1623–62) was said to have accused his compatriot of wishing to dispense with God, retaining him only because he was needed to get the cosmic ball rolling.[14] At the University of Cambridge, the Platonist philosophers Henry More (1614–87) and Ralph Cudworth (1617–88), initially keen advocates of Cartesian philosophy, also eventually came to wonder whether Descartes's emphasis on divine immutability might give some unintended encouragement to atheists.[15] Descartes's mechanical philosophy, which extended even to the suggestion that human bodies were mere machines, was also cause for alarm to those who thought his mechanism might eventually be extended to human minds. Ridding the world of all souls, excepting only those of human beings, seemed a dangerous precedent. (This is somewhat ironic, given that since the twentieth century, the charge has been the opposite one, that Descartes foisted on posterity a quasi-religious doctrine of the human mind.) Compounding these concerns was Descartes's explicit rejection of final causes, notwithstanding that fact that Descartes has offered theological reasons for this rejection. The intentions of God, he insisted, were simply inaccessible to finite human minds and it was presumptuous to pretend that we could know them. It necessarily followed that the customary search for final causes in natural philosophy was 'useless'.[16]

[13] For Descartes's religious orthodoxy, see John Cottingham, 'The Desecularization of Descartes', in *The Persistence of the Sacred in Modern Thought*, ed. Nathan A. Jacobs and Chris L. Firestone (Notre Dame, IN: University of Notre Dame Press, 2012), pp. 15–37.

[14] Blaise Pascal, *Pensées*, trans. A. J. Krailsheimer (Harmondsworth: Penguin, 1966), p. 355.

[15] Henry More, *An Explanation of the Grand Mystery of Godliness* (London, 1660), p. 214; Ralph Cudworth, *The True Intellectual System of the Universe* (London, 1678), p. 54; John Keill, *An Examination of Dr Burnet's Theory of the Earth* (Oxford, 1698), p. 19.

[16] Descartes, 'Meditations', in *Philosophical Writings*, vol. 2, p. 39. Descartes did allow a place for final causes in ethics: Descartes, 'Objections and Replies', in *Philosophical Writings*, vol. 2, p. 258.

A number of English natural philosophers, responding to these concerns, opted for a different way of involving God in natural philosophy, bringing in theological speculations at the end of the philosophical process rather than the beginning.

GOD AND THE ENDS OF NATURAL PHILOSOPHY

It is significant that the 'General Scholium' in which Newton's remarks about God and natural philosophy appear begins with a devastating critique of Descartes's natural philosophy. At this time, the Cartesian world system—a heliocentric model that had the planets mechanically moved in their orbits by whirlpools of invisible, ethereal matter—was virtually the sole competitor to the Newtonian theory. In the space of a few pages, Newton enumerates the putative deficiencies of Descartes's cosmology: his hypothesis is factually erroneous and inconsistent with the observed motions of the planets; it is methodologically unsound on account of its reliance upon hypotheses and deductions from hypothetical assumptions; it is metaphysically problematic because of its axiomatic assumption of the mechanical nature of all motion; and, finally, it is theologically dangerous because of its denial of final causes. From Newton's perspective, Descartes's whole way of proceeding had got things the wrong way around: we do not arrive at conclusions about the laws of nature from assumptions about God's nature; rather, from the laws of nature, discovered through observation, we can arrive at conclusions about God.

The difference between Descartes and Newton on laws of nature and how they relate to their divine source is further clarified in the preface to the second edition of the *Principia* (1713), which was penned by Newton's collaborator Roger Cotes (1682–1716). Here Cotes explains that the world must have arisen 'from the perfectly free will of God, who provides and governs all things'. From God, he continues, 'have all the laws that are called laws of nature come, in which many traces of the highest wisdom and counsel certainly appear, but no traces of necessity'.[17] Here, then, lies the great difference between the two approaches. Newton has accepted that there are laws of nature but proposes that these are freely chosen by God, and these choices attest to the wisdom of their divine source. Descartes, however, simply deduces the laws from the divine nature, and it follows that the world was the result of a kind of necessity rather than free choice. From Newton's perspective, not only is this poor philosophy, since it relies purely on human speculation rather than observation, but it is

[17] Cotes, preface to Newton, *Principia mathematica*, 2nd edn, p. 397.

poor theology. Descartes's God, Newton concludes, is virtually indistinguishable from 'fate and nature'.[18]

We find a similar position set out in the last of the 'Queries' appended to the later editions of the *Opticks*. Here Newton restates his position about how to treat God in natural philosophy. The correct manner of proceeding in the investigation of nature 'consists in making Experiments and Observations, and in drawing general conclusions from them by Induction'. He continues:

> And if natural Philosophy in all its Parts, by pursuing this Method, shall at length be perfected, the Bounds of Moral Philosophy will also be enlarged. For so far as we can know by natural Philosophy what is the first Cause, what Power he has over us, and what Benefits we receive from him, so far our Duty towards him, as well as that towards one another, will appear to us by the Light of Nature.[19]

Natural philosophy is described here as something separate from theology and moral philosophy, but nonetheless as having important implications for these disciplines. In fact, Newton implies that false theology and idolatry go hand in hand with erroneous natural philosophy.[20]

Newton's arguments to this point simply appear to be a more or less standard rehearsal of the argument from design: natural philosophy is about God in the sense that it points to the idea of a divine designer of the physical world and its living inhabitants. As he wrote in the *Opticks*, 'Such a wonderful Uniformity in the Planetary System [and] the uniformity in the Bodies of Animals... can be the effect of nothing else than the Wisdom and Skill of a powerful ever-living Agent'.[21] Laws of nature, along with the ingenious contrivance of animal bodies, point to the wisdom of a divine designer. Such reference to design in nature was commonplace in much of the natural philosophy and natural history of the period.

So far, then, we have two ways in which God might be involved in early modern natural philosophy. For Descartes, God's existence is established by rational arguments quite independent of natural philosophy, and God is then used as a premise for a new kind of scientific inquiry based on the idea of laws of nature. Natural philosophy is not used to provide evidence for God. Rather, as an enterprise, its very coherence depends on there being a God. The alternative is to begin not with rational arguments but with observations. Careful scrutiny of the operations of nature reveals a remarkable intelligibility to the world that can be expressed in terms of mathematical laws. The laws underpinning nature's operations can be understood as 'traces' of God's wisdom, as Roger Cotes put it, and thus natural philosophy provides evidence of a powerful

[18] Newton, *Principia*, pp. 939–43. It is a little more complicated than this, since in places Descartes does seem to have a place for God's choices (e.g. 'Principles of Philosophy', in *Philosophical Writings*, vol. 1, pp. 256–57).

[19] Newton, *Opticks*, p. 405.

[20] Newton, *Opticks*, p. 406.

[21] Newton, *Opticks*, pp. 402–3.

and wise God. This latter position we have identified with Newton, and indeed it was a very common approach to natural philosophy in the seventeenth century.

All that said, Newton's position turns out to be a little more complicated than this, and he and a number of his followers go well beyond simply giving God a role as the designer of the universe and author of its governing laws. Closer inspection of his remarks reveals some additional possibilities for relating God and natural philosophy.

CONTINUAL MIRACLES?

Among his more speculative comments in Query 31 of the *Opticks*, Newton makes reference to 'some inconsiderable Irregularities... which may have arisen from the mutual attractions of Comets and Planets upon one another, and which will be apt to increase, till this System wants a Reformation'.[22] Newton alludes here to the problem of the stability of the solar system. The law of universal gravitation offers a precise solution to the motion of a single planet revolving about the sun (leaving aside what we would now understand as relativistic considerations). However, the solar system consists of the sun and a number of planets, some with their own satellites, each of which exerts gravitational attraction on all the others. The question then arises as to whether the cumulative effect of the various gravitational forces will in time introduce a fatal instability, leading planets to collide with each other or the sun, or to stray into the far reaches of space. (Readers may be reassured by the fact that while the solar system is inherently chaotic, computer modelling suggests that it is highly probable that the planets will survive until the death of our sun in about five billion years.) Added to this is the problem of the fixed stars, which, although somewhat distant from the solar system and perhaps each other, ought to be exerting a continual attraction upon one another. Why were they not, as a consequence of their mutual attractions, destined to congregate together at the centre of the universe?

This quandary was put to Newton by the classicist and theologian Richard Bentley (1662–1742). Bentley would subsequently be nominated by Newton for the Boyle Lectureship, and he was eventually to become master of Trinity College, Cambridge, where Newton held the Lucasian chair of mathematics. In 1693 Bentley wrote to Newton asking why the stars in infinite space 'should maintain an equilibrium, and not convene together?' Bentley himself proposed to Newton that such an equilibrium 'could not be preserved but by ye power of God'.[23] Newton seems to have acceded to Bentley's explanation, for in remarks

[22] Newton, *Opticks*, p. 402.

[23] 'Richard Bentley to Newton', 1693; The Newton Project, accessed 25 January 2013, http://www.newtonproject.sussex.ac.uk/view/texts/normalized/THEM00257.

recorded by the mathematician David Gregory in 1694 he suggested that only the direct exercise of divine power would avert this disagreeable convention:

> [Newton says] that a continual miracle is needed to prevent the Sun and the fixed stars from rushing together through gravity: that the great eccentricity in Comets in directions both different from and contrary to the planets indicates a divine hand: and implies that the Comets are destined for a use other than that of the planets. The Satellites of Jupiter and Saturn can take the places of Earth, Venus, Mars if they are destroyed, and be held in reserve for a new Creation.[24]

A similar solution suggested itself for the irregularities of the solar system. The agent of the required 'reformation' of the solar system, to which Newton refers in the passage from the *Opticks*, is God. The Deity, Newton suggests, 'is more able by his Will to move the Bodies within his boundless uniform Sensorium, and thereby to form and reform the Parts of the Universe, than we are by our Will to move the Parts of our own Bodies.'[25] These discussions suggest a rather different role for God than being a premise of natural philosophy or being merely the conclusion to which natural philosophy points. God is here directly involved as a causal variable in order to maintain the stability of the cosmic system (and, at the same time, to maintain the coherence of the Newtonian theory).

It is tempting to think that the invocation of God in this way will be fatal to any rational account of the operations of nature. But it need not be, and in this case, arguably, was not. Provided that God's activity is reckoned to be uniform, there is little difference between attributing the regularity or lawfulness of events to powers inherent in nature or directly to God. In other words, if a miracle is a *continual* miracle, irrespective of the fact that the causes of the events in question are attributed directly to God, it might still be incorporated within a natural philosophy geared towards description in terms of mathematical laws. Newton himself seemed to hold that all events are equally works of God, and that human observers label some events miracles only because they are unusual: 'If they [miracles] should happen constantly according to certain laws impressed upon the nature of things, they would no longer be called miracles, but might be considered in philosophy as part of the phenomena of nature notwithstanding that the cause of their causes might be unknown to us.'[26] Gravity was a clear case in point. Newton admitted that he did not know the cause of gravity, although at times he entertained the possibility that it

[24] Memoranda by Gregory, 1694, in *The Correspondence of Isaac Newton*, ed. H.W. Turnbull et al. (Cambridge: Cambridge University Press, 1959–77), vol. 3, p. 336. According to present science, the expansion of the universe, following the 'big bang', keeps the stars moving away from each other. However, if the total mass of the universe is sufficient, all of the stars at some future time, will in fact 'convene', to use Bentley's term, in the scenario known as 'the big crunch'.

[25] Newton, *Opticks*, p. 403.

[26] Newton, *Sir Isaac Newton Theological Manuscripts*, ed. H. McLachlan (Liverpool: Liverpool University Press, 1950), pp. 17–18. On Newton and miracles, see Peter Harrison, 'Newtonian Science, Miracles, and the Laws of Nature', *Journal of the History of Ideas* 56 (1995): pp. 531–53.

might simply be God's direct action. Newtonians such as Bentley and the philosopher-theologian Samuel Clarke (1675–1729) were less circumspect, and unambiguously identified gravity with God's activity.[27]

One of the implications of this approach is that there is no genuine difference between natural and supernatural events. Clarke, the most philosophically acute defender of the Newtonian system in the early eighteenth century, thus announced that 'the course of nature, truly and properly speaking, is nothing else but the will of God producing certain effects in a continued, regular, constant, and uniform manner; which course or manner of acting, being in every moment perfectly arbitrary, is as easy to be altered at any time as to be preserved'.[28] English natural philosophers such as Robert Boyle (1627–91), Nehemiah Grew (1641–1712), and William Whiston (1667–1752) had set out similar positions.[29] On this view, which receives its most lucid and rigorous treatment in Clarke, there is no independent and subsistent 'nature', and the subject matter of natural philosophy is really the modalities of God's activity in the world. Natural philosophy, on this understanding, *is* ultimately about God.

Like Descartes's cosmology before it, the Newtonian system also attracted trenchant criticism. German philosopher Gottfried Wilhelm Leibniz (1646–1716), whose views of Newton were already jaundiced on account of a long-standing dispute over who had invented infinitesimal calculus, attacked on three fronts. First, Newton's God seemed rather like an incompetent workman, compelled to intervene to mend his handiwork from time to time. Surely, Leibniz reasoned, a competent creator could have made a world that did not require periodic 'reformations'. Second, Leibniz objected to the Newtonian understanding of miracles. If all of nature were miraculous, then traditional miracles were in fact no different from other events. This meant, in turn, that they could play no role in supporting the truths of the Christian religion. Finally, Leibniz commented that Newton's remarks about God moving objects in his 'boundless uniform sensorium' suggested that God was extended in space, and that the world was his body. To him, this sounded dangerously close to the pantheism of Spinoza. These objections were articulated in an extended correspondence between Leibniz and Clarke, who acted as Newton's spokesman in the debate.[30]

[27] Richard Bentley, *The Works of Richard Bentley, D.D.*, vol. 3: *Theological Writings*, ed. Alexander Dyce (London: Macpherson, 1838), pp. 74–5. Cf. Memoranda by Gregory, 21 December 1705, in Newton, *Theological Manuscripts*, p. 30.

[28] Samuel Clarke, *A Discourse Concerning the Unchangeable Obligations of Natural Religion and the Truth and Certainty of the Christian Revelation*, in *The Works of Samuel Clarke, D.D.*, vol. 2 (London: John and Paul Knapton, 1738), pp. 697–8.

[29] Robert Boyle, *A Disquisition about the Final Causes of Natural Things* (London, 1688), pp. 91, 96; Nehemiah Grew, *Cosmologia Sacra* (London, 1701), pp. 31, 87; William Whiston, *A Vindication of the New Theory of the Earth* (London, 1698), pp. 211, 219.

[30] *The Leibniz-Clarke Correspondence*, ed. H. G. Alexander (Manchester: Manchester University Press, 1998). The basic points of dispute are set out in Leibniz's first letter, p. 11–12.

At the outset, Leibniz makes it clear that it is the theological implications of Newtonianism that most concern him, although these are often inseparable from the scientific issues. For example, the disagreement over whether space is absolute (Newton) or relative (Leibniz) is connected to Newton's association of space with God's omnipresence. These connections were explored in detail and with great sophistication by Clarke in his Boyle Lectures. Clarke suggests that for God to act in the universe he must be substantially present everywhere, for otherwise he would need to act at a distance (a solution that all parties sought to avoid). Clarke ends up with the position that just as 'natural' causation is a mode of divine activity, so space and time are similarly attributes or, more correctly, modes of God.[31] Again, then, natural philosophy is in some ultimate sense, necessarily about God.

PHYSICO-THEOLOGY

On the question of miracles and the possibility of divine interventions to keep the cosmos in order, modern readers may feel some sympathy with Leibniz's position. Not only could this strategy be seen as impugning the wisdom and foresight of God (Leibniz's point), but for a natural philosopher to hold in reserve the possibility of invoking God's direct action on the occasion of encountering some explanatory difficulty seems inimical to the conduct of rational enquiry. That said, it is not entirely clear that Newton did in fact adopt this 'God-of-the-gaps' strategy in which explanatory lacunae in an otherwise naturalistic account are bridged by supernatural explanation. For a start, as we have already seen, the position adopted by Newton, Clarke, and a number of other English natural philosophers is that all causation is divine causation, and that talk of God's *intervention* is really just our way of describing an unusual, as opposed to a usual, act of God. God's usual activity could be described in terms of laws of nature, and this was the province of natural philosophy. But this did not prevent God from acting in ways that, from a human perspective at least, seemed to contravene laws of nature as they were currently understood. In principle, there was a way of incorporating all divine acts within a rational explanatory framework, by adding to natural philosophy additional information found in sacred texts. That hybrid activity was originally referred to as 'physico-theology'.

The correspondence may also be found on the Newton Project webpages, http://www.newtonproject.sussex.ac.uk/prism.php?id=135&name=39, accessed January 2013. For commentary, see Ezio Vailati, *Leibniz and Clarke: A Study of their Correspondence* (New York: Oxford University Press, 1997).

[31] Clarke, *Works*, vol. 4, p. 758. God's relation to time also means a different understanding of divine immutability from that held by Descartes. See Clarke, *Works*, vol. 3, p. 897.

Contemporary philosophers of religion, taking their lead from the Prussian philosopher Immanuel Kant (1724–1804), sometimes speak of 'the physico-theological argument', by which they mean the argument for God's existence from apparent design in the natural world. In its original seventeenth-century English context, however, 'physico-theological' typically referred to something quite different, namely, a kind of explanation that brought together natural philosophy with biblical accounts of the history of the world.[32] One of the shortcomings of early modern natural philosophy was its limited capacity to shed light on the origins of the earth, the geological changes that it had undergone in the past, and those that would befall it in the future. Newton, for example, insisted that the scope of natural philosophy extended only to 'the present frame of nature' and not the creation of the world, nor its eventual destruction.[33] It was generally agreed, however, that the biblical account, correctly interpreted, could be informative about these matters, and a number of attempts were made over the course of the seventeenth century to combine the resources of natural philosophy with those of sacred history. Newton's references to the possible destruction of the earth and to a new creation represent such a combination. The prospect of a destabilized solar system which his natural philosophy allowed for seemed to square with biblical prophecies regarding the destruction of the world by fire and the creation of a new heaven and a new earth. The explanatory scope of natural philosophy could thus be supplemented with additional information from the scriptures in order to provide a complete account of the history of the cosmos.

Previous attempts of this kind had been made from about the middle of the seventeenth century. Henry More had thought that destruction of the world by fire prophesied in scripture (2 Pet. 3:6–7) would take place in accordance with Cartesian philosophy, as the earth was drawn further and further into the solar vortex and eventually incinerated.[34] Thomas Burnet's *Telluris theoria sacra* (*Sacred History of the Earth*, 1681) provided a more sustained application of Cartesian philosophy to events of sacred history, explaining how Noah's flood could be accounted for in terms of elements of Descartes's cosmogony. A second part of the work that appeared in 1689 sought to explain the demise of the present earth in equally naturalistic terms. Burnet here proposed that natural causes alone would be sufficient to bring about the final conflagration of the earth, although he did allow that their combination and timing was the result of

[32] On the category 'physico-theology', see Peter Harrison, 'Physico-theology and the Mixed Sciences: The Role of Theology in Early Modern Natural Philosophy', in *The Science of Nature in the Seventeenth Century*, ed. Peter Anstey and John Schuster (Dordrecht: Springer, 2005), pp. 165–83.

[33] Newton, 'Draft Versions of "The Queries"', MS Add. 3970.3, 243r, Cambridge University Library, in Newton Project, accessed 25 January 2013, http://www.newtonproject.sussex.ac.uk/view/texts/normalized/NATP00055.

[34] Henry More, *An Explanation of the Grand Mystery of Godliness* (London, 1660), p. 240.

'a particular providence'.[35] Newton was an admirer of Burnet's work, in spite of its Cartesian framework, and his own speculations about the eventual fate of the solar system adopted a similar principle to Burnet's: a sequence of 'natural' events would coincide with a foreordained providential plan foreshadowed in biblical prophecies.

Newton did not take the trouble to apply his own natural philosophy to pivotal events of cosmic history in any sustained way. That task was taken up by Newton's successor in the Lucasian chair at Cambridge, William Whiston (1667–1752). In the *Principia*, Newton had demonstrated how his law of gravitation applied to the path of a bright comet that had appeared in 1680. Newton's protégé, Whiston, along with his better known contemporary Edmund Halley (1656–1742), became an early advocate of the periodicity of comets, and in *A New Theory of the Earth* (1696) he set out an ingenious scheme in which the Genesis flood was occasioned by the earth's near collision with a comet. Whiston also surmised that the end of the world would be wrought by the transit of a comet, possibly the same one that had originally visited the deluge upon the earth. Like Burnet before him, Whiston worked on the assumption that the moral state of the world and the history of the cosmos were intimately connected, and he consistently relied upon the general principle that what might look like a divine interposition into the natural order may in fact be the outcome of a providentially planned concatenation of natural causes.[36] The postulation of lengthy orbital periods of comets made it possible to bring together natural philosophy with the history of the world, which, before this, had been solely the province of sacred history.

One implication of these hybrid activities, in spite of their religious motivations, was that the scope of natural philosophy could potentially expand to incorporate events that had previously been categorized as miraculous, or as owing to God's immediate and direct activity. Pious natural philosophers were happy to push 'naturalistic' explanation as far as they could. Burnet, for example, cited what he called 'the first rule of miracles', according to which 'we must not flie to miracles, where Man and nature are sufficient'.[37] Others such as Grew and Whiston agreed that many events described in the Old Testament as miraculous were now susceptible to natural philosophical explanations.[38] 'Naturalistic' explanations were thus to be prosecuted as far as possible. Robert Boyle even essayed some physico-theological theories about how the resurrection

[35] Thomas Burnet, *The Theory of the Earth... the Last Two Books* (London, 1690), pp. 253, 271.

[36] See Whiston, *Vindication*, p. 31.

[37] Burnet, *Theory of the Earth... Last Two Books*, p. 281. This 'rule' goes back to Augustine, and had been recently restated in Benedict Pererius's influential commentary on Genesis, *Commentariorvm et dispvtationum in Genesin* (Cologne, 1601), p. 8. Cf. Burnet, *A Review of the Theory of the Earth* (London, 1690), p. 44; Leibniz, *Philosophical Essays*, 2.A.4, ed. and trans. Roger Ariew and Daniel Garber (Indianapolis, IN: Hackett Publishing Company, 1989), p. 254.

[38] Grew, *Cosmologia sacra*, pp. 195–203; Whiston, *A New Theory of the Earth* (London, 1696), pp. 218, 361.

of bodies might be effected (although he hedged his bets by conceding that this would not happen according to the ordinary course of nature).[39] These efforts were not motivated by a naturalism that was in any way sceptical about the involvement of the Deity in the natural world. Rather the idea was to locate natural philosophy within a broader explanatory framework that took as its domain the whole range of divine activity. Boyle, for example, explained that the Gospel sets out the mystery of human redemption, while mechanical natural philosophy seeks to derive all the phenomena of nature from matter and motion. But he argued that these two branches of learning were components of a 'more general theory of things' for which our sources are both the scriptures and the light of nature. Theology and natural philosophy, he writes, 'are the subordinate parts of the Universal Hypothesis, whose Objects I conceive to be *the Nature, Counsels, and Works of God, as far as they are discoverable by us* (for I say not to us) *in this Life*'.[40] The increasing explanatory scope of natural philosophy results in no diminution in the gamut of God's activities, representing only a change in how, from a human perspective, those activities are understood. The expanding boundaries of science portend no corresponding contraction of the boundaries of theology, but merely a change, from the human perspective, in how God's activities are understood.

GOD AND THE NATURAL PHILOSOPHER

Up until this point we have treated natural philosophy as being akin to a modern discipline with its own discrete subject matter and modes of explanation, and we have explored ways in which God may or may not have been part of the subject matter or, indeed, part of the explanatory framework. When we consider, however, that natural philosophy in the early modern period was a *philosophical* enterprise, and that philosophy at the time was still concerned to some extent with the traditional goal of shaping the person of the philosopher, we come to another way in which God may have been thought to be involved in the whole process—not in the formal understanding of the scope of the 'discipline' but in the lives of its practitioners.

One straightforward way in which God was thought to be involved with natural philosophers as individuals was as a direct source of motivation or inspiration. On St Martin's Eve in November 1619, for example, the young Descartes experienced a series of dreams that would determine the future course of his life. After an eventful day during which, on his own account, he had discovered 'the foundations of a wonderful science', the twenty-four year-old

39 Boyle, *Some Physico-Theological Considerations about the Possibility of the Resurrection* (London, 1675), pp. 3, 29.

40 Boyle, *The Excellency of Theology* (London, 1674), pp. 51–2.

Descartes, 'full of enthusiasm', retired for the evening. As he slept fitfully in his stove-heated room he experienced three dreams involving, among other things, a foreign melon or fruit, an evil demon, and an encyclopedia. In the third dream he laid his hand upon a book that contained the Seventh Ode of Ausonius, which begins with the words '*Quod vitae sectabor iter*'—What road in life shall I follow? Upon wakening concluded that his visions 'could have come only from on high'. He interpreted them as portending that he would found a new and comprehensive mathematical science, and piously vowed to the Virgin that he would visit her shrine at Loretto.[41]

Descartes was not unusual amongst his contemporaries in taking seriously the possibility of divine communication through dreams, or in his reliance upon divine inspiration to assist in important life decisions. Such communications could range from a general inspiration to pursue a career in natural philosophy, to the conveying of quite specific scientific truths to the individual. On the first point, it was not uncommon for natural philosophers to regard themselves as 'priests of nature', specially commissioned to learn more of God through the study of the creation. The astronomer Johannes Kepler (1571–1630) had originally intended to pursue a clerical career, but eventually came to the view that astronomy was theologically relevant too: 'I wished to be a theologian; for a long time I was troubled, but now see how God is also praised through my work in astronomy.'[42] The world, he wrote, was 'the temple of God' and 'the book of nature', and to contemplate the world was thus to worship God and act as a priest.[43] Robert Boyle also drew upon the commonplace of nature as God's temple to argue for the idea of the philosopher as a priest of nature. Natural philosophy was thus 'philosophical worship of God'.[44] On the second point, God might directly inspire particular individuals with truths about the natural world. Robert Boyle noted that followers of the Swiss medical reformer Paracelsus (1493–1541) and Flemish chemist Jan van Helmont (1579–1644) held that God 'discloses to men the great mystery of chemistry by good angels, or by nocturnal visions'. Boyle was himself reluctant to go that far, but he nonetheless stated his belief that God did promote men's proficiency in the study of nature,

[41] The report of the dream is provided by Descartes's first biographer, Adrien Baillet, *La Vie de M. Des-Cartes*, vol. 1 (1691), pp. 80–6, in Charles Adam and Paul Tannery (eds), *Oeuvres de Descartes*, rev. edn., vol. 10 (Paris: J. Vrin/C.N.R.S., 1964–76), pp. 180–5. See also Descartes, 'Early Writings', in *Philosophical Writings*, vol. 1, p. 4.

[42] Johannes Kepler, *Gesammelte Werke* (Munich, 1937–45), pp. 13, 40.

[43] Kepler, *Mysterium Cosmographicum*, trans. Alistair M. Duncan (Norwalk, CT: Abaris, 1999), p. 53; Kepler, *Gesammelte Werke*, pp. 7, 25. On the priest of nature motif, see Harold Fisch, 'The Scientist as Priest: A Note on Robert Boyle's Natural Theology', *Isis* 44 (1953): pp. 252–65.

[44] Boyle, *Usefulness of Experimental Natural Philosophy*, in *The Works of the Honourable Robert Boyle*, vol. 2, 2nd edn, ed. Thomas Birch, (London, 1772), pp. 32, 63. Cf. Thomas Sprat, *History of the Royal Society* (London, 1667), pp. 349–50; Joseph Glanvill, *Philosophia pia* (London, 1671), pp. 6, 12, 141–2. On the venerable idea of the world as God's temple, see Peter Harrison, *The Bible, Protestantism and the Rise of Natural Science* (Cambridge: Cambridge University Press, 1998), pp. 198–9.

usually through the communication of 'happy and pregnant hints', and that he did so much more than was commonly acknowledged.[45]

Related to the notion of God communicating truths to the individual was the traditional idea that for the philosopher to be the recipient of higher forms of knowledge, a certain work had first to be performed on the self. Ancient and medieval discussions of the divisions of the sciences had often been premised on the idea that the study of nature provided a preliminary stage that led on to the more elevated moral or theological sciences. Aristotle's longstanding division of the theoretical sciences into natural philosophy, mathematics, and theology was understood as reflecting not just the discrete subject matter of the sciences but also the order in which they were to be studied. Natural philosophical pursuits are thus a kind of groundwork preparing the way for the more abstract activities of mathematics, which led in turn to the most abstract disciplines of all: theology or metaphysics. This was not just a matter of obtaining the necessary prerequisites for the higher theoretical disciplines, moreover, but of undergoing a process of cognitive and spiritual formation. The Neoplatonist philosopher Simplicius (490–560), for example, wrote that natural philosophy assisted in the acquisition of moral virtues, that it was a ladder leading to knowledge of God, and that it incited acts of kindness towards God.[46] These stages were also reflected in Christian appropriations of classical philosophical practices. Cappadocian theologian Evagrius of Pontus (*c.*345–399) taught that Christianity consisted of the three stages of praxis, physics, and theology. By 'praxis' he meant a preliminary purification that made possible 'physics' or the contemplation of nature. This, in turn, led to theology—contemplation of God.[47]

These traditional conceptions of the goals of philosophy continued to inform, to some extent at least, early modern understandings of the discipline. Descartes's *Meditations* (the clue is in the title) are suggestive of a set of spiritual exercises that call for a cognitive purgation that is then followed by a kind of mental restructuring.[48] Something quite similar was also claimed for experimental

[45] Boyle, *Usefulness*, in *Works*, vol. 2, p. 371. There were biblical precedents for this. See Gen. 31:11–12; Philo, *On Dreams* 1, p. 23; Origen, *Letter to Africanus*, p. 10. Boyle also thought that the Bible itself offered 'nobler hints of natural philosophy than men are yet perhaps aware of', in *Usefulness of Natural Philosophy*, p. 19.

[46] Simplicius, *In Physica*, in *Commentaria in Aristotelem Graeca*, vol. 9, ed. H. Diels (Berlin: Reimer, 1882), 4.17–5.21, quoted in Remi Brague, *Wisdom of the World* (Chicago: University of Chicago Press, 2004), p. 116.

[47] Hadot, *What is Ancient Philosophy?*, trans. Michael Chase (Cambridge, MA: Harvard University Press, 2002), pp. 249–50.

[48] Matthew L. Jones, 'Descartes's Geometry as Spiritual Exercise', *Critical Inquiry* 28 (2000): pp. 40–71, esp. 58; Jones, *The Good Life in the Scientific Revolution: Descartes, Pascal, Leibniz, and the Cultivation of Virtue* (Chicago: University of Chicago Press, 2006), p. 43. See also John Cottingham, 'Descartes as Sage: Spiritual *askesis* in Cartesian Philosophy', in *The Philosopher in Early Modern Europe*, ed. Conal Condren, Stephen Gaukroger, and Ian Hunter (Cambridge: Cambridge University Press, 2006), pp. 182–201; Peter Harrison, 'Francis Bacon, Natural Philosophy and the Cultivation of the Mind', *Perspectives on Science* 20 (2012): pp. 139–58. Cf. Sorana Corneanu, *Regimens of the Mind: Boyle, Locke and the Early Modern* Cultura Animi *Tradition* (Chicago: University of Chicago Press, 2011).

natural philosophy. Bishop Thomas Sprat, who wrote the first history of the Royal Society in 1667, could accordingly suggest that the pursuit of experimental philosophy would accomplish the same goals as moral philosophy. The performance of experiments was said to promote a kind of spiritual mortification, purgation, and repentance.[49] Robert Boyle agreed that the formal study of nature would excite 'true Sentiments both of Devotion and of particular Vertues'.[50] Another defender of the Royal Society, Joseph Glanvill, pointed out that Boyle himself personified the formative qualities of the practice of natural philosophy, combining in his person 'the *profoundest* insight into *philosophy* and *Nature*, and the most *devout, affectionate* Sense of *God* and of *Religion*'.[51] Not merely, then, did experimental natural philosophy provide evidence of God's existence in some abstract way. Its practice was also thought to instil virtues and sentiments of piety in the person of the natural philosopher. It involved a kind of habituation that brought with it an immediate sense of the Deity. Boyle lecturer William Derham could accordingly declare in the Preface of his *Physico-theology* (1754) that 'nothing tended more to cultivate true religion and Piety in a Man's mind, than a thorough skill in [natural] Philosophy'.[52]

God played diverse roles in early modern natural philosophy—underpinning the basic premise that there were laws of nature, representing the logical conclusion to which natural philosophy pointed, inspiring certain individuals in their scientific pursuits, and being the object of an enterprise that some understood to be inherently religious. The proper question seems not to be *whether* God was involved in the practice of natural philosophy but, rather, *how* he was involved. Yet we may still ask whether it was then possible to conceive of a natural philosophy in which God played no significant role. The answer is a qualified yes, although it was usually thought that such natural philosophy would be deficient in some way. Ancient Epicureanism provided the chief model, and one of the reasons that early modern natural philosophers found it necessary to be explicit about their theological commitments was the need to distinguish their basic assumptions from those of the Epicureans. This was all the more so because most of the early moderns had adopted a matter theory that was essentially Epicurean. Thomas Hobbes (1588–1679) was generally thought to represent the modern face of this ancient philosophy, along with its atheistic and materialistic implications. Had it not been for the spectre of 'Hobbism' (and we can probably add 'Spinozism') the key players may not have expended so much effort in demonstrating how God fitted into their various schemes. Even so, the religious elements that we encounter do seem to play a substantive role in

[49] Sprat, *History*, pp. 341, 367. Cf. Joseph Glanvill, *Plus ultra* (London, 1668), p. 23.

[50] Boyle, *Disquisition*, p. 88.

[51] Glanvill, *Plus ultra*, p. 23.

[52] William Derham, *Physico-theology; or, A Discourse on the Being and Attributes of God* (London, 1754), p. vii.

providing the presuppositions and goals of natural philosophy, and to the extent that it would be difficult to imagine it working without them.[53]

It is also significant that the seventeenth century also sees the beginning of the end of the notion of natural philosophy as a genuinely *philosophical* enterprise, concerned with the moral and religious formation of the philosopher. While we have noted the persistence of this ideal, even its chief proponents wanted more from natural philosophy than this. Francis Bacon's (1561–1626) influential plan for the renovation of the sciences was motivated by the conviction that natural philosophy should be directed towards the mastery of nature, rather than self-mastery.[54] This new focus, Bacon had insisted, would give rise to a natural philosophy that would issue in material benefits for all. While this vision was originally justified in theological terms, the Baconian emphasis on practical utility ultimately led natural philosophy away from the more elevated goals of traditional philosophy and from the theological concerns that had once played a foundational role.

Finally, there were elements of both Cartesianism and Newtonianism that could be accommodated within a naturalistic framework with relative ease, and eventually they would be. Descartes's 'laws of nature' could be understood literally as laws *of nature* rather than of God. On this understanding of nature's regularities, it is possible to engage in natural philosophy without seeking further explanation of why there are lawful regularities to begin with. The Newtonian scheme, for its part, effectively abolished the distinction between natural and supernatural. While the intention had been to assert the ubiquity of divine power and to 'supernaturalize' the whole cosmic order, this approach was equally susceptible to a purely naturalistic reading, on the condition that the incessant and regular operations of God were reconceptualized simply as 'nature'. Ironically, this was not so far removed from Spinoza's monist 'God-or-nature' understanding of reality. Also significant was Newton's professed agnosticism about the cause of gravity. Once natural philosophy abandoned the quest for causal explanations and resorted instead to mathematical description, it became easier to remove God from the picture and simply 'do the maths'. This was precisely the scenario that Leibniz feared when he wrote presciently in 1669 that with the recent improvements in mathematics, chemistry, and anatomy, mechanical explanations could be offered 'for most of the things which the ancients referred only to the Creator or to some kind (I know not what) of incorporeal forms'. As a consequence, he rued, it became possible for those philosophers who neglected the foundations of their enterprise 'to try to save or explain natural phenomena... without assuming God'.[55]

[53] Conversely, as an aside, among the least fruitful natural philosophies of the period (judged by the standards of the present) are those of Hobbes and Spinoza, which at the time were closely associated with atheism.

[54] Stephen Gaukroger, *Francis Bacon and the Transformation of Early-Modern Philosophy* (Cambridge: Cambridge University Press, 2001).

[55] Leibniz, 'Confession of Nature against Atheists', in *Philosophical Papers and Letters*, 2nd edn, ed. Leroy E. Loemker (Dordrecht: Kluwer, 1989), pp. 109–10.

4

Between Isaac Newton and Enlightenment Newtonianism

The 'God Question' in the Eighteenth Century

J. B. Shank

Almost a century after Newton died, Pierre-Simon Laplace (1749–1827) asserted what many would call the quintessential Newtonian claim when he responded to the Emperor Napoleon Bonaparte that he had no need for the 'hypothesis' of God in his cosmological system. The Laplacian conception of the universe found in works like *Exposition du système du monde* (1796) and *Traité de mécanique céleste* (1799) does indeed operate without any reference whatsoever to a divine principle. His physical science is also rightly described as completely 'naturalistic' in the sense set out in the introduction to this volume—'a rejection of supernatural or spiritual explanations and a focus on what is explicable in terms of natural causes, forces, and laws'. Yet while Laplace was indeed a naturalist, a professed atheist, and a self-proclaimed Newtonian, nothing is more erroneous than making his late eighteenth-century 'Newtonian science' the verbatim echo of the new astronomy and physics initiated by Isaac Newton in his celebrated *Principia mathematica* (1687).

A frustratingly pervasive (and still for many persuasive) conception of eighteenth-century history nevertheless treats Laplace's naturalistic science and atheism as a direct consequence of Newton's pioneering work in the *Principia*. In this story, Newton is the herald of an uncompromising naturalism, and the increasing centrality of naturalistic approaches in the sciences of the eighteenth-century Enlightenment is imagined to be a direct consequence of his work. Within this frame, Alexander Pope's famous line that God let Newton be and all was light is treated as a persuasive historical thesis. The Enlightenment *philosophes* certainly came to conceive of their science in these terms, making Newtonian astronomy and physics the source for their new and more fully naturalistic approach to physical science. They also made 'Newtonianism', defined in this manner, their creed, deploying it as an imagined solvent for priestly power and religious superstition. Laplace rallied under this Enlightenment banner, articulating his scientific debt to Newton and his *philosophe* teachers with enthusiasm.

The architect Étienne-Louis Boullée was also of this party, and when he chose during the bloody Year II of the first French Republic to make a monumental cenotaph for Isaac Newton the centrepiece of his utopian enlightened city (Figure 4.1) he articulated a widely shared understanding of 'Newtonianism' around 1800, one that held it to be synonymous with scientific naturalism and the perceived overcoming through it of revealed religion, faith, and priestly authority.[1]

Contemporary textbooks on Western civilization rehearse the same story, and popular audiences are even more drawn to the eighteenth-century 'Just So' stories that make Newton the heroic founding father of secular scientific modernity. History, however, is not so simple, and complicating this Enlightenment understanding is the actual character of the man and the science that is claimed as the agent for this world-historical naturalizing revolution. In fact, the actual Isaac Newton bears almost no resemblance at all to the man which Enlightenment stories of the birth of scientific naturalism make him out to have been. Newton was never anything but a devout (if heretical) Christian,

Figure 4.1. Étienne-Louis Boullée, imagined cenotaph for Isaac Newton, interior view with orrery above tomb. Bibliothèque Nationale de France.

[1] J. B. Shank, *The Newton Wars and the Beginning of the French Enlightenment* (Chicago: University of Chicago Press, 2008), pp. 1–3.

and as we have seen from the previous chapter, he did not marginalize these religious views in his natural philosophy, but rather made them absolutely central to it. Moreover, not only was Newton's natural philosophy fully Christian (at least according to his understanding of it), he also pursued biblical and theological scholarship at least as vigorously as he pursued mathematics and physics. Newton's writings in what we would today call natural science form a minuscule pile when compared to those devoted to what we would today call religion and theology. And if we add to the religious pile the many overtly religious preoccupations to be found in the already small stack of texts supposedly devoted to natural science, the lie of calling Newton anything but a deeply religious natural philosopher becomes patently obvious.[2]

Yet Newton's historical legacy remains caught today in a historiographical 'two-body' problem produced by the gap between his actual scientific work and the naturalizing account of it created by the eighteenth-century Enlightenment. This chapter has no solution to offer to this historiographical disjuncture, but what it proposes is a reconstruction of the exceedingly crooked and even discontinuous path that actually connected Newton's natural philosophy as developed in the decades before 1700 with the very different, and indeed much more fully naturalistic, Laplacian science that he and his Enlightenment colleagues came to call 'Newtonianism' after 1800.

My analysis will move in two stages. First, I will show that Newton's actual achievements in his *Principia*, namely his theory of planetary motion according to the law of universal gravitation, were not as naturalizing as Enlightenment narratives would lead us to believe. While his astronomical science may have had the unintended consequence of opening the door to a collapse of the God–nature distinction, as Newton himself viewed it, it was a Christian natural philosophy that was to a large degree continuous with ancient understandings of the relationship between God and nature. Second, I will move across the divide separating Isaac Newton and his *Principia* from the eighteenth-century Enlightenment Newtonianism said to be derived from it in order to examine how a new and more fully naturalist cosmology that increasingly came to be set against revealed religion did come to be associated with Newton's name and legacy, but despite his very different intentions. My chapter will not therefore trace a genealogy of how Newton produced Laplacian naturalism, but will follow the historical passage across the gap that separates them, a journey that should force us to look beyond Newton's own scientific accomplishments when trying to account for the modern secular and naturalistic assumptions bound up in what is still called the 'Newtonian scientific worldview'.

[2] James E. Force and Richard H. Popkin (eds), *Newton and Religion: Context, Nature and Influence* (Dordrecht: Kluwer, 1999); Frank E. Manuel, *The Religion of Isaac Newton* (Oxford: Clarendon Press, 1974).

THE COMPLEX NATURALISM OF CHRISTIAN COSMOLOGY

Any universal claim for the intrinsic opposition of naturalistic science with religion in the Western tradition must contend with the complicated entanglement of each in the development of European scientific thought before 1700. Newton's late-seventeenth-century science of terrestrial and celestial mechanics was developed in a manner that was consistent, in many respects, with preceding understandings of the interplay between natural philosophy and theistic conceptions of the natural order. To a degree, his science marks not the revolutionary overturning of a theistically informed natural philosophy but perhaps its twilight moment in the history of Western thought. Indeed, if anything, the Newtonian achievement was to forge a more intimate connection between theology and science than medieval thinkers who, as Michael Shank's chapter has argued, set up institutional structures that sought to preserve a distance between the natural philosophy of the arts faculties and the doctrinal subject matter of the theology faculties.

The Ancient Greeks who charted the initial paths of Western science tied God and nature together when laying the foundations for their new science. By reducing the action of *physis* (nature) to unifying principles—fire, water, atoms of matter, and so on—and by claiming a universal rationality, or *logos*, as the immanent principle organizing the physical world, Greek science without question displaced divine explanations through a new naturalistic account of the world. But the divine was not eliminated altogether as a consequence of this change. Socrates's famous pronouncement that a *daemon*, or divine spirit, gave him the inner certainty of his convictions is one illustration, and while his challenge to the ruling Olympian myths of Athens was certainly a clash between a new philosophical mindset and traditional religion, his was not a naturalist campaign against religion itself so much as a fight against a certain conception of it. Aristotle likewise reinforced the same philosophical understanding when he traced the chain of physical causation that unified the heavens and the earth back to a first cause, or unmoved mover, who sustained the rationality of the cosmos. In all these ways, and as Daryn Lehoux's chapter has illustrated in more detail, Greek science was newly naturalistic and at the same time still inescapably anchored in notions of the divine and supernatural.

With the inception of Christianity, elements of this Greek inheritance were preserved in the writings of the earlier Christian thinkers. God was understood to be the creator of the cosmos, its guiding spirit (or *logos*), and the end toward which everything was moving. Jesus was likewise understood as divinity made flesh, a claim that was both supportive and disruptive of Christian natural philosophy. It encouraged Christian science by creating an avenue for joining Greek notions of an immanent natural order, one revealed through mathematical reasoning and empirical observation, with Christian understandings of a divinely created and predestined cosmos. Here, the equation of God with Jesus (and both with the *arche* and *logos* of the universe) that we encounter in the gospel of John, provided a catalyst for Christian scientific inquiries into the structure and principles of the material world.

Not everything was reconcilable. Aristotle's purely philosophical deduction that a creation *ex nihilo* was logically impossible, and that the world was thus eternal with no clear beginning or end, was a perpetual source of controversy, and Christ's miracles remained immune to rational criticism for centuries because of their status as truths given through unimpeachable sacred testimony. Yet these limits largely constituted the margins of what during the Middle Ages was an unproblematic compatibility between Christian religious orthodoxy and natural scientific understandings of the cosmos. Within this frame, Ptolemaic astronomy could be made the natural scientific basis of the Christian cosmos without generating even a bit of heat, and Aquinas was canonized a saint for showing how everything from the motion of moving bodies on earth to the miraculous transubstantiation of the bread and wine in the Eucharist was conceivable through the principles of Aristotelian physics. The Renaissance period inherited this tradition and in most ways merely extended it into the sixteenth and seventeenth centuries.

Newton's late-seventeenth-century cosmological science is a case in point, since he in no way broke with this ancient tripartite understanding of nature, God, and human scientific understanding. In fact, he actually aspired in his science to rejoin the ancients with the moderns so as to recapture the perfect wisdom of Christian natural philosophy, which he believed was found at the first moments when Christianity and ancient Greek science came together. To call him a modernizing scientific revolutionary, therefore, is to misrepresent completely who he was.

Two images separated by almost half a millennium illustrate well the ancient continuities present in the scientific mindset of Newton's day. The first (Figure 4.2), from the thirteenth century, shows Jesus-God as the creator of the cosmos, which he holds as a distinct body separate from himself, and regulates using mathematical rationalism as symbolized by his compass.

Central to the Christian orthodoxy of this conception is the clear distinction between the three realms of God (the Jesus figure), nature (the orb he manipulates), and mathematical rationalism (his tool as depicted by the compass) even as each is contained within the integral order of the world as depicted. Moving forward to the sixteenth century we see in Figure 4.3 the same fundamental tripartite divisions in a much more modern image of Christ as saviour of the world.

Jesus-God is again depicted separately from, and in control of, his celestial creation, and while the geometrical compass has now been embedded into the cartographic grid visible in the terrestrial globe that Jesus holds, naturalistic mathematical science is still present as a geographic structure of God's creation and as a tool appropriate for establishing human knowledge of the divine creation.

Figure 4.2. *God as Architect/Builder/Geometer/Craftsman, The Frontispiece of Bible Moralisée* (mid-thirteenth century), ÖNB Vienna, Codex Vindobonensis 2554, f.1 verso. Wiki-Commons.

This dualistic arrangement, which isolates God from nature while attaching him through science to the cosmos he created, also supports the fundamental Christian theology of the image because Jesus-God remains the sovereign and providential saviour distinct from, and master over, nature even as the suggestion is made that mathematical science is a divine, though humanly accessible, tool that connects us with the order of the universe.

This same ancient Christian cosmological foundation undergirds Newton's approach to cosmological science. His achievements in naturalistic science, which were many, are best understood as contributions to the further sustenance of this Christian natural philosophical understanding, not as its revolutionary overturning. That said, a collapse of the God–nature distinction that had anchored pre-modern Western cosmology for over two millennia did occur in the wake of Newton's pioneering new astronomy, and out of this collapse emerged our modern understanding of naturalistic cum atheistical science. Newton's 'naturalistic' science proved to be a catalyst for these transformations, but it was in spite of, rather than because of, his intentions.

Figure 4.3. *Christ Salvator Mundi*, artist unknown, lower Rhine, 1537–45. Getty Images.

In Chapter 3, Peter Harrison discussed some of the many seventeenth-century natural philosophical developments provoked by this increasingly unstable philosophical-theological terrain, one where the line separating the divine from the natural was coming to be called into question. One seventeenth-century response he did not explore in detail, however, was that of Benedict Spinoza (1632–77). Spinoza pushed to the most radical extreme the increasingly fraught division separating God and nature, going so far as to question the division itself as a pernicious fiction that natural philosophy needed to overcome. His monist concept of God/nature accordingly evaporated Christian natural philosophy by reducing all being to a single, rational world system driven by inexorable natural laws. His *Ethics* also argued for a notion of the good life that had no need for Christ or Christian redemption, since the good for Spinoza was nothing more than life led in accordance with the inexorable laws of a mathematically rational nature. In Spinozism, miracles have no reality, becoming purely natural phenomena whose apparent mystery is an illusion created by erroneous or incomplete rational understanding. God as saviour, and grace as the purposive intervention of the Deity to redeem sinners, is also declared a superstitious fiction imposed on a credulous populace by priestly authorities. Divine providence is likewise reduced to the deterministic outcomes of natural laws operating according to inviolable geometric logics.

Spinoza, in short, laid the foundation for a fully naturalist cosmology that eliminated God and Christian understandings altogether. But those living in the last decades of the seventeenth century did not experience a sudden and revolutionary reversal as a result of Spinoza's work. Spinozism was certainly present and influential, but for the vast majority of Europeans around 1700 the label represented less a creed or a rallying cry than a marker of the abominable limits across which natural philosophy should not transgress. The label 'Spinozist', in fact, was rarely used as a term of self-description and was more commonly deployed as an epithet against overly freethinking enemies. In this respect, its actual meaning in eighteenth-century learned discourse strongly resembles the way that the word 'communist' was used to stain any and all enemies of the United States in the political discourse of the Cold War.

None of the most important astronomical and cosmological thinkers of the eighteenth century were radically Spinozist in outlook, and it is a mistake, therefore, to see the modern ejection of the Christian God from natural science as a direct outcome of the spread of Spinozism. The claim is even more dubious when attached to Isaac Newton. Newton was hard at work on his monumental *Principia* when the first controversies surrounding Spinoza's work began to percolate, and while these battles provided an important context for the reception of his ideas they had no discernible influence on the development of his science. Overall, Newton found appalling any challenge to the God–nature duality foundational to Christian natural philosophy, and his *Principia* was not a rebellion against this understanding in the manner of Spinoza so much as a radically innovative effort to re-establish the traditional understanding on a new and better scientific footing.

To cut right to the heart of the matter, consider Newton's theory of universal gravitation as presented and defended in his epochal 1687 treatise. His argument derives from key innovations in the new naturalist mechanics and astronomy of the seventeenth century. From Kepler, he took the regular mathematical laws of planetary motion, especially the claim that planets move in elliptical orbits with the sun at one focus, sweeping equal areas in equal time during their motions. From Galileo and Huygens, he took the mathematical approach to mechanics that allowed him to treat bodies in motion as problems of pure geometry. These sources produced Book I of the *Principia*, where Newton used rigorous geometry to show (among other things) that planetary bodies moving in elliptical orbits according to Kepler's laws obey an inverse square law where the impetus of their motion, or the force animating it from the centre point, is inversely proportional to the square of the distance of the body from its orbital centre. Staying true to the pure mathematical approach of Book I, Newton showed that other laws also obtain mathematically. Book I, therefore, does not prove the inverse square law of universal gravitation as a physical law of nature, but provides a mathematical foundation for such a proof.

Newton needed to test his mathematical hypotheses empirically in order to demonstrate his law of universal gravitation as a physical property of nature, and this he accomplished in Book III of the *Principia*, subtitled 'The System of the World'. The new mathematical mechanics of the seventeenth century was again a central influence. By using experiments with pendulums and other complicated and innovative methods of inductive empirical reasoning, Newton clarified the difference between mass and weight, and established the quantitative law of their relation. He then set up a brilliant thought experiment—perhaps the most luminous moment in a book full of stunning bursts of insight—that physically confirmed the inverse square law as an empirical and mathematical property of observed phenomena. In this experiment, Newton imagined the moon to be a material body like any other, and then used two equally valid methods to calculate the rate of its 'fall to the earth'. The first was based on the mechanics of Galileo and Huygens, and it worked by imagining the moon as a massive body falling from the sky like a large stone. The second treated the moon as a planetary body moving according to Kepler's laws. In both cases, Newton found that the inverse square law derived in Book I determined the motive force that produced the actual observed motions of the moon, and that the result was identical to within a hundredth of a decimal point no matter which of his two methods he used.

The implications of this almost exact quantitative match are immense, and it is upon this rock of near perfect quantitative agreement that Newton built the epistemological foundation of his theory of universal gravitation. If the motive force driving the moon can be identically calculated using the quantitative laws of terrestrial mechanics and those of celestial mechanics, and if the results obtained each way are the same (or seemingly so with only an exceedingly small margin of difference), then there must be no difference between the two frames, and hence only one set of universal laws of motion applicable to pebbles on earth no less than planets in the heavens. Moreover, if one set of laws operates everywhere and at all times, then we no longer need to sustain the fundamental distinction between the earthly and the heavenly spheres that has been fundamental to ancient and medieval natural philosophy.

Newton also challenged the rationality that had sustained previous cosmological understandings when he made his inverse square law the basis of his new universal cosmological system. Why exactly does this of all laws hold the cosmos together, and what exactly is the force that appears to pull (or is it push?) matter in this as opposed to any other direction? Such questions would become a central theme of Newton's eighteenth-century reception, yet Newton famously refused to answer them, at least not in any direct or decisive way. When asked to explain the force of gravity that makes the inverse square law work, Newton famously proclaimed, '*hypotheses non fingo*', (I feign no hypotheses).[3] It was enough, he contended, that his natural philosophy described the motions of all bodies, both terrestrial and celestial, with stunning quantitative accuracy, and while he held out hope that an explanation of the gravitational force animating this elegant system would be found some day, he also refused to find any challenge to the validity of his system in his failure to offer such an explanation himself.

In many respects, Newton's refusal to speculate about the nature of gravity beyond his brilliant empirical-quantitative demonstration for its universality is his most significant act in the making of modern science. With this stance, physical science is defined in a newly limited way as empirico-quantitative description alone, while whole domains of what used be included within it in the broader category of natural philosophy are demoted as unnecessary for scientific explanation. Among these is religion, since to follow Newton strictly there is no reason, if our goal is solely empirico-quantitative scientific understanding, to inquire into who made the cosmos or toward what end. What science does is describe the natural phenomena we see with great quantitative and empirical precision, and if we want the answers to other questions about the cosmos, such as who made it, toward what end, and for what purpose, then we need to use other methods. As Newton summed up the matter: 'it is enough, that gravity does really exist, and act according to the laws which we have explained, and abundantly serves to account for all the motions of the celestial bodies, and of our sea.'[4] This statement opened the door to Laplace and his naturalistically optional 'God Hypothesis'.

Newton himself, however, was not intentionally stepping through the door of modern atheism by taking this stand, even if he was clearly trying to negotiate the relationship between science, naturalism, and religion in new ways. In fact, Newton's refusal to offer a full account of the nature of gravity was not designed to divorce natural science from Christianity but to make them newly connected, albeit according to his own idiosyncratic understanding of their proper relationship. For Newton, empirico-mathematical accounts of natural phenomena show us how God's creation works, and it is appropriate for us to pursue these inquiries since they reveal to us God's rationality and omnipotence. But Newton's God was not synonymous with his creation, and no error was so great in his mind as equating the two, as Spinoza had famously done in his God/nature monism.

Newton's hostility to this conflation was manifest in numerous ways, including some that we might call religious, while others were clearly scientific even by our modern secular understanding of the term. In the religious group were Newton's writings on biblical prophecy, and his work in the early modern science

[3] Isaac Newton, 'General scholium', *Philosophiae naturalis principia mathematica*, 2nd edn (London, 1713).

[4] Newton, 'General scholium'.

of chronology. Nature, in Newton's conception, was regular, harmonious, and mathematically lawful, but it was also the creation of an omnipotent, providential God who brought his son into the world as a saviour and redeemer, and who promised to return to judge the suffering faithful on the day of the apocalypse. Scripture revealed this intention, and its authority was absolute. Natural science could also play a role in helping humans to understand the pathways of providential history, however, and Newton's intellectual labours (which he would have called his scientific work) therefore included intensive inquiries into biblical prophecy, especially the Book of Daniel, where he searched for clues revealing the millennial prognostications that he knew to be contained in these holy books. His studies also extended to serious work in chronology, including a posthumously published treatise on the topic. Many still dismiss this work as a distraction from Newton's proper natural science, but it is more accurately described as an integral part of his fundamental project of understanding the full providential character of the mathematical God so elegantly revealed in the mathematico-empirical system described in the *Principia*.[5]

Newton also made the distinction between the mathematical God and the providential God clear in the *Principia*. Central to the Spinozist demonstration of a singular God/nature monism was the claim that each was beholden to a single natural law of geometric reason. Newton's *Principia* resembled Spinoza's *Ethics* in its deductive, geometrical approach, but it was decisively at odds with it in resisting any notion that the operations of nature could be grounded in geometric necessity. The complex, and ultimately analogous, structure of the empirico-mathematical argument for universal gravitation in the book was in fact religiously motivated in a way, since this method of reasoning worked epistemologically to keep mathematical necessity separate from the natural behaviours of the material world. Other aspects of the *Principia* also reveal Newton's hostility toward mathematically deterministic thinking.

Central to Spinoza's monist cosmos is the claim that a law of necessity drives God/nature toward rationally determined ends. If gravity were to be construed in these Spinozist terms, then the planetary system would be nothing more than a great clockwork moving in pure mathematical regularity without either a starting or an ending point. Also absent from this view would be any guiding hand of judgement, since for Spinoza ethics involves aligning oneself with the deterministic rationality of the cosmos, a stance that has no need for a Christian redeemer saving sinful man from his fallen self through divine grace.

Newton rejected all aspects of such a view. His convictions about providential history have already been noted, but he was similarly clear about his antipathy toward any deterministic or necessitarian understanding of the universe.

[5] See, e.g., Rob Iliffe, *Priest of Nature: The Religious Worlds of Isaac Newton* (Oxford: Oxford University Press, 2017); Steffen Ducheyne, *'The Main Business of Natural Philosophy': Isaac Newton's Natural Philosophical Methodology* (Dordrecht: Springer, 2012), pp. 264–9.

He went so far as to include in the *Principia* a demonstration that used empirical evidence to show that the planetary system was not stable and self-sustaining, and would require divine adjustment to return it to its regular course. Gottfried Wilhelm von Leibniz famously mocked this argument, accusing Newton's God of being such a bad watchmaker that he could not build a planetary system that did not need repeated repairs.[6] For Newton, however, God was not God unless he was continually sustaining the cosmos in its regular operations while at the same time retaining the capacity to intervene when he chose to do so. Without such an interventionist God, how were miracles possible, and by what mechanism were the sinners delivered of their sins? Certainly not through necessary geometric laws, Newton contended.

To avoid any possible confusion about his motives, Newton added a 'General Scholium' to the second edition of the *Principia,* published in 1713, which renders impossible any Spinozist interpretation of his work. In this text, gravity is explicitly distanced from any conception of it as a universal natural force, present inherently in all matter, which acts as a universal motive agent of the world. To describe gravity this way was to step onto the slippery slope leading directly to materialism and atheism, and Newton refused any such slips with respect to his natural science. Instead, for Newton, God is the ruler of all things, 'not as the world soul but as the lord of all', as 'Lord God *Pantokrator* [all-powerful]' a being 'eternal, infinite and absolutely perfect', but also 'a being with dominion'. 'God is one and the same always and everywhere', and he is omnipresent 'not only *virtually* but *substantially*. In him all things contain and move'. Putting the pieces together, Newton's Pantokrator was a 'Lord God' in the full sense, a being who is present to all change and also the lord over it, who governs through his will with providential wisdom.[7]

Many close to Newton also believed that he harboured a deeply divine and Christian (to him) understanding of gravity that he refused to state publicly for fear of its heterodox implications. Newton's young assistant and would-be protégé Fatio de Duillier was one who reported this suspicion, writing that, 'he would often seem to think that Gravity had its Foundation only in the arbitrary Will of God'.[8] If gravity were simply a naturalized term for the action of the omnipotent and omnipresent divine will of the Lord Pantokrator, then Newton would best be described as a kind of Christian pantheist who made nature the theatre within which the will and wisdom of the universal Christian God were

[6] Samuel Clarke, *A Collection of Papers, Which Passed Between the Late Learned Mr. Leibnitz, and Dr. Clarke, In the Years 1715 and 1716* (London, 1717), p. 3.

[7] The quotations are taken from Isaac Newton, 'The General Scholium', in Isaac Newton, *The Principia: Mathematical Principles of Natural Philosophy*, ed. and trans. I. Bernard Cohen and Anne Whitman (Berkeley, CA: University of California Press, 1999), pp. 940–41.

[8] The comment is cited in Richard Westfall, *Never at Rest: A Biography of Isaac Newton* (Cambridge: Cambridge University Press, 1980), p. 495, and Westfall notes that it 'is found on a sheet among his papers written after 1701': p. 495, n. 82.

performed. Newton's definition of absolute space in the 'General Scholium' as 'God's sensorium' generated similar suspicions on account of the implication that space is part of God even as it is a distinct arena in which he perceives and acts. Each of these conceptions avoided Spinoza's monist materialism, but at the cost of redefining the traditional distinctions between God and nature in ways that were equally heterodox with respect to the traditional Christian God–nature duality. In his 'General Scholium', Newton accordingly attempted to eradicate precisely these fears by defining God as a being distinct from nature, but nevertheless the omnipotent governor of it. For him, God ruled over the world as a 'Pantokrator', or judge and overlord, seeking to bring about the righteous ends he intended for his creatures. That God's rule was overwhelmingly rational was revealed in the mathematical elegance of the *Principia*, but Newton also insisted that in his conception the divine Pantokrator was an omnipotent being distinct from the mathematically ordered universe he created, and a free, sovereign master over it.

Leibniz once again revelled in boxing Newton into awkward logical corners through the use of these attempts at theological clarification. If bodies move in an absolute and infinite space, yet God is not identified with this absolute space, where precisely, he inquired, is God?[9] Despite the logical gymnastics that such critiques compelled, Newton refused to allow his science to be reduced to an all-consuming naturalism. Mathematics and empirical physics were always for him new ways of grounding traditional Christian natural philosophy, and for this reason, to see Newton as actively promoting the move toward anti-religious scientific naturalism in the West is to misunderstand completely his actual historical work.

THE RECEPTION OF NEWTON AND THE MOVE TOWARD FULLY NATURALIST SCIENCE

Newton did not deliberately push cosmology in a purely naturalistic direction, and he in fact worked vigorously to stem precisely this tide. Yet in spite of his explicit and vigorous efforts to produce a different outcome, the reception of his work led either to an increasing identification of God with nature or to the idea of a nature whose operations could be fully accounted for simply by referring to its own inherent laws. It also led to the increasing condensation of all these modern ideas into the single label 'Newtonianism' that located, however inaccurately, the historical source for this naturalizing turn in the work of Isaac Newton.

[9] Clarke, *A Collection of Papers*, p. 189.

A key impetus toward this newly naturalist understanding of Newtonianism was the Spinozist reading of his theory of universal gravitation, which took hold in the very first decade of the *Principia*'s reception. The Irish radical John Toland exemplified this strand of eighteenth-century 'Newtonianism'. Paying little attention to the complex rigour of the *Principia*—which, admittedly, few were equipped to handle in any case—Toland seized on the claim that a universal attractive force responsible for driving earthly and celestial bodies existed in matter, and then conflated this idea, contrary to Newton's stated wishes, with Spinoza's conception of active, monist matter. This made Newtonian gravitational attraction the central animating force driving the God/nature unity in Spinoza's system. Combining these quasi-Newtonian formulations with other provocative ideas drawn from Spinoza that had no Newtonian pedigree whatsoever—the natural character of republican government, for example, and the fictional nature of the Christian miracles—Toland went on to produce an influential radical stew that he circulated under the name of 'Newtonianism'.[10]

Toland and Newton were contemporaries, and during the final decades of his life Newton made strenuous, if often indirect, efforts to thwart this very development. As already noted, his 'General Scholium' published in the 1713 second edition of the *Principia* was one attempt to demonstrate the Christian foundations of Newtonian natural philosophy. But even more important were the annual series of Boyle Lectures, begun in 1692, that offered a systematic commentary on the relation between natural science and religion. Robert Boyle, the distinguished chemist and founding member of the Royal Society of London, had endowed the programme, asking that a series of sermons be preached each year in churches throughout London emphasizing the harmony between natural science and Christianity. Newton was neither a cleric nor the sort of intellectual who would have been asked to deliver these sermons, but those who did often consulted closely with him, especially Richard Bentley and Samuel Clarke, who delivered the Boyle Lectures in 1692 and 1704–5 respectively. Each targeted what they saw as the new atheism brewing in the materialist and naturalist arguments of people like Toland. Each also used his public sermons (which were later published as widely read books) to systematically defend an orthodox reading of Newton's *Principia* that explicitly distanced its science from interpretations like those offered by Toland.

Bentley's correspondence with Newton, as he went about preparing his Boyle Lectures, is today one of the best sources available for examining Newton's religious understanding of his own work. Peter Harrison's chapter summarizes well the key issues. But since these letters remained unpublished until much later, and since Newton kept his own, voluminous religious writings secret,

[10] John Toland, *Letters to Serena* (London, 1704), Letters 4 and 5; Jeffrey R. Wigelsworth, 'Lockean Essences, Political Posturing, and John Toland's Reading of Newton's *Principia*', *Canadian Journal of History* 38 (2003): pp. 521–35.

for fear (well-founded) that his exceedingly heterodox religious views would cause him difficulties, they had no direct impact on contemporary opinion. Only in the less-than-limpid sentences included in the 'General Scholium' did Newton offer a direct intervention into the religious and philosophical debates triggered by the theory of universal gravitation. In the absence of direct participation in these debates it was left to Bentley, Clarke, and the other Boyle Lecturers to articulate the orthodox Christian Newtonianism of Newton's own texts from the London pulpits and in the texts of the lectures which they published.

This orthodox Newtonian discourse made clear the traditional Christian nature of his science, especially the insistence on maintaining the God–nature distinction. Spinozist Newtonianism, by contrast, challenged this very divide, and the debate was further complicated when Leibniz found himself accused by the English Newtonians of plagiarizing Newton's infinitesimal calculus in 1709. The so-called priority dispute that ensued was a public intellectual fight of titanic proportions, and while it began with questions about the calculus, it quickly escalated into a violently heated battle about the orthodox foundations of natural philosophy.

Leibniz initiated the escalation in 1717 when he sent a brief letter from Hanover to his patron, Princess Caroline, who was in London at the time with another of his patrons, the newly installed Hanoverian King George I. His letter purported to instruct the princess about the dangerous tendencies of English natural philosophy.[11] Leibniz had Newton in his sights, and in his analysis, Newton's *Principia* was positioned as a source for the newly aggressive Spinozism and deism percolating in Europe. Samuel Clarke, with Newton's encouragement, immediately took up the pen in response, offering Leibniz the full arsenal of orthodox Newtonian rebuttals which had been forged over the previous decade. Leibniz had not written to produce a gentlemanly discussion, however, so in response, he channelled all of his immense intellectual powers toward a scathing rebuttal of the orthodox Newtonian position. Leibniz was no Spinozist however, so the result, contained in five increasingly lengthy and heated letters that were published in English in 1719, and then in a widely read French translation a year later, was less a victory for either side than a cosmological battle royal that ultimately illuminated the Spinozist position as much as either opponent's supposed antidote to it.

Meanwhile, between 1704 and 1715 Newton also published the various editions of his *Opticks*, books that contained concluding queries speculating more freely about the nature of matter and force and their interaction in the physical world. Despite Newton's insistence that his queries were only hypothetical conjectures lacking the rigour of actual scientific demonstrations, many of them could be read, and were read, as essays supporting the naturalistic conviction that gravity was a universal active agent, present in all matter, which served as the motive

[11] Clarke, *A Collection of Papers*, pp. 3–5.

force behind everything from chemical affinities among substances to the attractions and repulsions of magnetic and electric bodies. Newton's actual intentions in publishing the queries to the *Opticks* were irrelevant to these judgements, since, once printed, the queries became new evidence that Newton held matter to be active by nature and to be governed by rational mathematical laws. The distance between this understanding and Spinoza's geometrically demonstrated active matter materialism was often hard to sustain, and it further shrank, at least outside England, when John Locke's *Essay Concerning Human Understanding* further suggested the possibility of a materialist understanding of human consciousness. By the time of Newton's death in 1727, he and Locke were becoming the twin giants of English philosophy, and to the extent that each could be read as supporting a naturalistic materialism (Locke's actual position, like Newton's, was much more complicated), their imagined affinity supported the conflation with Spinozist understandings of both sets of views.

After Newton's death a new generation came of age with no reason to care about his original intentions, and the founding debates of the first three decades of the eighteenth century began to take on a life of their own. At one end of the spectrum were clerics who developed a general notion of 'subversive natural philosophy', which encompassed all of these naturalistic arguments, be they Spinozist, Newtonian, Leibnizian, Lockean, or some eclectic mix. Louis-Bertrand Castel (1688–1757), the only French Jesuit ever made a Fellow of the Royal Society of London, well illustrates the point. From 1720 to 1745 he combined work as a mathematical teacher at the prestigious Parisian Collège Louis-le-Grand (of which Voltaire and d'Alembert were alumni) with work as the science editor of the widely read Jesuit monthly the *Journal de Trévoux*. Learned journals were the main theatres of Enlightenment scientific debate, and Castel was a firebrand who kept the Jesuits at the forefront of contemporary intellectual discussion while also ensuring the presence of Catholic arguments in this space. He saw Newton's theory of universal gravitation as a classic misapplication of mathematics to natural philosophy, and while he had nothing but praise for the mathematico-empirical description of Newton's system of world, he also made clear that it was the most dangerous sort of error to derive physical and metaphysical understandings from mere mathematico-empirical descriptions.[12] Other clerics advanced the same cause in different ways, including the French Cardinal Melchior de Polignac, who used a Latin poem (quickly translated into French), modelled on Lucretius's *De rerum natura*, titled *L'anti-Lucrèce*, to pre-empt the new Epicureanism he saw brewing in Newtonian natural philosophy.

By 1740, arguments like these were immersing the European scientific community in a massive 'Newton war' that pitted defenders of the materialist

[12] Louis Bertrand Castel, *Le vrai système de physique générale de M. Isaac Newton exposé et analysé en parallèle avec celui de Descartes, à la portée du commun des physiciens* (Paris, 1743).

and anti-Christian readings of Newton against the moderate defenders of Newtonian orthodoxy and the clerical opponents of each.[13] In the wake of this contestation, many returned to Newton's now-famous declaration that his was an empirico-mathematical account of gravity alone, and not a physical or metaphysical explanation of anything. From this position, these moderate Newtonians worked to create a space for a natural science free of the contestatory distractions of philosophy and religion. The Royal Academy of Sciences in Paris had been founded in 1666 with a pledge to avoid philosophical and religious sectarianism, and under the direction of a new generation of mathematicians that came of age after 1740—Alexis Clairaut, Jean le Rond d'Alembert, Pierre Louis Maupertuis, and the Comte de Buffon, to name only the luminaries—a practice of Newtonian celestial and terrestrial mechanics was established that explicitly drew epistemological lines between the pursuit of empirico-mathematical description on the one hand and enquiries into its metaphysical or theological intentions on the other. Voltaire also advocated for precisely this separation in his defence of Newtonianism, adding a section to the second, authoritative edition of his *Éléments de la Philosophie de Newton* called 'La Métaphysique de Newton', which argued that the special strength of Newtonian science lay in its explicit rejection of metaphysics altogether. These French scientists and *philosophes* were Laplace's teachers, and it was from them that he formed the idea of a Newtonian science detached from religion, and a celestial mechanics where God could be a hypothesis, entertained or not as one preferred.

Yet this developing positivist comfort with the epistemological distinction between natural science on the one hand and metaphysics and religion on the other did not become dogma all at once. For some, science without metaphysical explanation was no science at all, and many continued to push for a more unified, natural philosophical understanding of universal gravitation, including those that explicitly asked the God question. One interesting case was Voltaire's partner, the Marquise du Châtelet (1706–49), who never accepted that natural science could be empirical mathematical description alone. Following Leibniz, she searched for the rationality that explained the inverse square law, and she was supported in her work by her other partner, Pierre Louis Maupertuis (1698–1759), who also sought to discern the rational divinity at work in the Newtonian cosmological system. For Châtelet, this involved returning to Leibniz's principle of sufficient reason and trying to derive the laws of mechanics deductively from first principles. For Maupertuis, the search led to the articulation of universal principles, such as his Least Action Principle, that posited a higher teleological reason for the regularity of the universe. Each worked to find a new naturalist reason for the cosmos comparable to that of the providential God in Christian natural philosophy, but each also conceived of

[13] For more details, see Shank, *The Newton Wars*.

their organizing *logos* as a natural reason free from any attachment to scriptural revelation or Christian doctrine. In this way, they also pushed toward a more fully naturalized Newtonianism.

Others shared in this rationalist naturalizing quest, and another approach was to try to answer Newton's call for a physical-mechanical explanation of the force of gravity that his science suggested must exist. For many, gravity could only be explained rationally by conceiving of it as an effect produced by some fundamental mechanical action or process. Both at the end of his 'General Scholium' and in some of his queries in his *Opticks*, Newton had suggested such a possibility, hypothesizing that gravity could result from an all-pervasive ether that generated natural phenomena such as corporeal heaviness or bodily attraction across seemingly empty space through the vibrations and impacts of its material action. To some, the ether became the Holy Grail that, if discovered, would resolve all mysteries of gravity once its mechanical principles were fully understood. For others, it was just Spinozism by another name, since the rational action of an all-pervasive, determinist, and active ether was hard to distinguish from Spinoza's active God/nature monism.

Another strategy, adopted by the young Immanuel Kant (1724–1804), among others, involved saving the God–nature distinction by moving God ever further away from the actual operations of the cosmos. This deistic option allowed for a fully quantitative and empirical accounting of the cosmos down to the minutest detail, while stopping short of the Spinozist claim that God was inseparable from, and inherent within, the nature that was the object of natural science. It also saved a providential and judgemental God working to accomplish a transcendental cosmological purpose, even if it located the source and action of this teleology entirely in the initial moments of creation. In this way, Kant retained God as the mathematical architect of the cosmos, and preserved Christian morality as the universal human morality, while evacuating the actual substance of the Christian theology that anchored this ethics—namely, the idea of an omnipresent and purposive God who operates through judgement and the delivery of miraculous grace. Newton had predicted just this outcome if the cosmos were ever reduced to determined mathematical regularities alone, and Kant's naturalistic Newtonianism was in many ways the realization of the very conception he abhorred.

Alongside all these various efforts to save God while increasingly removing him from the nature he was said to have created were numerous Spinozist challenges that simply erased the duality altogether. Maupertuis's search for a universal rationality underlying terrestrial and celestial mechanics eventually gave way in the 1750s to a more aggressive materialism grounded in the Spinozistic claim that a universal active force, or spirit, must animate all things. In his *Essai de Cosmologie*, Maupertuis found God (or the organizing *logos* of nature) in a combination of mathematical rationality and the fecund material activity empirically evident in nature. Central to his argument was a union of universal gravitation, as revealed in the motions of physical bodies, with the force of life observed in the growth and activity of living organisms. A new naturalistic and materialist science of life further began to take shape after 1750 as Newtonians with materialist leanings extended the idea of universal laws of matter and motion to the living world. Denis Diderot played a vigorous role in this movement, and he was joined later by other thinkers, including Erasmus Darwin in England, the German Romantics, and, in Italy, Luigi Galvani, who first showed that animal life and electricity were intertwined. In this way, the stage was set for Mary Shelley's suggestion in *Frankenstein*, first published in 1818, that humans were capable of performing the ultimate divine act: generating human life itself, thanks to the natural science of 'Enlightenment moderns' like the 'Newtonians'.

Politics also played a role, for when the French revolutionary state used the authority of natural law to place the Christian church under its control, and then attempted to supplant Christianity altogether by instituting a Cult of the Supreme Being as a new national religion, the revolutionaries were acting upon the new conflation of God and nature that had been introduced via Spinoza, and then catalysed, whether unwittingly or not, by the reception of Newtonian science. Boullée's monumental cenotaph honouring Newton (Figure 4.1), produced at precisely this moment of radical political upheaval in France, is a striking illustration of the emerging secular understanding of Newton that was part and parcel of this new revolutionary worldview. Treating his name as synonymous with the natural motions of the stars, which this planetarium-like monument puts on display, the cenotaph honours Newton as a God-like prophet of secular science, one who has helped humanity to progress by leading it out of the superstitious fog of religion and into the light of the true natural science. Newton would have found this description of himself and his work repellent, but by 1800 it did not matter: Boullée's understanding of Newton's importance and legacy was becoming the mainstream view all across Europe.

William Blake (1757–1827) also articulated this understanding, albeit in a critical, anti-Newtonian way, when he pictured Newton as the 'Divine Geometer' in a painting from 1795, later reworked and reissued in 1805 (Figure 4.4). We have already seen the reconfiguration of the God-nature-human relationship in other early modern European paintings, and Blake's image brings this series to a close, along with the argument of this chapter.

In the painting, the mortal, though superhuman, Newton is depicted as the divine architect incarnate using mathematical science, not as a human tool that mirrors God's rationality but as an incarnation of natural reason itself and the human mastery over the universe. Jesus-God is nowhere in evidence, and Newton is not only the master of the creation that unrolls before him, he is the beginning and end of this cosmos. In short, the Christian God–nature distinction has completely departed, replaced by a natural world that is wholly constituted

Figure 4.4. William Blake, *Newton*.
Source: Tate Images.

and ruled by mathematical reason, and a world available to complete human mastery so long as the tools of secular science are used.

Blake found the naturalist cosmos assumed in this Enlightenment understanding dispiriting, but Laplace was energized by it. When he reported to the new emperor Napoleon with ardent revolutionary and atheist fervour that he saw no need for a God hypothesis when explaining the mechanism of the heavens, he was articulating the 'Newtonian' views of many other moderns of his day. Ever since, people have echoed Laplace's formulation and conflated it with Newton's own views. But the two were not identical. Newton's 'Newtonianism' and the one which came to be held by many across Europe by the end of the eighteenth century were two very different things; in trying to disentangle them from one another, this chapter has worked to disrupt the simple story that asserts their equation and then explains it as an inevitable outcome produced by some imagined rational progress of scientific knowledge. Newton's *Principia* does not need to be defended as a major step forward in the development of astronomy and physics, but by showing how Newton's pioneering work was anything but a direct agent in the Enlightenment's isolation of cosmological science from God and religion, we have seen how science and naturalism are separate things that historically have not developed in lock-step with one another.

5

God and the Uniformity of Nature

The Case of Nineteenth-Century Physics

Matthew Stanley

Newton had something for everyone. The natural philosophers of the time noticed that he was perfectly comfortable discussing the possible religious meaning and origin of many parts of his new theory of nature. His 'General Scholium' told readers that 'much concerning God... does certainly belong to Natural Philosophy' and freely considered the interaction of God with the physical world.[1] And yet his rules of reasoning seemed to restrict physical explanation to natural causes. As natural philosophy gave way to professional physics in the nineteenth century, his universe continued to be used to justify both theistic and naturalistic approaches to nature. Wherever you were on the spectrum of belief, his work had something for you.

Newton was deliberately ambiguous about the deep meaning of his physics, seen in his refusal to say he understood the cause of gravity. This created a conceptual space that allowed physics to be framed as either a naturalistic or theistic enterprise. Indeed, representatives of both groups seized on Newton as the exemplar of their way of doing science. While virtually no physicist promoted direct divine intervention in the universe, many felt validated by Newton in thinking of their basic conceptual tools as fundamentally religious entities. Looking back from the twenty-first century, their practice of physics appears to adhere to scientific naturalism as we understand it. Their explanations of phenomena invoked only natural processes, laws, and events, with no recourse to direct supernatural intervention or causation. However, to the historical actors themselves, the practice of physics was closely bound with the divine. This practice, which I will here call 'theistic physics', embraced natural laws, energy, the unity of nature, and causality as religious concepts throughout the nineteenth

[1] Quoted in I. Bernard Cohen and Richard Westfall (eds), *Newton* (New York: Norton, 1995), p. 342.

century. Causality was also the pivot point to a fully naturalistic approach to physics that took hold in the twentieth century, and which eventually claimed the laws of nature as its own. Despite this, physics remained a fertile ground for a new wave of non-naturalistic approaches at the end of that century. The central practices of physics were flexible enough to be justified and motivated in both the naturalistic and theistic worldviews.

LAWS

Physics in its modern sense emerged at the beginning of the nineteenth century.[2] The most essential concept for the practice of the discipline was that of natural laws: the idea that events and phenomena happen according to certain fixed patterns, usually of a mathematical form.[3] Newton's influence on shaping this concept was enormous, and his argument that laws were the result of divine creation was widely accepted. The Danish physicist Hans Christian Oersted (1777–1851), combining Christianity with Schelling's *Naturphilosophie*, elegantly described this view:

> The progress of discovery continually produces fresh evidence that Nature acts according to eternal laws, and that these laws are constituted as the mandates of an infinite perfect reason; so that the friend of Nature lives in a constant rational contemplation of the Omnipresent Divinity.... The laws of Nature are the thoughts of Nature; and these are the thoughts of God.[4]

An important exception to this general view was revolutionary France, where natural laws were often considered to be completely independent of God and supportive of an atheistic worldview. This threat to the standard theistic interpretation of natural laws stimulated British men of science to articulate explicitly the deep links they saw between these foundations of scientific practice and a Christian worldview. It was Victorian Britain that most loudly argued for the necessity of theism for physics.

The 1830 *Preliminary Discourse* by the authoritative man of science John Herschel (1792–1871) was hugely influential along these lines. It was a formative text for a generation or two of scientists and helped articulate scientific orthodoxy in the early nineteenth century. Herschel stressed that the regularity of natural phenomena was an essential part of the universe. Those phenomena that

[2] Iwan Rhys Morus, *When Physics Became King* (Chicago: University of Chicago Press, 2005).

[3] Peter Harrison, 'The Development of the Concept of the Laws of Nature', in *Creation: Law and Probability*, ed. Fraser Watts (Aldershot: Ashgate, 2008), pp. 13–36. There was enduring controversy over precisely what natural laws were, and whether they were descriptive or active in nature, throughout this period. The subtleties of this debate cannot be addressed here.

[4] Quoted in Baden Powell, *Essays on the Spirit of the Inductive Philosophy, the Plurality of Worlds, and the Philosophy of Creation* (London: Longman, Brown, Green, and Longmans, 1855), p. 113.

'happen uniformly and invariably' could be considered as laws of nature, which were both the goal and proper focus of attention for the natural philosopher.[5] He rejected the idea that these laws, created by the 'Divine Author of the universe', could change or be interrupted. Their uniformity came from 'the constant exercise of his direct power in maintaining the system of nature, or the ultimate emanation of every energy which material agents exert from his immediate will, acting in conformity with his own laws'.[6] The laws of nature were stable only because of God's constant and ubiquitous action. If matter and energy were left to their own devices, the universe would be a place of chaos. The orderliness of natural phenomena could only be explained, Herschel argued, if God ensured that it was so.

The Cambridge mathematician William Whewell (1794–1866) developed these arguments further in his contribution to the Bridgewater Treatises. Like Herschel, he saw the laws of physics as created by God. He explicitly quoted Newton to support his claim that God's universal agency was 'the only origin of any efficient force'.[7] He further argued for the possibility of learning of the Deity's 'intelligence and benevolence, superintendence and foresight' through the study of those laws, and that the minds best suited for doing science were those capable of seeing the evidence for a creator.[8] Whewell emphasized a kind of natural theology particularly compatible with physics. Rather than pointing to complexity, as William Paley did, he put forward simplicity and precision as the mark of design. The law of gravity was of the simplest mathematical form that allowed for stable planetary orbits; the laws of heat were of precisely the form that allowed for the formation of clouds and rain. Whewell stressed that these laws could have had many different forms, thus making the case that their actual form was the result of deliberate choices on the part of God.[9] The possibility that physics could have been otherwise showed that the laws had been designed for the benefit of humanity.

ENERGY

Beyond the broad concept of natural laws, many physicists also saw their explanatory entities as having a religious character. Particularly notable among these concepts was energy (sometimes referred to as force in the early part of

[5] John F. W. Herschel, *A Preliminary Discourse on the Study of Natural Philosophy* (London: Longman, Rees, Orme, Brown, and Green, and J. Taylor, 1830), p. 119.

[6] Herschel, *Preliminary Discourse*, p. 137.

[7] William Whewell, *Astronomy and General Physics Considered with Reference to Natural Theology* (London: W. Pickering, 1836), p. 362; Matthew Stanley, 'A Modern Natural Theology?', *Journal of Faith and Science Exchange* 3 (1999).

[8] Whewell, *Astronomy and General Physics*, pp. 4, 304.

[9] Whewell, *Astronomy and General Physics*, chaps 9–10.

the nineteenth century). Energy was an immensely useful idea with properties seen by many investigators as having clearly theistic origins. Most commonly, it was argued that energy's conservation and convertibility—the idea that the quantity of energy always remained the same despite changed appearances—were the result of solely divine agency. The physicists Michael Faraday (1791–1867) and James Joule (1818–89) believed that energy could be used and manipulated by man, but its creation or destruction was something completely other, apparently reserved only for the Deity's direct creative acts.[10] It was, therefore, an agency designed and placed by God into the physical world. While Faraday and Joule disagreed about the details of energy conversion, they both agreed with the dominant interpretation that those processes showed its religious character.

Another property of energy seen as having religious significance was the second law of thermodynamics. This law, articulated mid-century, was described in several different ways. For our purposes, it postulated the continual decrease in energy available for physical processes (natural or artificial). This implied that a system left to itself will run out of useful energy in a finite time. William Thomson (1824–1907), later Lord Kelvin, was one of the major figures involved in articulating and interpreting the law. He immediately drew religious conclusions from it, most notably that it implied both a discrete beginning and end to the universe, which Thomson happily associated with the Christian creation and eschaton. This sense that God literally built his cosmic history into the laws of nature was both amazing and obvious to Christian physicists, and thermodynamic arguments for the reality of God's creation were widespread well into the twentieth century. Even further than this broadly theistic sense of creation, Thomson saw the second law as being deeply consonant with scripture. Writing with his collaborator P. G. Tait (1831–1901), he declared that

> we have the sober scientific certainty that heavens and earth shall "wax old as doth a garment" [Ps. 102:26]; and that this slow progress must gradually, by natural agencies which we see going on under fixed laws, bring about circumstances in which "the elements shall melt with fervent heat" [2 Pet. 3:10].[11]

This was only one example of the general sense among religious physicists at the time that God carried out plans and actions through natural laws. They did not welcome the idea of God interrupting the natural order to achieve certain results, instead envisioning his agency coming through far-sighted uses of those laws.[12]

[10] Geoffrey Cantor, *Michael Faraday: Sandemanian and Scientist* (London: Palgrave, 1993); Crosbie Smith, *The Science of Energy* (Chicago: University of Chicago Press, 1998), p. 63.

[11] Crosbie Smith and M. Norton Wise, *Energy and Empire: A Biographical Study of Lord Kelvin* (Cambridge: Cambridge University Press, 1989), p. 535.

[12] For theological perspectives on this, see Baden Powell, *The Connexion of Natural and Divine Truth* (London: John Parker, 1838); Frederick Temple, *The Relations Between Religion and Science* (London: Macmillan, 1884).

Despite his deep belief in the reality of divine action, Thomson demanded that physicists restrict themselves to explanations in terms of natural laws: 'If a probable solution [to any scientific problem], consistent with the ordinary course of nature, can be found, we must not invoke an abnormal act of creative power.'[13]

UNITY

This stress on the continuity and uniformity of divine action through natural laws led to an important conclusion: that the unity of nature itself was an indication of God's design. James Clerk Maxwell (1831–79) famously addressed the British Association for the Advancement of Science on the uniformity of molecular spectra throughout the universe. He argued that hydrogen in distant stars, or liberated from rocks buried since time immemorial, was identical to that in the Cavendish laboratory. This incredible uniformity among matter scattered through space and time indicated the hand of a divine manufacturer: uniformity could only be explained through God.[14] Maxwell powerfully linked the unity and order of nature not just with divine creation itself but also with the role of humans in that creation. In a letter to a bishop, he suggested that we can see 'wisdom and power' in the uniformity of natural laws just as effectively as in the beneficial adaptations of living creatures: 'uniformity, accuracy, symmetry, consistency, and continuity of plan are as important attributes as the contrivance of the special utility of each individual thing.'[15]

This deep-seated faith in the divinely ordained unity of the natural world had a critical influence on Maxwell's work in physics. In a playful essay, he wondered why we should think that nature is organized like a book (with each chapter leading to the next, with a unified argument) instead of like a magazine (with each article standing on its own). This was the difference between a natural order in which laws were connected and interrelated versus one in which every natural process worked on its own terms.[16] But Maxwell was convinced that nature was like a book, because he knew that the author of that book had designed the universe with unity, consistency, and

[13] Joe Burchfield, *Lord Kelvin and the Age of the Earth* (New York: Science History Publications, 1975), p. 48.

[14] James Clerk Maxwell, 'Address to the Mathematical and Physical Sections of the British Association' (1870); repr. in *Scientific Papers of James Clerk Maxwell*, ed. W. D. Niven, vol. 2 (Cambridge: Cambridge University Press, 1890), p. 224.

[15] Maxwell to Charles John Ellicott, Bishop of Gloucester and Bristol, 22 November 1876, in *Scientific Letters and Papers of James Clerk Maxwell*, ed. P. M. Harman, vol. 3 (Cambridge: Cambridge University Press, 2002), p. 417.

[16] Maxwell, 'Are There Real Analogies in Nature?', in *Scientific Letters and Papers*, vol. 1, pp. 381–2.

simplicity. Physicists, he wrote, needed to understand that 'the laws of nature are not mere arbitrary and unconnected decisions of Supreme Power, but that they form essential parts of one universal system, in which infinite Power serves only to reveal unsearchable Wisdom and eternal Truth'.[17] This principle was an essential part of Maxwell's famous work uniting electricity, magnetism, and light.[18]

Decades earlier, Whewell had made similar arguments for the religious significance of the interrelationship of the laws of physics. He contended that in order for humans to see, there had to be a perfect harmony among the laws governing light, air, and the ether. This complex interaction of simple laws for complex ends was held to be a clear indicator of divine intent:

> The contemplation of the material universe exhibits God to us as the author of the laws of material nature; bringing before us a wonderful spectacle, in the simplicity, the comprehensiveness, the mutual adaptation of these laws, and in the vast variety of harmonious and beneficial effects produced by their mutual bearing and combined operation.[19]

Whewell wrote that the ether was one of the best examples of this interconnection and unity. Its role as a link among many different areas of physics showed its divine purpose.

Many other physicists also saw the unifying aspect of the ether as having supernatural significance.[20] Balfour Stewart (1828–87) and P. G. Tait's famous *The Unseen Universe* argued that rigid deductions from the laws of physics resulted in a realization that the ether was a connector to a higher, invisible reality. This invisible reality was a spiritual realm in which human souls would live on forever very much as described by Christian doctrine. The idea that the ether provided unity to matter and spirit just as it did to light and electricity became quite popular. Oliver Lodge (1851–1940) continued to promote this view well into the twentieth century. Integrating his experiences with spiritualism, Lodge constructed a detailed system of how the ether functioned as the medium for interaction between mind and matter. After death, the mind would live on in vibrations of the ether. The ether itself, therefore, was 'the primary instrument of Mind, the vehicle of Soul, the habitation of Spirit. Truly it may be called the living garment of God'.[21] As justification for these sorts of speculation, he again pointed back to Newton.

[17] Maxwell, 'Inaugural Lecture at King's College London', in *Scientific Letters and Papers*, vol. 1, p. 670. 'Unsearchable Wisdom' is probably a reference to the *Westminster Confession of Faith* (1717), chap. 5, article 4.

[18] Matthew Stanley, 'By Design: James Clerk Maxwell and the Evangelical Unification of Science', *British Journal for the History of Science* 45, no. 1 (2012): pp. 57–73.

[19] Whewell, *Astronomy and General Physics*, p. 251.

[20] Geoffrey Cantor, 'The Theological Significance of Ethers', in *Conceptions of Ether*, ed. Geoffrey Cantor (Cambridge: Cambridge University Press, 1981).

[21] Oliver Lodge, *Ether and Reality* (London: Hodder and Stoughton, 1925), p. 179.

CAUSALITY

Shared belief in unity did not mean the theists and the naturalists always got along. An assault on the long-entrenched tradition of theistic physics came in the second half of the nineteenth century as a new group of scientific naturalists sought to replace it with their own worldview. This group—the X-Club and its allies—largely agreed with the theists on basic principles of using natural laws, the value of hypotheses, and so on. However, they argued that these principles were incompatible with a traditional theological worldview and they were heavily critical of integrating religious and scientific thought (despite their methodological similarity to Maxwell and others). One of this group's leaders was John Tyndall (1820–93), the Irish physicist whose fiery lectures and writing helped establish the very notion of naturalistic physics.

Tyndall stressed a particular facet of the uniformity of natural laws: causality. If the laws of physics were truly uniform, he argued, then every event was fully and solely caused by previous events. He declared that the set of energy laws, for instance, 'binds nature fast in fate'.[22] This left no room for supernatural intervention:

> Now, as science demands the radical extirpation of caprice and the absolute reliance upon law in nature, there grew with the growth of scientific notions a desire and determination to sweep from the field of theory this mob of gods and demons, and to place natural phenomena on a basis more congruent with themselves.[23]

The same uniformity that was obviously divine to Maxwell was, to Tyndall, the reason that divine action was impossible. Tyndall insisted that the deterministic causality of physics allowed him to 'wrest from theology, the entire domain of cosmological theory'.[24] According to him, science, as a complete scheme of the universe, could have no interaction with religion other than accepting its surrender. Before the advent of science, Tyndall said the unlearned masses had no option other than filling the world with 'witchcraft, and magic, and miracles, and special providences'. This appeal to the unseen was a natural human behaviour, but foolish and dangerous nonetheless. The power of physics would simply squeeze the world until nothing else remained: 'the law of gravitation crushes the simple worshippers of Ottery St. Mary, while singing their hymns, just as surely as if they were engaged in a midnight brawl.'[25]

Tyndall was merciless in drawing this line during the debates of the early 1870s over the efficacy of prayer.[26] He said that once science had demonstrated

[22] John Tyndall, 'Address', in *Report of the Forty-fourth Meeting of the British Association for the Advancement of Science; Held at Belfast in August 1874* (London: John Murray, 1875), p. lxxxviii.

[23] Tyndall, 'Address', p. lxvii.

[24] Tyndall, 'Address', p. xcv.

[25] Tyndall, *Fragments of Science for Unscientific People*, 4th edn (New York: D. Appleton and Company, 1875), pp. 67, 49.

[26] Robert Bruce Mullin, 'Science, Miracles, and the Prayer-Gauge Debate', in *When Science and Christianity Meet*, ed. Ronald L. Numbers and David C. Lindberg (Chicago: University of Chicago Press, 2003).

the uniformity of nature, the age of miracles was past. There was no indication that physics was ever suspended, and therefore there was no possibility of miracles. The only way out, he said, was to retort: 'How do you know that a uniform experience will continue uniform? You tell me that the sun has risen for six thousand years: that is no proof that it will rise tomorrow; within the next twelve hours it may be puffed out by the Almighty.'[27] He said someone attacking uniformity in this way could barely function in the normal world. They had no reason to believe that Jack and the beanstalk was not a true story, since perhaps the natural laws governing bean growth had been suspended at some time. The rhetorical move here was a clear one: someone who believed in the miracles of the Bible or in God answering a prayer for their sick child was no different than someone who believed in fairy tales.

The precise role of causality in physics was a subject of significant controversy at exactly this time, however. The introduction of sophisticated statistical methods led to deep questioning about how deterministic nature truly was. The increasing application of physics to living things raised painful questions about the nature of human consciousness and the soul. Physicists could predict the motion of comets for centuries to come—if the soul was simply a by-product of the material brain, surely they could predict the behaviour of the soul as well? Tyndall and the scientific naturalists concluded that this was obviously the case. But Christian physicists were reluctant to accept this and found themselves in a difficult position. They had no desire to give up the uniformity of physical law (which they saw as divinely ordained) or the possibility of human free will (which they saw as essential to Christianity). Many physicists saw these questions as well within their area of expertise and developed sophisticated systems to allow both uniformity and free will. G. G. Stokes and others argued that this could be done with a model where the soul acted as a steering force rather than an energetic source. Maxwell even incorporated these ideas into his famous 'demon' thought experiment in which a conscious being could circumvent the second law of thermodynamics.[28]

These issues were still alive and controversial in the early twentieth century when quantum physics appeared. The quantum world's apparent reliance on probability rather than determinism seemed to provide a window of action for human free will. Physicists such as Arthur Holly Compton (1892–1962) took a straightforward approach to this, deciding that the soul must act on the subatomic level, coaxing an indecisive quantum particle to take one path rather than another.[29] These conscious influences would need to be magnified by some

[27] Tyndall, *Fragments*, p. 409.

[28] David B. Wilson, 'A Physicist's Alternative to Materialism: The Religious Thought of George Gabriel Stokes', in *Energy and Entropy*, ed. Patrick Brantlinger (Bloomington, IN: Indiana University Press, 1989); Matthew Stanley, 'The Pointsman: Maxwell's Demon, Victorian Free Will, and the Boundaries of Science', *Journal of the History of Ideas* 69, no. 3 (2008).

[29] Edward B. Davis, 'Prophet of Science: Arthur Holly Compton on Science, Freedom, Religion, and Morality', *Perspectives on Science and Christian Faith* 61 (2009): pp. 175–90.

process in the brain, but could at least in principle allow for purely mental influences in the physical world. The astrophysicist A. S. Eddington (1882–1944) used a slightly different strategy, arguing that the new uncertainty did not suggest a particular mechanism for free will, but rather that quantum physics had removed our prejudice in favour of determinism.[30] In a quantum world, he said, the burden of proof was placed on the determinist to show that free will could not function, not the other way around.

These particular claims about the relation of scientific and religious thought were distinctive of the early twentieth century as the high water mark of liberal theology.[31] Liberal religious thinkers were far less interested than their Victorian predecessors in finding proofs of God or connections to scripture. Instead they emphasized personal religious experience as the root of true religion. So Eddington, a Quaker, was focused on the problem of defending the validity and authenticity of spiritual experience, and ignored questions of creation or design completely. His God did not make himself known through the material world, rather preferring to connect with believers by individual inner contact. The indeterminate properties of physical entities only provided a space for religious experience to function on its own terms.

Questions of causality continued to be fruitful ground for non-naturalistic thinking throughout the twentieth century. In addition to liberal Christian physicists arguing along the lines of Eddington, the 1960s American counterculture brought completely novel approaches. Some physicists calling themselves the 'Fundamental Fysiks Group' grew dissatisfied with the contemporary utilitarian, non-philosophical approach to quantum theory. They began drawing on New Age, parapsychological, and Eastern religious thought to better understand the strange causal behavior of the quantum world, particularly Bell's theorem and non-locality. Fritjof Capra's famous book *The Tao of Physics* started life as a physics text stimulated by these physicists. But more importantly, the Fysiks group used their readings of Buddhism, Hinduism, and Taoism to develop innovative approaches to quantum causality and entanglement. While non-Western religions have rather different understandings of what 'supernatural' might mean, it is clear that their religious and spiritual approaches deeply influenced this physics. This was not merely after-hours speculation: the group's interests directly shaped the projects undertaken and experiments designed by these physicists, and eventually led to innovations such as quantum cryptography.[32]

Not all twentieth-century physicists were quick to embrace indeterminism, however. Many religious figures wanted to maintain the traditional sense that God could be revealed through the regularity of the laws of physics. The mathematical physicist James Jeans (1877–1946) wrote a series of books celebrating the universal

[30] Matthew Stanley, *Practical Mystic: Religion, Science, and A. S. Eddington* (Chicago: University of Chicago Press, 2007), chap. 6.

[31] Peter Bowler, *Reconciling Science and Religion* (Chicago: University of Chicago Press, 2001).

[32] David Kaiser, *How the Hippies Saved Physics* (New York: W. W. Norton, 2012).

reach and unbroken action of natural law. He specifically emphasized the mathematical character of these laws. Jeans concluded that 'from the intrinsic evidence of his creation, the Great Architect of the Universe now begins to appear as a pure mathematician.'[33] This led to Bertrand Russell's famous quip that 'Eddington deduces religion from the fact that atoms do not obey the laws of mathematics. Jeans deduces it from the fact that they do.'[34] While the jab was intended to mock the inconsistency of religious thought, it actually highlights the crucial issue of the variety of religious thought in play at this time. Eddington, representing liberal religion, conceived of the relationship between science and religion quite differently from Jeans, a much more conservative Christian thinker.

Jeans's position that physical laws were designed by God would have been quite familiar to Whewell or Herschel, and indeed this similarity was probably the result of direct intellectual continuity through the tradition of Cambridge physics. Jeans was a member of the last generation of physicists to be educated in the British tradition of theistic science that was the norm during the Victorian period. The direct reading of the laws of physics as creations of godly design, and the thinking of the practice of physics in reverent terms, was a standard part of science education until the waning years of the nineteenth century.

It was only late in this period that Tyndall and the other scientific naturalists found success in their manoeuvres to control the British educational system. Their winning strategy in the end was to place themselves in charge of the examinations system that qualified one to teach science in the new state schools created by the 1870 Elementary Education Act. This allowed the introduction of a purely naturalistic physics as the norm, replacing theistic physics but maintaining the previous tradition's stress on the lawful order of nature.[35] This was not a silver bullet for Tyndall and his allies, but theistic physics did become less and less common in British and American universities. With their allies in charge, the X-Club could be assured that the next generation of science students was trained in a naturalistic perspective, which would then be passed on to its students.

It might seem that overthrowing a centuries-old tradition such as theistic science would require a dramatic revolution, but in fact it was surprisingly smooth. This was because the basic methodological values of theistic and naturalistic physics (such as the uniformity of nature) were actually quite similar. So the practices and methods of theistic physics could often be imported into

[33] James Jeans, *Mysterious Universe* (1930; 2nd edn, Cambridge: Cambridge University Press, 1931), p. 122.

[34] Bertrand Russell, *The Scientific Outlook* (London: George Allen and Unwin, 1931), p. 108.

[35] James Elwick, 'Economies of Scale: Evolutionary Naturalists and the Victorian Examination System', in *Victorian Scientific Naturalism: Community, Identity, Continuity*, ed. Gowan Dawson and Bernard Lightman (Chicago: University of Chicago Press, forthcoming); Matthew Stanley, 'The Uniformity of Natural Laws in Victorian Britain: Naturalism, Theism, and Scientific Practice', *Zygon* 46 (2011); Stanley, *Huxley's Church and Maxwell's Demon: From Theistic Science to Naturalistic Science* (Chicago: University of Chicago Press, 2014).

naturalistic work with simple relabelling, or sometimes without comment at all. For example, after Faraday's death, Tyndall reimagined that deeply religious physicist as an exemplar of scientific naturalism. Faraday's scientific values that theistic physicists saw as clearly theological—unified laws, tentative claims, freedom of thought—were recast by Tyndall as purely naturalistic.[36] In this way, the agreement between theistic and naturalistic physics provided the grounds for the transition between the two. Because the new physics shared so much methodology with the old, no methodological disruptions were necessary. The practice and teaching of physics gradually shifted from one framework to the other, and a generation after Tyndall's death, naturalism had become the 'normal' way to think about physics. Orthodox Christian understandings of the laws of physics would never again be a standard part of the mainstream academic community.

EINSTEIN'S LEGACY AND THE NEW ANTHROPOCENTRISM

Perhaps the best known modern association of physics and religion came at the end of this transition, from the pen of no less than Albert Einstein (1879–1955). Einstein would often wax poetic about the spiritual elements of scientific progress: 'I want to know how God created this world. I'm not interested in this or that phenomenon, in the spectrum of this or that element. I want to know His thoughts, the rest are details.'[37] For him, as with the Victorians, the laws of nature were the expression of God in the world. Complete causality was the trademark of Einstein's God ('God does not play dice').[38] However, his ideas were far from the orthodox Christianity of Whewell—Einstein drew his theistic physics from Spinoza, not the Bible, and rejected both a personal God and most of organized religion.[39]

While Einstein's musings on physics and religion were popular among liberal religionists of his time, the lasting legacy of imagining physics as God's thoughts ended up being quite divergent. The late physicist Stephen Hawking (1942–2018) famously stated that a unified theory of physics would be the only way to know 'the mind of God.'[40] This gesture toward a religious frame for physics

[36] For example, compare Tyndall, *Faraday as a Discoverer* (New York: D. Appleton and Co., 1890) to Maxwell, 'Faraday', in *Scientific Papers*, vol. 2. For a full discussion of this strategy, see Stanley, *Huxley's Church*, chap. 7.

[37] Max Jammer, *Einstein and Religion* (Princeton, NJ: Princeton University Press, 1999), p. 123.

[38] Albert Einstein, 'Science and Religion', in *Ideas and Opinions* (New York: Crown Publishers, 1954), pp. 46–8.

[39] Matthew Stanley, 'Einstein and a Personal God', in *Galileo Goes to Jail and Other Myths About Science and Religion*, ed. Ronald Numbers (Cambridge, MA: Harvard University Press, 2009).

[40] Stephen Hawking, *A Brief History of Time* (New York: Bantam, 1988), p. 175.

no doubt helped Hawking's book sales, but his actual theories were decidedly naturalistic. Indeed, the cosmology presented in *A Brief History of Time* had no indication of or need for a creator of any kind.

In a similar vein, Leon Lederman (1922–) invoked Einstein satirically to dub the Higgs boson 'the God particle'.[41] The key to knowing the mind of Einstein's God was, to Lederman, hidden in this particle. With tongue in cheek, Lederman presented passages from the 'Very New Testament' of this Einsteinian God. Once a superconducting particle collider replaced the Tower of Babel, the world would regain its original unity. Einstein's reverent meditations became only a way to demonstrate the superiority of scientific understandings of the world.

Steven Weinberg (1933–) dispensed with even this literary offering to theistic physics. He acknowledged the temptation of the 'irresistible metaphor to speak of the final laws of nature in terms of the mind of God'.[42] But he completely rejected any religious meaning to physics. He relentlessly argued that physics could only be done naturalistically, and vigorously asserted that the laws of physics carried 'a chilling impersonality'.[43] Completely divorcing himself from any claim that scientific practice could lead to religious insight, Weinberg wrote that the 'more the universe seems comprehensible, the more it seems pointless'.[44] He did not make the case that science could replace religion per se. Instead, he simply wanted to show that religion had completely retreated before science, leaving no trace of theological truth behind.

However, this naturalization of Einstein was not universal. Parallel to Lederman, Hawking, and Weinberg were those who saw themselves as carrying on Einstein's quest for God's thoughts in the original spiritual sense. Paul Davies (1946–) wrote in *The Mind of God* that it was essential to consider the laws of physics to be the result of a creator.[45] The intelligibility and mathematical structure of the universe, he argued, could only be explained through divine forethought. While the arguments of Davies and his cohort might look similar to those of the Victorian theistic physics tradition, they are only distantly related in an intellectual inheritance sense. Most of these physicists are not even aware of that earlier tradition.

Rather, this group was largely stimulated by developments in cosmology in the 1970s and 1980s. It had become increasingly clear that many of the properties of the universe were determined by a handful of physical constants. It was noted that if these constants were even slightly different, then various features of our universe—atoms, galaxies, planets, carbon-based life—would be impossible. This realization of the so-called 'fine-tuning' of nature for human life came to be

41 Leon Lederman, *The God Particle* (Boston, MA: Houghton Mifflin, 1993), pp. 22–4.

42 Steven Weinberg, *Dreams of Final Theory* (1992; repr. New York: Vintage Books, 1994), p. 242.

43 Weinberg, *Dreams of Final Theory*, p. 145.

44 Steven Weinberg, *The First Three Minutes: A Modern View of the Origin of the Universe* (1977; repr. New York: Basic Books, 1988), p. 154.

45 Paul Davies, *The Mind of God* (New York: Simon and Schuster, 1992). Interestingly, the title is drawn from Hawking's formulation, not Einstein's original.

called the anthropic principle.[46] Trying to explain why these constants of nature should have the values they do, rather than some others, has become a significant branch of theoretical physics. However, this mystery has also given rise to design arguments based on fine-tuning. The so-called strong anthropic principle (SAP) asserts that the universe *must* have the physical properties necessary for the appearance of life at some point, and lends itself directly to arguments for a creator deity working through the laws of physics. Even beyond these natural theological claims are those physicists making the case that the SAP is a scientific principle with genuine explanatory power. For example, they point to Fred Hoyle's use of anthropic reasoning to predict a particular carbon resonance in stellar nucleosynthesis. These moves are, of course, controversial—especially given the Intelligent Design community's interest in SAP. But fine-tuning remains enough of a mystery that cosmologists are willing to postulate an infinite number of universes to explain it, so it is unlikely that strict methodological rules regarding it will be agreed upon any time soon.

This use of fine-tuning is only the latest example of a practice that has been pervasive since the development of modern physics. Physicists' disciplinary caution regarding explicitly discussing the essential causes and deep meaning behind phenomena ('how,' not 'why') had an unexpected side-effect. This acceptance of uncertainty allowed the basic categories of physics to be read easily in terms of theological secondary causes. These explanatory concepts and physical entities (such as causality or the ether) were thus understood in religious terms. The theistic physicists did not invoke direct divine intervention. But neither did they discard their religious understandings at the laboratory door: in the nineteenth century, there was plenty of God-talk among physicists. Maxwell's and Thomson's work *looks* perfectly naturalistic to modern eyes, but they would *not* have agreed that their work was naturalistic. They practised their physics as deeply theistic. The puzzle is that the basic concepts of physics could be framed either naturalistically or theistically, with very similar methodological practices either way.

This is not to say that physics today is not naturalistic. It is generally taught, practised, and interpreted without reference to religious categories. But this is not a result of a methodological victory of naturalistic physics over theistic physics. Instead, it came from the intervention of Tyndall and his allies in the educational system, interrupting the intellectual tradition of theistic physics. Thinking non-naturalistically became something that individual physicists had to discover and defend for themselves. By the mid-twentieth century, naturalistic physics had become the default.

46 The most robust formulation of the reasoning behind the anthropic principle remains John Barrow and Frank Tipler, *The Anthropic Cosmological Principle* (Oxford: Oxford University Press, 1988).

However, the ease of thinking about the fundamentals of physics in religious terms has not changed. It is relatively straightforward to envision a deity working through natural laws. Religious physicists rarely have to ponder the recalcitrant issues relating to the soul or human uniqueness that might complicate matters for their colleagues in the life and social sciences.[47] Similarly, it is not an accident that many of the most influential contemporary reconcilers of science and religion—such as Ian Barbour or John Polkinghorne—are or have been physicists. Despite the naturalistic consensus of the twenty-first century, physics retains the robust combination of scope, power, and deliberate uncertainty that allows for a close productive relationship between religious belief and apparently naturalistic scientific practice.

[47] James Leuba's 1914 survey showed that physical scientists were much more likely to be religious than those working in the life or social sciences. However, the 1996 reproduction of that survey found this correlation no longer to be the case. James H. Leuba, *The Belief in God and Immortality: A Psychological, Anthropological and Statistical Study* (1916; repr. Chicago: Open Court Publishing Company, 1921), p. 278; Edward J. Larson and Larry Witham, 'Scientists Are Still Keeping the Faith', *Nature* 386, no. 6624 (1997): pp. 435–6; Larson and Witham, 'Leading Scientists Still Reject God', *Nature* 394, no. 6691 (1998): p. 313.

6

Chemistry with and without God

John Hedley Brooke

The assumption that scientific progress has been a potent force in the corrosion of religious belief has been well entrenched in the culture of modern science. It is endemic in secular critiques of religion where a simple contrast between the 'natural' and the 'supernatural' is commonplace and where avowal of belief in supernatural agents is often presented as a defining characteristic of religions.[1] Given the scientist's preoccupation with natural causes it is easy to believe that gains in scientific knowledge must mean losses for religion. That is how anthropologist Anthony Wallace saw matters in 1966, predicting that 'belief in supernatural powers is doomed to die out, all over the world, as a result of the increasing adequacy and diffusion of scientific knowledge'.[2] This ill-fated prophecy was noted by Ronald Numbers in one of the many fine essays in which he has challenged simplistic models of the impact of science on religious sensibility. Numbers observed that 'contrary to such wishful prophecies, supernaturalism not only persisted but flourished. Instead of becoming more rational and liberal, world religions in the late twentieth century became more fideistic and militant'.[3]

There are many sciences, even more religions, and the relations between them have been constructed in numerous ways. In some contexts, new scientific knowledge has promised gains for a religion, in other contexts a loss.[4] Different sciences have carried different implications for religious belief, even today when God-talk is more common among physicists than evolutionary biologists. Simplicity and elegance in the mathematical formulation of physical laws, indicative to some of a transcendent intelligence, have stood in sharp contrast

[1] Daniel C. Dennett, *Breaking the Spell* (London: Penguin, 2007), pp. 9, 24–6.

[2] Anthony F. C. Wallace, *Religion: An Anthropological View* (New York: Random House, 1966), pp. 264–5.

[3] Ronald L. Numbers, *Science and Christianity in Pulpit and Pew* (New York: Oxford University Press, 2007), p. 129.

[4] John Hedley Brooke, *Science and Religion: Some Historical Perspectives* (Cambridge: Cambridge University Press, 1991).

to the random and seemingly disorganized processes at work in evolution. What, then, might we expect from the distinctive science of chemistry?

The first President of the American Chemical Society, John William Draper (1811–82), was the same John Draper who, in his *History of the Conflict between Religion and Science* (1875), launched an influential diatribe against the Catholic Church, posing an antithesis between scientific naturalism and clerical obscurantism. As it happens, chemistry did not loom large in Draper's story. He found enough ammunition in the Church's response to Copernican astronomy, to the nineteenth-century extension of the age of the Earth, and to what he called the 'great theory of Evolution'.[5] Chemistry, however, was and is the science that deals par excellence in the manipulation and transformation of matter. When Draper's contemporary John Tyndall (1820–93) gave his historical account of how science had restored to matter capacities traditionally ascribed to spirit, the embodiment of atomism in chemical theory had a high profile. Tyndall's notorious Presidential Address at the 1874 Belfast meeting of the British Association, where he proclaimed that science would 'wrest from theology the entire domain of cosmological theory', transformed conflict between science and religion into a public spectacle having adverse consequences locally for the reception of Darwin's science. Arguing that matter had been much maligned, that 'the principle of every change resides in matter', Tyndall looked to chemistry for support: 'The atomic doctrine, in whole or part, was entertained by Bacon, Descartes, Hobbes, Locke, Newton, Boyle, and their successors, until the chemical law of multiple proportions enabled Dalton to confer upon it an entirely new significance.'[6] It was chemistry that had given unprecedented empirical support to a doctrine that, in antiquity, had expelled gods from the world.

Yet chemistry was not always on the side of quasi-materialist creeds. For the Romantic poet Samuel Taylor Coleridge and his chemical mentor Humphry Davy (1778–1829), chemical combination was paradigmatic for a holism in which wholes were more than the sum of their parts.[7] Think of the properties of water compared with those of hydrogen and oxygen. In earlier centuries, the study of chemical transformation even invited comparison with the spiritual transformation of the religious believer.[8] Because chemistry has been exploited by religious apologists, by defenders of spiritual realities, as well as by reductionists and materialists, I shall argue that it has been a fluid, ever-changing and ambivalent resource in religious controversies. In pursuing this argument, I hope to convey its distinctiveness as a science and the various ways in which it has been appropriated for polemical purposes.

[5] John W. Draper, *History of the Conflict between Religion and Science*, 18th edn (London: Kegan Paul, 1883), p. 247.

[6] John Tyndall, 'The Belfast Address', in *Fragments of Science*, 6th edn, vol. 2 (London: Longmans, Green, 1879), pp. 136–203, esp. pp. 163, 199.

[7] David M. Knight, *The Transcendental Part of Chemistry* (Folkestone: Dawson, 1978), pp. 63–6.

[8] Lawrence M. Principe, *The Secrets of Alchemy* (Chicago: University of Chicago Press, 2013), pp. 190–2.

THE NON-NATURALNESS OF NATURALISM

When Draper marshalled his evidence for recurring conflict between science and religion, he conveniently structured his anticlericalism by positing just two models for the governance of the world. It was governance either by Providence or by law. The former was 'maintained by the priesthood', the latter substantiated by the sciences.[9] His dichotomy survives today and captures much that resonates with popular understandings of how the sciences may erode religious authority and commitment. There is a simple choice. As Tyndall insisted in his Belfast Address, 'two courses and two only are possible. Either let us open our doors freely to the conception of creative acts, or abandoning them, let us radically change our notions of Matter.'[10]

These had not been the only options during preceding centuries when the sciences had gathered momentum in Europe. There were leading European scientists and philosophers before Tyndall's generation for whom the natural and the supernatural were not mutually exclusive categories. Even Charles Darwin (1809–82) had once spoken of a 'Creator' who 'creates by laws', confiding to Asa Gray in May 1860 that he was 'inclined to look at everything as resulting from designed laws', with the details left to chance.[11] Darwin had firmly rejected Christianity by then, but his willingness still to countenance design in the laws of nature suggests that even he did not believe that advancing naturalism entailed disbelief in a transcendent power. There had been a long tradition in which the very concept of *laws* of nature had been understood to presuppose a legislator, whose prescience in ensuring that the laws would combine to guarantee a viable world had impressed religious apologists. Interested only in extreme positions, Draper excluded mediating theologies in which laws, and their sustenance, were themselves expressions of Providence. As Numbers has underlined, 'scientific naturalism of the methodological kind could—and did—coexist with orthodox Christianity'.[12] The principle that only natural causes were permissible in scientific explanation did not commit one to an ontological naturalism in which 'nature' constitutes all there is.

A methodological naturalism could coexist, as it did for Isaac Newton (1642–1727), with the proposition that natural causes were God's instruments

[9] Draper, *Conflict*, xxi. [10] Tyndall, 'Belfast Address', p. 191.

[11] John Hedley Brooke, 'The Relations between Darwin's Religion and his Science', in *Darwinism and Divinity*, ed. John Durant (Oxford: Blackwell, 1985), pp. 40–75, esp. pp. 46–7; also Brooke, '"Laws Impressed on Matter by the Creator"?: The *Origin* and the Question of Religion', in *The Cambridge Companion to the 'Origin of Species'*, ed. Michael Ruse and Robert J. Richards (Cambridge: Cambridge University Press, 2009), pp. 256–74; Darwin to Asa Gray, 22 May 1860, in *The Correspondence of Charles Darwin*, vol. 8, ed. Frederick Burkhardt et al. (Cambridge: Cambridge University Press, 1993), pp. 223–4.

[12] Ronald L. Numbers, 'Science without God: Natural Laws and Christian Beliefs', in *When Science and Christianity Meet*, ed. David C. Lindberg and Ronald L. Numbers (Chicago: University of Chicago Press, 2003), pp. 265–85, esp. p. 284.

in the governance of the world. Comets, according to Newton, had a divinely scripted role in restabilizing the solar system. Newton and one of the great chemists of the seventeenth century, Robert Boyle (1627–91), both thought of God's action in the world as analogous to the action of persons when voluntarily moving their limbs. It was possible for Boyle to provide chemistry with a theoretical base, grounded in a mechanical understanding of chemical reactions, while simultaneously resisting language (such as 'nature abhors a vacuum') that conferred autonomous power on 'nature'. Exasperated by many ambiguities in references to 'nature', Boyle even proposed that the word be scrapped.[13] His was a naturalism without nature. Moreover, from Boyle's time to that of Darwin and beyond, the natural world was often interpreted as the work of a divine artist or craftsman, so intricately designed that one could celebrate the wisdom and ingenuity of its Creator.[14] For Boyle, the cathedral clock of Strasbourg served as a model for the universe: to learn how it worked in no way nullified the fact that it had been designed and made by an external intelligence. Likewise for Newton, the beauty and mathematical elegance of the solar system testified to a Creator 'very well skilled in mechanics and geometry'.[15]

Such interpenetration of scientific and religious motifs strongly suggests that until the theistic connotations of the 'law' metaphor ceased to be compelling, the discovery of more scientific laws was unlikely to be sufficient in itself for the desacralization of nature. Even a science-without-God (in which theistic reference was excluded from the technical literature of the science) was not necessarily as powerful an agent of secularization as other social and political roots of religious disenchantment. There is an important distinction between the desacralization *of* science and secularization *by* science.[16] Critics of religion have often imagined a division of labour between God and nature in which God's share was simply redistributed to nature. Yet such a division is precisely what a full-blown Christian theism had rejected. Writing at the end of the nineteenth century, the Oxford High Church theologian Aubrey Moore (1848–90) tried to recapture that earlier theology of nature in which nature and supernature were not at odds. For the Christian theologian, Moore insisted, the facts of

[13] Boyle, *A Free Enquiry into the Vulgarly Receiv'd Notion of Nature* (1686), sec. 2, 4, in *Selected Philosophical Papers of Robert Boyle*, ed. M. A. Stewart (Manchester: Manchester University Press, 1979), pp. 176–91; Michael Hunter and Edward B. Davis, 'The Making of Robert Boyle's *Free Enquiry into the Vulgarly Receiv'd Notion of Nature* (1686)', *Early Science and Medicine* 1 (1996): pp. 204–71.

[14] Brooke, *Science and Religion*, pp. 192–225; John Brooke and Geoffrey Cantor, *Reconstructing Nature: The Engagement of Science and Religion* (Edinburgh: T&T Clark, 1998), pp. 141–243; Alister E. McGrath, *Darwinism and the Divine: Evolutionary Thought and Natural Theology* (Chichester: Wiley-Blackwell, 2011), pp. 49–142.

[15] Newton to Richard Bentley, 10 December 1692, in *Newton's Philosophy of Nature*, ed. H. S. Thayer (New York: Hafner, 1953), pp. 46–50.

[16] John Hedley Brooke, 'Science and Secularization', in *The Cambridge Companion to Science and Religion*, ed. Peter Harrison (Cambridge: Cambridge University Press, 2010), pp. 103–23.

nature are the acts of God.[17] From this perspective it was even possible to claim that Darwin's science had marked a gain for religion in that it discouraged dependence on a naive concept of divine intervention that, for Moore, smacked of deism.

Even this brief sampling of philosophical positions suggests that there is no self-evidently 'natural' nature of naturalism. How it has been interpreted has depended on prior philosophical assumptions and other cultural preconceptions, including those derived from religious and indeed anti-religious beliefs. With that richness of possibility in mind, I now turn to chemistry and to historical examples of ways in which it once lent support to religious ideals. It was a science that could also prove embarrassing to religious orthodoxies.

CHEMISTRY IN COMBINATION WITH RELIGION

The roots of chemistry lie in alchemy and, as the best modern scholarship has shown, no clear rupture separated them.[18] The combination of alchemy with religion was often so intimate that, in some historical interpretations, alchemical practices have been seen as quintessentially religious, denuded of any proto-chemical significance. While this reduction of empirical practices to religious categories is now seen as a distortion, the connections that were made in late medieval and early modern Europe between alchemical and spiritual goals were real and diverse. For the French Franciscan friar, John of Rupescissa (b. *c.*1310), writing from prison in the fourteenth century, knowledge of alchemy and the power of the 'philosophers' stone' was vital to the security of the Church, threatened as he believed it was, by an imminent return of the Antichrist.[19]

The alchemical dream of turning base metals into gold had motives that could be noble as well as base: a discourse of purification was common both to the experimental study of metals and the enrichment of a spiritual life. Religious leaders critical of astrology could warm to alchemy, as when Martin Luther (1483–1546) declared that he liked it very well:

> I like it not only for the profits it brings in melting metals, in decocting, preparing, extracting, and distilling herbs...; I like it also for the sake of the allegory and secret signification, which is exceedingly fine, touching the resurrection of the dead at the last day.[20]

[17] Richard England, 'Natural Selection, Teleology, and the Logos: From Darwin to the Oxford Neo-Darwinists, 1859–1909', *Osiris* 16 (2001): pp. 270–87.

[18] William R. Newman, *Atoms and Alchemy* (Chicago: University of Chicago Press, 2006); William R. Newman and Lawrence M. Principe, *Alchemy Tried in the Fire* (Chicago: University of Chicago Press, 2002).

[19] Principe, *The Secrets of Alchemy*, pp. 63–6.

[20] Quoted from the 'Table Talk' of Martin Luther in Stanton J. Linden, 'Alchemy and Eschatology in Seventeenth Century Poetry', *Ambix* 31 (1984): pp. 102–24.

As the earthly alchemist purified through fire, leaving the dregs at the bottom of the furnace, the divine alchemist at the Day of Judgement would separate all things through fire, the righteous from the ungodly. To put it crudely, chemistry in its early alchemical phases could provide gains for religion, as it provided a distillation of metaphors that permeated devotional literature. A seventeenth-century bishop, Jean-Pierre Camus (1584–1652), exhorted his hearers to 'put all our good and bad thoughts, affections, passions, vices, and virtues all mixed together into the alembic of our understanding'. This was then to be placed 'upon the memory and recollection of the eternal fire as if upon a furnace, and we shall see some marvelous subtle effects'. This 'fiery cogitation' would 'separate the confused elements, the hullabaloo of ambition, the earth of greed and lust, the winds of vanity, the waters of covetousness, the air of presumptions'. Now then, he asked, is this not a 'fine chymistry'?[21]

In the depiction of alchemical pursuits, chemical and religious practices were often conjoined, as in a striking plate from Heinrich Khunrath's *Amphitheatrum Sapientiae Aeternae* (1595), where the alchemist is shown at prayer, his oratorium and laboratorium placed side by side.[22] To achieve a propitious spiritual state was often seen as a prerequisite of experimental success. Parallels would be drawn between the creation of the soul by God and the creation of the philosophers' stone by the alchemist. For Pierre-Jean Fabre (1588–1658), author of *Alchymista Christianus* (1632), alchemy was a virtuous activity for orthodox Christians, a means of expressing and confirming theological truth.[23] There was certainly two-way traffic in the complex relations between alchemy and religion, but the chemical knowledge that accrued was almost always described as a gift from God. This was knowledge that might be guarded secretively and which was also understood to be 'holy'.

In his bid to reform medical practice, the sixteenth-century iconoclast Paracelsus (*c.*1493–1541) claimed to be making chemistry an even holier science by deflecting alchemists from the pursuit of gold to the preparation and purification of chemical remedies for diseases, which he believed were caused by external agents rather than by an internal imbalance of humours.[24] Critical of physicians who too often absolved themselves by describing particular diseases as incurable, Paracelsus wedded chemistry to medicine by insisting that God in his mercy had provided remedies that did, however, require chemical skills to extract and purify them from their substrates. Paracelsus was not averse to using metals as medicines, with the consequence that not all his patients recovered; but the long marriage that ensued between chemistry and

[21] Principe, *The Secrets of Alchemy*, p. 191.
[22] Brooke, *Science and Religion*, p. 68.
[23] Principe, *The Secrets of Alchemy*, pp. 201–2.
[24] Walter Pagel, *Paracelsus: An Introduction to Philosophical Medicine in the Era of the Renaissance* (Basel: Karger, 1958); Owen Hannaway, *The Chemists and the Word* (Baltimore, MD: Johns Hopkins University Press, 1975).

medicine gave the science a distinctive role in contexts where the imperative to heal was often associated with Christ's example as healer.

Chemistry, even when deprived of allegorical richness, could still be presented as a science replete with religious connotations. For Paracelsus and his followers, the Genesis creation narrative was understood to be describing a chemical process, as the divine chemist separated the elements from primordial water. Chemistry had value as an adjunct to biblical exegesis. There was even a sense in which the toil and sweat of the chemist were tokens of his place in a fallen world, to the redemption of which his science might contribute. For later chemists, too, chemistry could not be sacrilegious if, in its pharmacological guise, it promised the relief of suffering. The medical value of the various gases he identified were uppermost in the mind of Joseph Priestley (1733–1804) as, late in the eighteenth century, he speculated on their utility. He wondered whether water infused with 'fixed air' (carbon dioxide as it is known today) might constitute a cure for scurvy.[25] Briefly, in Bristol, the medical virtues of various gases were instantiated in the Pneumatic Institution of another Unitarian, Thomas Beddoes (1760–1808).[26] Despite his religious radicalism, to which we shall return, Priestley also shows how chemistry could be supportive of a doctrine of Providence. In a letter from America, written in April 1800, he looked back on a career in which one of his primary objects had been to join (natural) philosophy to Christianity, from which it had been 'too much separated'.[27]

A belief in divine Providence both regulated and was reinforced by Priestley's chemical research. The belief that nature is a designed system had led him to investigate mechanisms of restoration that he believed Providence must have provided, especially for the replenishment of air fouled by breathing. On the basis of a sustained inquiry, he eventually concluded that vegetation was the key. He did not achieve a complete theory of photosynthesis, but when presented with the Royal Society's Copley Medal, he was congratulated by the President, Sir John Pringle, for having shown that 'no vegetable grows in vain'.[28] In the vitiation and restoration of air, Priestley found a chemical parable of the transformation of evil into good.

The advent of a chemistry without God, and a chemistry less dependent for its transmission on university medical faculties, is usually ascribed to the autonomy achieved by Antoine Lavoisier (1743–94) and his scientific contemporaries working in France just before and in the aftermath of the 1789 revolution,

[25] John Hedley Brooke, '"A Sower went Forth": Joseph Priestley and the Ministry of Reform', in *Motion Toward Perfection: The Achievement of Joseph Priestley*, ed. A. Truman Schwarz and John McEvoy (Boston, MA: Skinner House, 1990), pp. 21–56.

[26] Jan Golinski, *Science as Public Culture: Chemistry and Enlightenment in Britain, 1760–1820* (Cambridge: Cambridge University Press, 1972), pp. 153–87.

[27] Joseph Priestley to B. Lynde Oliver, 3 April 1800, in Robert E. Schofield, *A Scientific Autobiography of Joseph Priestley (1733–1804)* (Cambridge, MA: MIT Press, 1966), p. 302.

[28] F. W. Gibbs, *Joseph Priestley: Adventurer* (London: Nelson, 1965), p. 81.

when politically challenging secular ideals held sway. Having been associated, as a government employee, with the collection of taxes, Lavoisier was guillotined during the Terror. Before his life was cut short, his introduction of rigorous quantitative methods into the study of chemical reactions, with its ascription of particular importance to gains and losses of weight, had led to a new theory of combustion, in which metals ceased to be compounds of a calx with a principle of inflammability, phlogiston, becoming instead elements that combined with oxygen from the air when burning. The reform of chemical nomenclature that accompanied this chemical revolution produced an enduring redefinition of the language of chemistry.[29] God was not to be found in the textbooks of the nineteenth-century chemists who consolidated a science that would soon be yielding models and mechanisms for atomic combination.

Paradoxically, however, by the early years of the nineteenth century, chemistry was furnishing arguments which, by embarrassing materialist precepts, offered support for spiritual readings of nature. One of the ironies of the 'chemical revolution' was that the new theoretical system in which Lavoisier's concept of oxygen was embedded quickly became obsolete in one crucial respect. Lavoisier had chosen the name 'oxygen' (meaning acid-producer) to designate the gas that Priestley and others had isolated because, from its combination with elements such as carbon, nitrogen, and sulphur, acids were generated. The implication was that chemical elements could continue to be regarded as property-bearing 'principles', directly impressing their generative powers on the compounds to which they contributed. It soon emerged, however, that oxygen was not a *principle* of acidity in this monocausal sense. In a forceful critique of what in England was sometimes seen as a French bid for hegemony in the reformation of chemistry, Humphry Davy decomposed the alkalis that we know as sodium and potassium hydroxide, showing that they, too, contained oxygen—hence his quip that the principle of acidity of the French chemists could just as easily be dubbed the principle of alkalinity.[30] In several other respects, Davy showed that the relationship between elements and their compounds defied simple reduction. Even one and the same element, carbon, could exist in contrasting forms, charcoal and diamond. The combination of nitrogen with oxygen could produce laughing gas, but also the brown toxic fumes of nitrogen dioxide that were no laughing matter. In addition, Davy's electrochemical researches showed that the reactivity of a chemical agent could be changed simply by giving it a positive or negative charge. Chemical properties manifestly did not reside in material particles.[31] Active powers, such as invisible forces of electricity, must also be

[29] Ferdinando Abbri and Bernadette Bensaude-Vincent (eds), *Lavoisier in European Context: Negotiating a New Language for Chemistry* (Cambridge, MA: Science History Publications, 1995).

[30] David M. Knight, *Humphry Davy: Science and Power* (Oxford: Blackwell, 1992), pp. 80–7; John Hedley Brooke, *Thinking about Matter: Studies in the History of Chemical Philosophy* (Aldershot: Ashgate Variorum, 1995), chap. 3.

[31] Knight, *Humphry Davy*, pp. 63–72.

given their due, as Priestley had argued and as, later, would Michael Faraday (1791–1867). There were those, such as Benjamin Franklin (1706–90), who, in the eighteenth century, had argued that electrical powers inhered in matter. But for Davy and Faraday, there was more to nature than a materialist philosophy would allow.[32] Far from catalysing a loss of faith, Davy's chemistry was a friend to religious inference. In a lecture delivered at the Royal Institution in 1812, he made the connection himself:

> Active powers must be considered as belonging to matter; but it is not necessary to suppose them inherent in it. [Matter] may be regarded... as inert; and all effects produced upon it as flowing from the same original cause, which, as it is intelligent, must be divine.[33]

Faraday was religiously more conservative than Davy in the simple biblical faith he espoused as a member of a minority Christian sect, the Sandemanians. Davy's writing became more redolent of a pantheistic spirituality in which 'as poet, as philosopher, as sage' he asked of nature 'have I not worshipped thee with such a love as never mortal man before displayed?'[34] Contrasting though their religious positions were, both believed that chemistry, albeit indirectly, offered further solace to spirituality. It was receptive to models of the unity of nature that were in turn associated with the monotheistic belief that the universe is the product of a single self-consistent mind.

This metaphysical principle of the unity of nature had been fundamental to Newton's science and would find expression in Faraday's conviction that the various forces of nature (magnetic, electrical, chemical, gravitational, etc.) were ultimately one and interrelated.[35] Belief in the ultimate unity of *matter* had been a pervasive feature of alchemy, symbolized by the image of a snake devouring its tail.[36] It was a belief that had made the transmutation of metals a plausible hope and it was still underpinning Boyle's aspirations for transmutation in the second half of the seventeenth century. Committed to an ultimate unity of matter, despite an enormous range in the densities of different substances, Newton removed the seeming paradox by suggesting that matter is porous, that the amount of truly solid matter in the universe could be put in a nutshell.[37] This diminution of brute matter had itself been a gift to religious apologists who, like the first Boyle lecturer Richard Bentley (1662–1742), found it incredible that in a vast universe, where the proportion of space to matter was so great, atoms would collide at all, let alone produce an ordered world.[38]

[32] Geoffrey Cantor, *Michael Faraday: Sandemanian and Scientist* (London: Macmillan, 1991), pp. 178–81.

[33] Knight, *The Transcendental Part of Chemistry*, p. 68.

[34] Knight, *Humphry Davy*, p. 9.

[35] Cantor, *Michael Faraday*, pp. 171–2, 186–9, 246–8.

[36] Principe, *The Secrets of Alchemy*, pp. 24–6.

[37] Arnold W. Thackray, '"Matter in a Nutshell": Newton's *Opticks* and Eighteenth-Century Chemistry', *Ambix* 15 (1968): pp. 29–53.

[38] Brooke, *Science and Religion*, pp. 157–8.

A return to the unity of matter in the early nineteenth century was prompted by the fact that in Lavoisier's new system of chemistry, now fortified by John Dalton's allocation of a distinctive weight to the atom of each element, the number of the elements had proliferated. Davy, who was a critic of Dalton's atomism, ironically added to their number when, by electrolysis, he isolated sodium and potassium. But suppose the elements were complex. One could then reduce their number by positing a smaller number of subunits differently organized in each. Davy toyed with the idea that hydrogen might be the ultimate unit of matter, a hypothesis made famous when another English chemist, William Prout (1785–1850), observed that the atomic weights of the elements were remarkably close to integers. More refined measurements put a stop to Prout's hypothesis in its original form, but an aesthetic preference for unity and simplicity in nature continued to favour the compound nature of elements throughout the nineteenth century and in several cases had religious overtones. Prout was a physician, a chemist who discovered hydrochloric acid in the gastric juices, and an exponent of natural theology. In his *Bridgewater Treatise*, composed in the early 1830s, he marvelled at the ingenuity of a Creator who had constructed living systems from such refractory elements as carbon, hydrogen, oxygen, and nitrogen, three of which were invisible gases. One of those, nitrogen, was inert. Amidst the wonders of creation, Prout declared,

> it is perhaps difficult to say what is most wonderful; but we have often thought, that the Deity has displayed a greater stretch of power, in accommodating to such an extraordinary variety of changes, a material so unpromising and so refractory as charcoal, and in finally uniting it to the human mind; than was requisite for the creation of the human mind itself.[39]

At least in England, God was not yet expunged from chemical texts. Chemistry could still be on the side of the angels.

CHEMISTRY AND THE CORROSION OF BELIEF

I have been arguing that during the emergence of modern science there was nothing intrinsic to the pursuit of explanation by 'natural' causes that necessarily jeopardized religious belief. An acquaintance with the fine details of the structure of nature could generate awe and reverence, as when Boyle had marvelled at the skill and power of a Creator who had injected life into the minutest of mites. An enthusiasm for chemistry was sometimes akin to religious enthusiasm, as when Boyle referred to his laboratory as a kind of *Elysium*.[40] But this is only part of the

[39] William Prout, *Chemistry, Meteorology, and the Function of Digestion Considered with Reference to Natural Theology*, 2nd edn (London: Pickering, 1834), p. 22.

[40] Brooke and Cantor, *Reconstructing Nature*, p. 324.

story. Chemistry might have been on the side of the angels—sometimes—but it also had pretensions as a practical science that could make the chemist a dangerous and suspect figure, a member of that breed of scientist most likely to be accused of 'playing God'.

In an essay on the roots of secularization, Peter Burke perceptively remarked that scientists were often destructive 'despite themselves'.[41] Boyle affords an interesting example. His mechanical metaphors for a universe exquisitely designed were affirming of religious belief. At the same time a clockwork universe could be a gift to deists less respectful of scripture than Boyle himself. Indeed, the gifted naturalist John Ray (1627–1705) had to apologize to Boyle for having supposed him a deist before becoming better acquainted with his Christian writings. The kind of natural theology of which both Boyle and Ray were exponents could also be counterproductive, if exaggerated claims were made for 'proofs' of divine wisdom from the intricacies of nature. Though itself an exaggeration, it was famously quipped that no one doubted the existence of God until the Boyle lecturers undertook to prove it.

How might a history of chemistry slot into a broader narrative of threats to, and loss of, religious belief? Much of the appeal of chemistry has derived from the excitement that accompanies the manipulation of matter, from the expectation that something unexpected might be made, a novel resource for improving the world. The alchemists' claim that they were only accelerating a perfectly natural and subterranean process of ripening metals could be acceptable to religious observers such as the sixteenth-century Jesuit Martin del Rio. But when their seeming lust for power over nature was combined with secrecy, alchemists were open to accusations of diabolical practice.[42] I have argued elsewhere that chemistry was not, like astronomy or anatomy, a typically descriptive science.[43] It was a science concerned with process, as when Newton presented his elitist chemistry as an attempt to imitate the processes at work in the growth of vegetation. Moreover, chemists were not always satisfied with the imitation of nature. To this day they have aimed to transcend natural processes, for instance in the artificial production of substances unknown in nature.[44] Although chemistry could furnish arguments for a natural theology, as we have just seen in William Prout, its profile in Christian apologetics has been relatively low. It has tended to devalue the natural. A modern example of such devaluation would be allegations of 'stupid design' in a DNA-based template

[41] Peter Burke, 'Religion and Secularisation', in *The New Cambridge Modern History*, vol. 13, ed. Peter Burke (Cambridge: Cambridge University Press, 1979), p. 303.

[42] Martha Baldwin, 'Alchemy and the Society of Jesus in the Seventeenth Century: Strange Bedfellows?', *Ambix* 40 (1993): pp. 41–64.

[43] Brooke and Cantor, *Reconstructing Nature*, pp. 314–46.

[44] John Hedley Brooke, 'Overtaking Nature? The Changing Scope of Organic Chemistry in the Nineteenth Century', in *The Artificial and the Natural*, ed. Bernadette Bensaude-Vincent and William R. Newman (Cambridge, MA: MIT Press, 2007), pp. 275–92.

for living systems, from biochemists bent on improving it.[45] The world God created may have been 'very good', but chemists have presumed to make it better. As David Knight has nicely put it, chemistry is a science for a fallen world.[46]

A culture of improvement, central to Enlightenment ideologies, was not centred exclusively on chemistry. But it is rarely absent from chemical texts. The *Chemical Essays* of the Cambridge divine and Professor of Chemistry Richard Watson, published in the 1780s, celebrated a science that had already changed the world:

> It cannot be questioned, that the arts of dying, painting, brewing, distilling, tanning, of making glass, enamels, porcelane, artificial stone, common salt, sal-ammoniac, salt-petre, potash, sugar, and a great variety of others, have received much improvement from chemical inquiry, and are capable of receiving much more.... There are a great many combinations of metals which have never been made.[47]

Although Watson was happy to baptize the science, the transformative power of chemistry could easily have the science fiction appeal of redesigning nature to a human rather than divine specification. In such contexts, there surely was tension with conventional religious beliefs and a potential in the science to marginalize them. Two of the chemists discussed in the previous section, Davy and Prout, each on occasion referred to the chemist's godlike faculties.[48] The charge of 'playing God', often heard in today's reactions to biotechnology, continues to have, as one of its springs, a religious distrust of such hubris.[49]

In Priestley's combination of chemistry and religion, scientific progress provided both a model for and a means to social progress. Chemistry still belonged inside a providential scheme. But if his philosophy of nature is compared with that of Boyle, a century earlier, one can see a loss for religion. Whereas Boyle was fascinated by evidence for a world of spirits and even referred to God's communicating 'pregnant hints' as he conducted his experiments, Priestley developed a monistic understanding of the relationship between body and mind, in which reference to spirits was banished. The idea that God might exert a direct influence on the human mind had become, for Priestley, sheer superstition. He exorcised spirits from the language of chemistry and from the language of his radical religion.[50]

[45] Stephen Benner, quoted in Roberta Kwok, 'Chemical Biology: DNA's New Alphabet', *Nature: News Feature*, 21 November 2012, http://www.nature.com/news/chemical-biology-dna-s-new--alphabet-1.11863.

[46] David M. Knight, 'Chemical Sciences and Natural Theology', in *The Oxford Handbook of Natural Theology*, ed. Russell Re Manning (Oxford: Oxford University Press, 2013), pp. 434–48.

[47] Richard Watson, *Chemical Essays* (Dublin: Moncrieffe, 1786), p. 17.

[48] Humphry Davy, *Collected Works*, vol. 9, ed. John Davy (London: Smith Elder, 1840), p. 361; Prout, *Chemistry, Meteorology, and the Function of Digestion*, p. 7.

[49] Sven Wagner, *The Scientist as God* (Heidelberg: Universitätsverlag Winter, 2012).

[50] John Hedley Brooke, 'Joining Natural Philosophy to Christianity: The Case of Joseph Priestley', in *Heterodoxy in Early Modern Science and Religion*, ed. John Brooke and Ian Maclean (Oxford: Oxford University Press, 2005), pp. 319–36.

A hundred years after Priestley, chemistry was adopted for even more radical purposes when it was incorporated into a secular philosophy that had as an explicit goal not the reform of Christianity, but its displacement by a science-based culture. This 'positivist' philosophy originated in France where Auguste Comte (1798–1857) outlined a three-stage process in the cultural development of humanity. There had first been a religious phase when natural phenomena had been ascribed to the gods. This had been followed by a metaphysical phase in which speculative explanations had been regulated by philosophical abstractions that referred to concepts and entities beyond empirical proof. Now at last in a third, triumphant stage, a true science of nature was possible, in which facts and laws, established empirically, were the basis of human understanding. In this scenario, science could spell the demise of religion, though paradoxically Comte's positivism functioned for a while as a surrogate religion with its own churches, rituals, and celebrations (especially of the achievements of French scientists).[51]

This positivist philosophy, with its anti-Catholic thrust, found fertile ground in the secular ethos of France's Third Republic, which from 1870 had as one of its scientific gurus the chemist Marcellin Berthelot (1827–1907). Whereas Priestley, and many early nineteenth-century chemists, had believed that in all living things was a vital spark only explicable as the product of a divine creation, Berthelot relished the loss for religion that would occur if the relatively new science of organic chemistry could breach the barrier separating living things from the inanimate. Crucial to his campaign were the recent successes chemists had enjoyed in synthesizing organic compounds artificially. He himself had synthesized formic acid in November 1855—the first direct synthesis of an organic compound, in this case from carbon monoxide and steam, themselves immediate products of the direct union of their elements. To synthesize an organic compound was a long way from synthesizing the complex parts of a living creature, but for Berthelot this was the thin end of a wedge that would eventually demystify the processes occurring within them. He claimed 'without reservation' that the chemical forces governing organic matter were the same as those in the inorganic realm.[52]

In England, Berthelot's contemporary T. H. Huxley (1825–95) was also preaching lay sermons and advocating 'the physical basis of life'. Chemistry even reinforced his defence of biological evolution. For Huxley, the secret of life lay in what he called the 'protoplasm' of the cell, which in all its forms appeared to behave similarly when exposed to chemical reagents, electric shock, or heat. Crucially, in all living things there was an ultimate unity of composition manifested by the presence of the same four elements—carbon, hydrogen, oxygen, and nitrogen—in organic matter. In this unity, Huxley saw

[51] Brooke and Cantor, *Reconstructing Nature*, pp. 47–57.

[52] Brooke, *Thinking about Matter*, chap. 8.

evidence for common ancestry in a single evolutionary process.[53] Not perhaps the profoundest of arguments, but one that presaged later recourse to chemistry in reductionist accounts of living things. When, in 1953, Francis Crick and James Watson announced the double-helix structure for DNA, elucidating a mechanism for gene replication, it was claimed that the mystery of life had been solved. Crick's anti-religious sentiments are well known. He resigned his Fellowship at Churchill College, Cambridge when plans went ahead for the building of a chapel. His book *The Astonishing Hypothesis* (1994) was astonishing for its stark reductionist creed: 'The Astonishing Hypothesis is that "you", your joys and your sorrows, your memories and your ambitions, your sense of personal identity and free will, are in fact no more than the behavior of a vast assemblage of nerve cells and their associated molecules.'[54] Chemistry, in the service of a reductionist molecular biology, has definitely not been on the side of the angels. The 'no more than' in Crick's declaration would be offensive to most religious sensibilities.

THE AMBIVALENCE OF CHEMISTRY AS A CATALYST

The fact that chemistry, in different contexts, has catalysed both gains and losses for religion means there is no single story to be told about its relevance to religious authority, beliefs, and practices. There is a deeper reason for this, in that many developments in the science were not univocal in their implications. The atomism that was so important to John Tyndall's story of the emancipation of science from religious interference had been seen by Francis Bacon (1561–1626) in the seventeenth century as strongly supportive of a religious position, more supportive than the Aristotelian cosmology of his day. Bacon did not embrace the atomic doctrine of antiquity, but he found it inconceivable that a philosophy that so stressed the random motions and collisions of atoms could possibly account for the ordered world he experienced. In his essay 'Of Atheism', he protested that

> even that school which is most accused of atheism doth most demonstrate religion; that is, the school of Leucippus, and Democritus, and Epicurus—for it is a thousand times more credible that four mutable elements and one immutable fifth essence, duly and eternally placed, need no God, than that an army of infinite small portions, or seeds unplaced, should have produced this order and beauty without a divine marshal.[55]

[53] Brooke and Cantor, *Reconstructing Nature*, pp. 334–7.

[54] Francis Crick, *The Astonishing Hypothesis: The Scientific Search for the Soul* (New York: Simon and Schuster, 1995), p. 3.

[55] Francis Bacon, *Essays* (London: Dent, Everyman's Library, 1965), p. 49.

Conversely, chemical innovations advertised for their religious utility were interpreted quite differently by sceptics. The arguments for a divine marshal adduced by William Prout in his *Bridgewater Treatise* left Tyndall cold. In his preface to a posthumous fourth edition, Tyndall confessed that he would have thought more highly of Dr Prout had he not read his book. Arguments for the unity of nature and the unity of matter, which for Newton and Davy had religious connotations, were subsequently incorporated into stories about the progressive unification of science, in which religion was the loser.[56] In the service of both medicine and agriculture, chemistry was often seen as a virtuous science compatible with religious imperatives to relieve human want and suffering. Yet the consequences in both spheres could be debilitating for a religious life. A dependence on chemical remedies could, over time, diminish that sense of immediate dependence on Providence for one's continuing existence which, in earlier times, had been a feature of the spiritual life. The improvement of chemical fertilizers was seen by the great German chemist Justus von Liebig (1803–73) as a religious duty; and yet they have been identified as powerful agents of secularization in agrarian societies.[57]

The ambivalence of chemistry as a resource is particularly apparent in the relations between chemical innovation, vitalism, and religious belief. It has often been said that Friedrich Wöhler's (1800–82) artificial synthesis of urea in 1828 sounded the death knell for the vital forces that were considered peculiar to living things. But it was never that simple. Much of the excitement greeting his experiment sprang from the fact that it furnished an early, surprising example of isomerism. Wöhler had produced urea by combining potassium cyanate with ammonium chloride, leading to the conclusion that ammonium cyanate and urea were isomers. Those enamoured of vital forces would argue that the conditions under which Wöhler obtained urea could never obtain in a living organism, rendering the synthesis irrelevant to the question whether vital forces were operating *in vivo*. Wöhler had even begun with material of organic origin. Chemists well aware of his work, including J. J. Berzelius, Liebig, and Wöhler himself, continued to include vital forces in their scientific vocabulary.[58] If this were not complication enough, there was no simple correlation between religious belief and a commitment to vitalism. It was possible to challenge vital forces on religious grounds, since if they had the character of quasi-intelligent agencies they could be seen as detracting from God's sovereign relationship to nature.[59]

Scientific controversies are complicated and they have generally been more so when antagonists have claimed consequences for religion in their or their

[56] Steven Weinberg, *Dreams of a Final Theory: The Scientific Search for the Ultimate Laws of Nature* (New York: Vintage Books, 1994).

[57] Otto Sonntag, 'Religion and Science in the Thought of Liebig', *Ambix* 24 (1977): pp. 159–69; William H. Brock, *Justus von Liebig: The Chemical Gatekeeper* (Cambridge: Cambridge University Press, 1997), pp. 145–82; Peter Burke, 'Religion and Secularisation', p. 309.

[58] Brooke, *Thinking about Matter*, chap. 5.

[59] Brooke, *Thinking about Matter*, chap. 4.

opponent's position. What Frederick Gregory has said of the relations between the physical sciences and religion works for the chemical sciences too: 'Throughout the last two centuries in virtually all cases of interaction between physical science and religion, the diversity of opinion displayed has stemmed from the variety of assumptions that have been *brought to* the issues by the participants.'[60] This means we have to look even more closely at what the loss of religion might mean in relation to scientific advance.

CHEMISTS WITH OR WITHOUT GOD

A moment's reflection on what it might mean for an individual to lose their religious faith shows how difficult it can be to specify reasons for their loss. It is possible to reject conventional strands of religious belief and to retain others with greater vehemence as a consequence. Joseph Priestley rejected the Calvinist doctrines of predestination, of the 'fall', and of Christ's atonement, even of the Christian Trinity. And yet, in his Unitarian Christianity, other religious doctrines were resolutely upheld, especially that of bodily resurrection, which he considered essential for social control. He considered it immoral that all men should suffer divine judgment because of the sin of one man, Adam, and he was repelled by a concept of divine sovereignty that implied a predetermination on God's part of those who would be saved and those who would not. But he never lost his commitment to a God of love who wanted nothing less than human happiness.[61]

A second complication concerns the specificity of context in which a questioning of faith might occur. According to Robert Boyle, to be in London in the middle years of the seventeenth century would severely test one's faith because Christianity was being compromised by the proliferation of puritan sects during the interregnum, each claiming their own hotline to God.[62] The persecution of religious dissent by the Catholic Church in early eighteenth-century France triggered the hostility of Voltaire. In the renunciation of religion there can be many staging posts on the way to scepticism—from agnosticism on one or a few points of faith, to agnosticism on all matters pertaining to God-talk, to a self-conscious application of rationalist principles, to a scepticism that might eventuate in atheism, for which one might finally become an evangelist.

[60] Frederick Gregory, 'Intersections of Physical Science and Western Religion in the Nineteenth and Twentieth Centuries', in *The Cambridge History of Science*, vol. 5: *The Modern Physical and Mathematical Sciences*, ed. Mary Jo Nye (Cambridge: Cambridge University Press, 2003), p. 53.

[61] Brooke, 'Joining Natural Philosophy to Christianity', pp. 325–8.

[62] Michael Hunter, *Boyle: Between God and Science* (New Haven, CT: Yale University Press, 2009), pp. 84–5.

In his magisterial study of secularization in the West, Charles Taylor desisted from making the sciences a critical motor of change in such transformations. In his view, the underlying most powerful determinant of a secular mentality has been the change from a society in which 'it was virtually impossible not to believe in God, to one in which faith, even for the staunchest believer, is one human possibility among others.'[63] This invites a far broader analysis of the loss of faith than a focus on scientific naturalism alone would allow. There can be many twists to the story when individual cases are considered. In the nineteenth century, John Ruskin (1819–1900) finally departed from his evangelical faith not primarily because the geologists were chipping away at precious biblical verses, conscious though he was of their chiselling hammers. Rather, he found he could no longer condone evangelical censure of Catholic art by which he was deeply moved, notably by Veronese's *Queen of Sheba*, which had overwhelmed him just before he entered a Waldensian chapel where a 'little squeaking idiot was preaching to an audience of seventeen old women and three louts, that they were the only children of God in Turin'. He came out of the chapel 'a conclusively *un*-converted man'.[64] It is not difficult to imagine parallel cases where one might relinquish an evangelical affiliation not because one believed that it was incompatible with a particular science but because one grew tired of evangelical opprobrium levelled against that science. There are endless possibilities for why a faith might be damaged, attenuated, or lost.

In one of the most revealing surveys of reasons given by leading secularists for their loss of faith, Susan Budd examined the direct testimony of 150 unbelievers in the period 1850 to 1960. The implications of scientific knowledge were barely mentioned.[65]

Of far greater moment were moral objections to religious doctrines, the problem of suffering, the influence of radical texts such as Tom Paine's *Age of Reason*, even the study of the Bible itself, with its depiction of a vengeful anthropomorphic deity. In 1912, the president of the National Secular Society in Britain exclaimed that biblical stories of 'lust, adultery, incest and unnatural vice' were 'enough to raise blushes in a brothel'.[66] Higher criticism of the Bible raised further doubts. The unbecoming behaviour of clergy also took its toll. Renunciation of religious allegiance often accompanied migration from conservative to more radical politics. By the second half of the nineteenth century, there was no shortage of reasons why a chemist who had contributed to a chemistry without God might be a chemist without God.

[63] Charles Taylor, *A Secular Age* (Cambridge, MA: Harvard University Press, 2007), p. 3.

[64] Tim Hilton, *John Ruskin: The Early Years, 1819–1859* (New Haven, CT: Yale University Press, 1985), p. 254.

[65] Susan Budd, *Varieties of Unbelief: Atheists and Agnostics in English Society, 1850–1960* (London: Heinemann, 1977).

[66] Budd, *Varieties of Unbelief*, p. 109.

One who fits this description was the pioneer of organometallic chemistry Edward Frankland (1825–99). He is remembered as one who laid the foundations of valence theory and who, in common with the German chemist Hermann Kolbe (1818–84), advanced the chemistry of hydrocarbons. As a teacher of chemistry, he devised what became known as 'graphic' formulas, depicting the bonds between atoms in chemical molecules.[67] Frankland's spiritual trajectory and eventual loss of faith are particularly instructive. An Anglican in early life, he experienced a youthful evangelical conversion in a Congregational Church. During his evangelical phase, he devoted time to assisting the poor in London, in association with a large Independent chapel. During time spent with Kolbe in Marburg, he encountered historically based criticism of the Bible, which precipitated a reaction against his evangelical formation. Eventually finding a spiritual home among Unitarians, he finally lapsed into an acerbic agnosticism. When recalling reasons for his loss of faith he, like many of those examined by Budd, picked out the reading of Tom Paine. His critical remarks about religion were most vituperative when expressing solidarity with the X-Club, of which he became a member and which included Huxley, Tyndall, and other prominent freethinkers. In this context he would take a swipe at what he called two 'corollaries of religion': the idea that marriages are made in heaven and that Providence regulates the number of children, both 'among the most fertile sources of human misery'.[68] Frankland's biographer, Colin Russell, suggests that his eagerness to express solidarity with the secularist ethos of the X-Club may have triggered pronouncements that were not typical of his otherwise reticent stance on matters of belief. It is, however, apparent that anticlericalism more than chemistry catalysed his unbelief. When Huxley preached his lay sermon on protoplasm, Frankland congratulated him, predicting that it would 'frighten the parsons more than anything they have encountered for a long time'.[69] The power of the Catholic Church was a particular irritation.

Of the membership of the X-club it was said that Huxley was 'Xalted', Tyndall 'Xcentric', and Frankland the 'Xpert'. In 1881 his chemical expertise was invoked when asked to analyse 'holy water' from Mecca. Its nitrate concentration proved to be six times that in strong London sewage. When, ten years later, there were eleven thousand deaths in Mecca from cholera, Frankland was not surprised.[70] The guilt lay with decidedly *un*holy water. Chemistry could have been of service to religion, if only.... Chemistry without God had met God without chemistry.

It is tempting to make this episode my dénouement given what some might see as contemporary resonances. However, as Ronald Numbers has taught us,

[67] Colin A. Russell, *Edward Frankland: Chemistry, Controversy and Conspiracy in Victorian England* (Cambridge: Cambridge University Press, 1996).

[68] Russell, *Edward Frankland*, pp. 333–5. [69] Ibid., p. 335. [70] Ibid., pp. 393–5.

historical studies of science and religion should always spring a surprise. Here, it consists in the fact that Frankland's opposite number in Germany followed a strikingly different path. An early passion for chemistry had deflected Kolbe from following his father into the Lutheran ministry. At the time of his marriage, he was deflected further into a Reformed church. There is no evidence, however, that he ever relinquished a faith that was firmly grounded in scripture and in a natural theology to which chemistry contributed its testimony. Kolbe's chemistry gave him a standpoint from which to attack deviant forms of chemistry as well as deviant forms of religion. The heretics in chemistry were none other than the pioneers of structural theory, Frankland, Auguste Kekulé, and J. H. van 't Hoff. Kolbe was annoyed by the presumption he saw in their models of chemical structure, which he slated for their materialistic connotations. His objection was primarily epistemological. It would always be impossible, he maintained, to arrive at a notion of the spatial arrangement of atoms.[71]

Kolbe had many chemical achievements to his credit, including the first artificial synthesis of acetic acid. He had even promoted his subject as one that should be taught to students of theology. Chemistry would give them the apparatus for rebutting atheism—but not if the science was to be sabotaged by fanciful formulas derived from flights of imagination. Pictures of putative atomic arrangements had to be avoided, 'just as the Bible warns us from making a visual depiction of the Godhead'.[72] Posterity would not be on Kolbe's side. Nevertheless, his attack on the pretensions of Frankland and Kekulé shows that at a critical juncture in the history of science, a chemist with God still had something to say to a chemistry without.

[71] Alan J. Rocke, *The Quiet Revolution: Hermann Kolbe and the Science of Organic Chemistry* (Berkeley, CA: University of California Press, 1993), pp. 325–39.

[72] Rocke, *The Quiet Revolution*, p. 314.

7

Removing God from Biology

Michael Ruse

In his classic study of the scientific revolution, written in the middle of the last century, A. Rupert Hall offered this observation:

> No Christian could ultimately escape the implications of the fact that Aristotle's cosmos knew no Jehovah. Christianity taught him to see it as a divine artifact, rather than as a self-contained organism. The universe was subject to God's laws; its regularities and harmonies were divinely planned, its uniformity was a result of providential design. The ultimate mystery resided in God rather than in Nature, which could thus, by successive steps, be seen not as a self-sufficient Whole, but as a divinely organized machine in which was transacted the unique drama of the Fall and Redemption. If an omnipresent God was all spirit, it was the more easy to think of the physical universe as all matter; the intelligences, spirits and Forms of Aristotle were first debased, and then abandoned as unnecessary in a universe which contained nothing but God, human souls and matter.[1]

Despite being over half a century old, and notwithstanding more recent discussions about whether there even was a scientific revolution, this assessment has stood the test of time. At the heart of the great happenings in science that occurred in the sixteenth and seventeenth centuries was a change of root metaphor. Before Copernicus, the world was seen as an organism, or at least in organic terms. After Newton, the world was seen as a machine; the mechanistic era had started. No one claims that this was an immediate move to a Richard Dawkins kind of atheism. If anything, it was a move to a more directly Christianized view of creation. Machines have machine-makers with purposes. Automobiles are for transport; vibrators are for muscle tone. However, ideas (a bit like children) have their own tempo, and before long all that was left of the metaphor was a system bound by unchanging laws, simply going through the motions

[1] A. R. Hall, *The Scientific Revolution 1500–1800: The Formation of the Modern Scientific Attitude* (London: Longman, Green and Company, 1954), pp. xvi–xvii.

without end or purpose. In the words of another of the great historians of the revolution: 'The mechanism of the world picture led with irresistible consistency to the conception of God as a retired engineer, and from this to His complete elimination was only a step.'[2] In the language of philosophy, in the old-world picture, it was legitimate to think in terms of ends, of what Aristotle called 'final causes'. In the new world picture, the only allowable causes were 'efficient' or 'proximate'. Things took time. Newton, of all people, still invoked God to keep his system running smoothly. But by the beginning of the nineteenth century, all was functioning well. Famously, when questioned about God by Napoleon, the great French physicist Laplace replied: 'Sire, I have no need of that hypothesis.'

Mechanism. Is this the same as materialism? Frankly, I am not quite sure what materialism is. But I will assume that 'naturalism' is a term that applies here and use it interchangeably. More precisely, I will distinguish between 'methodological naturalism' and 'metaphysical naturalism', meaning by the former much that one would mean by 'mechanism'—a world understood as working by unbroken law, blind (that is unguided), and without interference by the Deity. A world understood without God. By the latter I mean the belief that methodological naturalism is all that there is. There is no God. Many people think that methodological naturalism leads automatically to metaphysical naturalism. Historically there may be some truth to this, and if you are a metaphysical naturalist you will be a methodological naturalist, but logically it does not follow that if you are a methodological naturalist you must be a metaphysical naturalist. Many people have been like the great chemist-philosopher Robert Boyle, who strove to be a methodological naturalist but who had no thoughts of being a metaphysical naturalist. In any case, my focus here is on methodological naturalism, which from now on I shall refer to as 'naturalism' without qualification.

BIOLOGY: THE EARLY YEARS

Physics and chemistry are all very well, but what about the life sciences? What about biology (to speak somewhat anachronistically)? This is the question I want to address, and I shall limit my discussion by concentrating on the side of biology that focuses on organisms at work and play, that which (today) makes evolution central. I shall ignore the more physiological side of biology—although I very much doubt that the story there would be so very different. I shall also set the discussion between two temporal limits. The earlier is around the middle of the seventeenth century. Descartes had argued that organisms are just machines, things belonging exclusively to the world of *res extensa*, and,

[2] E. J. Dijksterhuis, *The Mechanization of the World Picture* (Oxford: Oxford University Press, 1961), p. 491.

pushing the retired engineer line-of-thinking, denied that we should or could think of ends. 'When dealing with natural things, we will, then, never derive any explanations from the purposes which God or nature may have had in view when creating them <and we shall entirely banish from our philosophy the search for final causes>. For we should not be so arrogant as to suppose that we can share in God's plans.' Stick rather to efficient cause: 'starting from the divine attributes which by God's will we have some knowledge of, we shall see, with the aid of our God-given natural light, what conclusions should be drawn concerning those effects which are apparent to our senses.'[3]

Would-be methodological naturalist he may have been, but Boyle saw clearly that this will not do. Apart from the fact that we obviously do know the ends of the hand and the eye, there were significant difficulties in regarding organisms as things totally explicable in terms of blind law. I am not sure who first thought up Murphy's Law—if something can go wrong, it will go wrong—but everyone saw that blind law does not lead to complex, adaptive functioning. The hand and the eye demand something more. Here is Boyle again:

> For there are some things in nature so curiously contrived, and so exquisitely fitted for certain operations and uses, that it seems little less than blindness in him, that acknowledges, with the Cartesians, a most wise Author of things, not to conclude, that, though they may have been designed for other (and perhaps higher) uses, yet they were designed for this use. As he, that sees the admirable fabric of the coats, humors, and muscles of the eyes, and how excellently all the parts are adapted to the making up of an organ of vision, can scarce forbear to believe, that the Author of nature intended it should serve the animal to which it belongs, to see with.[4]

Boyle continued that 'the very supposition . . that a man's eyes were made by chance, argues, that they need have no relation to a designing agent; and the use, that a man makes of them, may be either casual too, or at least may be an effect of his knowledge, not of nature's.' But not only does this then take us away from the urge to dissect and to understand how the eye 'is as exquisitely fitted to be an organ of sight, as the best artificer in the world could have framed a little engine, purposely and mainly designed for the use of seeing'[5]—it also takes us away from the designing intelligence behind it.

Thus my earlier limit. In important respects, the life sciences were not purely naturalistic. The feeling is that, without bringing in God in some way, you simply could not get a full understanding of living phenomena. This was very much the view a hundred years later, when Immanuel Kant wrote in his *Critique of Judgement* that the biological sciences can never achieve the level of the physical

[3] R. Descartes (1644), *The Principles of Philosophy*, in *The Philosophical Works of Descartes*, ed. and trans. Elizabeth Haldane and G. R. T. Ross (New York: Dover Publications, 1955), p. 169. <> brackets mean added to French translation by Descartes [the work was originally published in Latin].

[4] Robert Boyle, *A Free Enquiry into the Vulgarly Received Notion of Nature* (1688), ed. E. B. Davis and M. Hunter (Cambridge: Cambridge University Press, 1996), p. 397.

[5] Boyle, *Free Enquiry*, p. 398.

sciences. Either you have to bring God into the science, which was *ganz verboten* for Kant, or you recognize that the science is limited, and then have to turn to theological explanation. Hence the conclusion: 'We can boldly say that it would be absurd for humans even to make such an attempt or to hope that there may yet arise a Newton who could make comprehensible even the generation of a blade of grass according to natural laws that no intention has ordered; rather, we must absolutely deny this insight to human beings.'[6]

BIOLOGY: TODAY

The later limit I set to my discussion is the present day, the beginning of the twenty-first century. Naturalism has conquered the life sciences. The extent to which biologists refuse to allow God any part in their work is starkly illustrated by the total rejection of so-called intelligent design theory, an explicitly God-infused program of explaining living beings. Give the floor to Richard Dawkins. He is promoting an explicitly atheistic agenda and outright metaphysical naturalism; but remembering that metaphysical naturalism implies methodological naturalism, no active biologist would disagree with his description of the living world as seen through the lens of biology.

> In a universe of blind physical forces and genetic replication, some people are going to get hurt, other people are going to get lucky, and you won't find any rhyme or reason in it, nor any justice. The universe we observe has precisely the properties we should expect if there is, at bottom, no design, no purpose, no evil and no good, nothing but blind, pitiless indifference. As that unhappy poet A. E. Housman put it:
>
> For Nature, heartless, witless Nature
> Will neither know nor care.
>
> DNA neither knows nor cares. DNA just is. And we dance to its music.[7]

These limits define my question. Three hundred and fifty years ago, naturalism in the life sciences seemed impossible. You had to bring in God to complete the task. Today, naturalism *über alles*. God is not needed. This is so, even if you believe in God. How did this happen? And there is a related question. Someone like Richard Dawkins undoubtedly thinks that the story has been one of warfare, of science set against religion, and of the triumph of science. One suspects that above his bed is an embroidered sampler quoting Thomas Henry Huxley: 'The cradle of every science is surrounded by dead theologians as that of

[6] Immanuel Kant, *Critique of Judgement* [1790], trans. J. H. Bernard (New York: Hafner Publishing, 1951), p. 270.

[7] Richard Dawkins, *A River Out of Eden* (New York: Basic Books, 1995), p. 133.

Hercules was with strangled serpents.'[8] But if the quotation that opened this essay has any truth, the situation may be more complex and interesting than that. Perhaps religion had a positive role to play, even if it did not quite turn out as people expected.

THE PROBLEM OF FINAL CAUSE

With good reason, Charles Darwin tends to be the central figure in these sorts of discussions, so let us start with the two centuries leading up to the publication of the *Origin of Species* in 1859. A number of facts stand out. First, no one cracked the problem of final cause: the design-like, end-directed nature of organisms. I hesitate yet again to quote Richard Dawkins, but he is right. Until Darwin, it was impossible to be an 'intellectually fulfilled atheist'.[9] Thinking in terms of philosophy and not biology, David Hume's *Dialogues Concerning Natural Religion* are devastating to ideas of final cause and underlying design by a deity. Yet at the end, Hume has to concede that perhaps there is something there after all. Philosophers talk in terms of inference to the best explanation. This is the methodology of Sherlock Holmes: '"You will not apply my precept", he said, shaking his head. "How often have I said to you that when you have eliminated the impossible, whatever remains, however improbable, must be the truth?"'[10] Blind law does not lead to functioning, adaptive complexity. There is no natural explanation of the hand and the eye. They must be the result of God's designing creation. There were those like Kant who said that this is not part of science. The nineteenth-century philosopher and historian of science William Whewell said that science says nothing but 'she points upwards'.[11] But in science or not, there is no naturalistic explanation.

SCIENCE AND THEOLOGY IN HAPPY HARMONY

Second, not everyone was worried about this. The very opposite, in fact. You had a research programme. You had what Thomas Kuhn calls a 'paradigm', where there are puzzles set and you can get on and solve them. Boyle pointed to this. Assume final cause and now set about finding it in action. This was

[8] Thomas Henry Huxley, 'The Origin of Species', in *Darwiniana: Essays by Thomas H. Huxley*, vol. 2, edited by Thomas Henry Huxley (London: Macmillan, 1893), p. 52.

[9] Richard Dawkins, *The Blind Watchmaker* (New York, NY: Norton, 1986), p. 6.

[10] This famous passage comes from *The Sign of the Four*, the second Sherlock Holmes novel, by Arthur Conan Doyle, first published in 1890.

[11] William Whewell, *The History of the Inductive Sciences*, vol. 3 (London: J. W. Parker, 1837), p. 588.

exactly the agenda of his contemporary, the clergyman-naturalist John Ray (1628–1705), especially in his *Wisdom of God, Manifested in the Words of Creation* (1691). An argument to adaptive complexity was stated clearly and unambiguously: 'Whatever is natural, beheld through [the microscope] appears exquisitely formed, and adorned with all imaginable Elegancy and Beauty. There are such inimitable gildings in the smallest Seeds of Plants, but especially in the parts of Animals, in the Head or Eye of a small Fly; Such accuracy, Order and Symmetry in the frame of the most minute Creatures, a Louse, for example, or a Mite, as no man were able to conceive without seeing of them.'[12] Everything that we humans do and produce is just crude and amateurish compared to what we find in nature. But that is just fine, because it gives us the opportunity to look at nature and to work out the functioning of organisms and their parts.

Ray was but one of a long line of parson-naturalists in Britain who did sterling work finding out about the ways of the organic world. One of the deservedly best known and important was William Kirby, co-author of *Introduction to Entomology, or Elements in the Natural History of Insects* (1815–28). This was the definitive work on the topic and a major influence on the Darwinians (especially Henry Walter Bates) in their work on insect wing colours and patterns. Thanks to this work, everyone knew full well that the colours of insects 'deceive, dazzle, alarm or annoy' their enemies. A prize example was the 'mimicry' of the Brazilian walking stick insect (*Phasma*), something that closely resembled the twigs on which it spent its living days. Although no full-blown theory was offered, it was clearly noted that it had a function, because the author 'has often been unable to distinguish it [the insect] from them [the twigs], and the birds probably often make the same mistake and pass it by'.[13]

I should add that the fact that so many naturalists were (Anglican) clergymen was not pure chance. The great thing is that you had more than just a scientific paradigm. You also had (and even for an Anglican clergyman this was no bad thing) a theological paradigm. You were working on the traditional argument to design! The living world was likened to a product of design. A machine implies an architect or an engineer, and so, likewise, inasmuch as the world of life is machine-like, it too implies a being, as much above us as the world of life is above our artefacts and creations. 'There is no greater, at least no more palpable and convincing argument of the Existence of a Deity', wrote John Ray, 'than the admirable Art and Wisdom that discovers itself in the Make and

[12] John Ray, *Wisdom of God, Manifested in the Works of Creation*, 7th edn (London: Samuel Smith, 1717), p. 58.

[13] William Kirby and William Spence, *An Introduction to Entomology: or Elements of the Natural History of Insects*, vol. 2 (London: Longman, Hurst, Reece, Orme, and Brown, 1818), p. 220.

Constitution, the Order and Disposition, the Ends and uses of all the parts and members of this stately fabrick of Heaven and Earth.'[14]

This made for a neat package, as the teleological way of thought in biology was tied back into the proof of the divine:

> That under one skin there should be such infinite variety of parts, variously mingled, hard with soft, fluid with fixt, solid with hollow, those in rest with those in motion:—all these so packed and thrust so close together, that there is no unnecessary vacuity in the whole body, and yet so far from clashing or interfering with one another, or hindering each other's motions, that they do all help and assist mutually on the other, all concur in one general end and design.[15]

Moreover, this is design which is of absolutely the top quality and so the same must be said of the intelligence behind it. This is the full-blooded Christian God, not some ethereal spirit worshipped by the heathen.

THE COMING OF EVOLUTION

The third point I want to make is that there was a worm in the bud. The scientific revolution did not necessarily imply inquiries into origins, but once you had got the universe up and running it was natural to start asking about how it all came about. At the same time discoveries in the life sciences and in geology particularly spurred people to ask about the early times in life history. Combine this with the beginnings of biblical criticism by the likes of Spinoza and the discoveries of other civilizations with origin stories very different from those of Genesis, and it is little wonder that by the beginning of the eighteenth century, people were wondering about how life came about and whether or not a natural story was plausible. There were many items that went into the pot—the Great Chain of Being, for instance—but the key notion was that of progress. The Greeks never really had such an idea—for them things were eternal and just kept repeating—but by the beginning of the eighteenth century (with reason called the start of the Age of the Enlightenment), people were starting to think that by their own efforts, through science and technology, through education and political reform, they could make for an ever-better state of society. Very quickly this became a theory of organic origins, as people read progress into life's history and then often turned around and used life's history to justify their beliefs in social progress! Erasmus Darwin, the grandfather of Charles Darwin, is the paradigm.

[14] Ray, *Wisdom of God*, p. 30.

[15] Ray, *Wisdom of God*, pp. 290–1.

Organic Life beneath the shoreless waves
Was born and nurs'd in Ocean's pearly caves;
First forms minute, unseen by spheric glass,
Move on the mud, or pierce the watery mass;
These, as successive generations bloom,
New powers acquire, and larger limbs assume;
Whence countless groups of vegetation spring,
And breathing realms of fin, and feet, and wing.

Thus the tall Oak, the giant of the wood,
Which bears Britannia's thunders on the flood;
The Whale, unmeasured monster of the main,
The lordly Lion, monarch of the plain,
The Eagle soaring in the realms of air,
Whose eye undazzled drinks the solar glare,
Imperious man, who rules the bestial crowd,
Of language, reason, and reflection proud,
With brow erect who scorns this earthy sod,
And styles himself the image of his God;
Arose from rudiments of form and sense,
An embryon point, or microscopic ens![16]

He made no bones about the way in which he tied his biology into his philosophy. This idea of organic progressive evolution 'is analogous to the improving excellence observable in every part of the creation ... such as the progressive increase of the wisdom and happiness of its inhabitants'.[17]

In many respects, cultural notions of progress and Christianity are rivals. Central to the latter is the idea—particularly conspicuous in Augustinian and Calvinist thought—that we ourselves can do nothing good without the assistance of divine grace. Self-help, central to progress, is arrogance. It is the sin that brought down Satan. Yet the story is a little more complex than this. There are reasons for saying that evolutionary theory is the bastard child of Christianity. It is different and then in the half-light you see the family resemblances. For a start, evolution (I will use this word, although it did not come into general use until the mid-nineteenth century) is a historical story of origins, analogous to (or substituting for) the historical story of origins found in the Bible. It is not a story of eternal uniformity as we find in Greek thought. It is (at least it was back then) a story that makes humankind as central as does Christianity. Erasmus Darwin's evolution does not lead up to warthogs. We are the focus of the drama. And also, at least again as it was back then, it is a story where God has the major role to play. Few if any of the pre-Darwinian evolutionists were

[16] Erasmus Darwin, *The Temple of Nature* (1803), canto 1, 11, lines 295–314.

[17] Erasmus Darwin, *Zoonomia; or, The Laws of Organic Life* (London: J. Johnson, 1794–6), p. 509.

traditional Christians, but they all (or nearly all) believed in God. They believed in the God who works through unbroken law, who built the machine and set it going, and who now sits back and watches the effects. In other words, a God for whom a naturalistic account of origins is support rather than refutation. We have the God of the deist, as opposed to the God of miraculous intervention, the God of the theist. A God, incidentally, who fits in nicely with the Industrial Revolution. A popular metaphor was of the deist's God being one who works through machines rather than by hand, as does the theist's God.

But there was still the problem of final cause, and, if anything, the coming of evolutionary ideas exacerbated it. How can blind law create such intricately functioning organisms? Basically, there were two answers. On the one hand, there were those who opted for some kind of (we today would call) guided evolution (or theistic evolution, although note that generally they were not theists). Perhaps the great French evolutionist Lamarck fell into this category. He certainly believed in some kinds of vital forces pushing organisms up the scale of nature, and these forces presumably came from God. He was also a 'Lamarckian' meaning that he (like a lot of other people) believed in the inheritance of acquired characteristics. Note, however, that even if this gives you some kind of organization through law, it is going to be limited. Darwin stressed that many features simply could not have a Lamarckian origin. The sterility of worker ants for instance. On the other hand, there were those who simply ignored the problem! They were ignorant of or indifferent to final cause. The pre-Darwinian and Scottish evolutionist Robert Chambers probably fell into this category. It is surely significant that he was not a trained scientist and so was ignorant of the work of those who had worked in the design paradigm. Had he done so, he would probably have thought very differently.

CHARLES DARWIN AND THE 'ORIGIN'

One who was trained within the design paradigm and who did think and work very differently was Charles Darwin, so let us move now to him. We begin with the basic facts. In the *Origin of Species*, Darwin made two separate, but related, claims. First, he offered a mechanism of change—natural selection. There is an ongoing struggle for existence, new undirected variations are always appearing in populations, and so we get a natural selecting of some forms over others. Given time this leads to permanent change. Second, Darwin argued that the history of organisms is that of an ever-branching tree from just a few forms, perhaps only one. To make this case he marshalled evidence from behaviour, paleontology, biogeography, morphology, systematics, embryology, and more. It is generally agreed that Darwin was a lot more successful in his second aim than his first. Almost overnight people accepted evolution. This is true, with some hesitation, even of religious people, with obvious exceptions. Very few accepted natural selection, at least in any significant fashion. Other mechanisms were favoured, including Lamarckism, evolution by large variations (saltationism), and a kind of general momentum (orthogenesis). Natural selection had to wait until the twentieth century and the development of an adequate theory of heredity. Once this was done, building on ideas first discerned by the Moravian monk Gregor Mendel, natural selection could come into its own. This it did in the 1930s, and there has been no looking back. Today's evolutionary theory is firmly Darwinian, meaning natural selection resting on a bed of genetics, initially Mendelian and now molecular. And as noted earlier, it is entirely naturalistic. God is out of the picture.

So what happened? Go back to Darwin. He cracked the final cause problem. Natural selection speaks not only to change but also to the end-directed nature of organisms, the hand and the eye, what Darwin called 'contrivances' and what today's biologists call 'adaptations'. Those organisms that do survive and reproduce, the 'fitter', have features not possessed by the losers, and over time these build up into full-blown end-directed characteristics like the hand and the eye. Murphy's Law is defeated! Blind law can lead to functioning organization. God is out. Darwin had no need of that hypothesis. He was the Newton of biology. (And didn't he just know it!) Richard Dawkins is right. Naturalism wins. Like Jacob and the angel, science wrestled with religion and, like Jacob, science won. Well, yes, but this isn't quite the entire story, and the story is a lot more interesting once you start to unpack it.

DARWIN'S SCIENCE AND THE GOD PROBLEM

First, Darwin did not see what he was doing as a simple victory of science over religion. As is well known, late in life, Darwin had moved to agnosticism. But when he wrote the *Origin*, he believed in a *God*! It was the deist's God, not the God of the Christians, but it was a God nevertheless. He made this completely clear in a letter written at the time to his great American supporter Asa Gray. 'I cannot anyhow be contented to view this wonderful universe and especially the nature of man, and to conclude that everything is the result of brute force. I am inclined to look at everything as resulting from designed laws, with the details, whether good or bad, left to the working out of what we may call chance.'[18] He continued:

> Certainly I agree with you that my views are not at all necessarily atheistical. The lightning kills a man, whether a good one or bad one, owing to the excessively complex action of natural laws,—a child (who may turn out an idiot) is born by action of even more complex laws,—and I can see no reason, why a man, or other

[18] Letter from Charles Darwin to Asa Gray, 22 May 1860, in *The Correspondence of Charles Darwin*, ed. F. Burkhardt et al. (Cambridge: Cambridge University Press, 1985-) vol. 8, p. 224.

> animal, may not have been aboriginally produced by other laws; and that all these laws may have been expressly designed by an omniscient Creator, who foresaw every future event and consequence.[19]

Second, in line with the sentiment in this letter, absolutely and entirely, Darwin had no place for God in his science. He was as hardline a methodological naturalist as it is possible to imagine. He thought that natural selection does the trick and that is an end to things. When people like the astronomer-philosopher John Herschel said that natural selection was the law of higgledy-piggledy, and that we have to have design, Darwin bluntly said that would be to take things outside of the realm of science.[20] End of argument. We are set firmly on the path to the present.

Third, and this is the really important point I want to make, naturalist though he may have been, Darwin's thinking was about as God-impregnated—Christian-God-impregnated—as it is possible for something to be. 'How do I love thee? Let me count the ways.'[21] Again and again, in the half-light, the resemblances shine through. It cannot be overemphasized that science is not a matter of simply describing and then explaining reality. You don't just go in there one day and look at the world and give what I like to call the police photographer's picture of reality—'Just the facts, ma'am, just the facts'—and then set about explaining. Darwin knew the score. Without a hypothesis, without a problem, without a question, one 'might as well go into a gravel pit and count the pebbles and describe the colours. How odd it is that anyone should not see that observation must be for or against some view if it is to be of any service.'[22] And as one comes up with one's answers, one puts them in the contexts one understands and can use. Above all, one reaches for metaphors that capture the experience, give it meaning, and help move forward to new insights and problems.

NATURAL SELECTION AND ITS CHRISTIAN CONTEXT

Go right to natural selection. It is there to solve a problem, namely that of change, but of change of a particular kind: change that produces adaptations and design-like features. Why did Darwin have this problem? Because he was thoroughly steeped in the Anglican natural theology of his day. As an undergraduate at

[19] Ibid.

[20] Darwin to Charles Lyell, 10 December 1859, in *The Correspondence of Charles Darwin*, vol. 7, p. 423.

[21] This is the first line of the well-known poem (Sonnet 43), with the same name, by Elizabeth Barrett Browning. First published in 1850 and read at every wedding ceremony over which I have officiated (in my role as a Notary Public).

[22] Darwin to Henry Fawcett, 18 September 1861, in *The Correspondence of Charles Darwin*, vol. 9, p. 269.

Cambridge, he had been force-fed the works of Archdeacon William Paley of Carlisle, and he knew all about functioning organization. He joked that he could have written out Paley by heart on the subject, except it wasn't a joke. Darwin's biggest supporter, Thomas Henry Huxley, had no such training and he was basically indifferent to adaptation. While Darwin was salivating over the challenges produced by butterfly and moth wing markings, Huxley was calmly denying that they had any significance at all. For Huxley, they just weren't a problem (whereas, given his metaphysical interests that segued into his reforming agenda for society, the simian origin of humankind was a huge matter of concern and interest.)

My point is that, for Darwin, the problem of final cause was a Christian-set problem. And his solution, natural selection, was a Christian-set solution! Look at how he introduced natural selection in the earliest full account of his theory. He had been talking about how humans select the organisms that they want and breed from them, thus creating new forms.

> Let us now suppose a Being with penetration sufficient to perceive differences in the outer and innermost organization quite imperceptible to man, and with fore-thought extending over future centuries to watch with unerring care and select for any object the offspring of an organism produced under the foregoing circumstances; I can see no conceivable reason why he could not form a new race (or several were he to separate the stock of the original organism and work on several islands) adapted to new ends. As we assume his discrimination, and his fore-thought, and his steadiness of object, to be incomparably greater than those qualities in man, so we may suppose the beauty and complications of the adaptations of the new races and their differences from the original stock to be greater than in the domestic races produced by man's agency: the ground-work of his labours we may aid by supposing that the external conditions of the volcanic island, from its continued emergence and the occasional introduction of new immigrants, vary; and thus to act on the reproductive system of the organism, on which he is at work, and so keep its organization somewhat plastic. With time enough, such a Being might rationally (without some unknown law opposed him) aim at almost any result.[23]

This is virtually identical reasoning to Paley, who, having shown how wonderful the eye is compared to the telescope, then refers the designing/creative process to what I like to call the 'Great Optician in the Sky'. It is true that, in the *Origin*, the Being is replaced by nature, but the argumentation is identical. The metaphor is right there.

> How fleeting are the wishes and efforts of man! how short his time! and consequently how poor will his products be, compared with those accumulated by nature during whole geological periods. Can we wonder, then, that nature's productions should be far "truer" in character than man's productions; that they should be infinitely

[23] Charles Darwin, *The Foundations of the Origin of Species: Two Essays Written in 1842 and 1844*, ed. Francis Darwin (Cambridge: Cambridge University Press, 1909), p. 86.

> better adapted to the most complex conditions of life, and should plainly bear the stamp of far higher workmanship?
>
> It may be said that natural selection is daily and hourly scrutinizing, throughout the world, every variation, even the slightest; rejecting that which is bad, preserving and adding up all that is good; silently and insensibly working, whenever and wherever opportunity offers, at the improvement of each organic being in relation to its organic and inorganic conditions of life.[24]

Let me make clear what I am arguing and what I am not arguing. I am saying that Darwin's thinking at this crucial point is set in a mould given to him by Christianity. (I would say the same of natural selection's co-discoverer Wallace; although it needs arguing and, with most historians of this topic, I would warn against a ready identification of Wallace's thinking with that of Darwin.) I am not saying that Darwin was surreptitiously breaking from naturalism and bringing in God. Some readers of the *Origin* thought he was doing so and indeed some readers were happy that he was doing so. With indignation, Darwin repudiated this suggestion.

> Others [critical of natural selection] have objected that the term selection implies conscious choice in the animals which become modified; and it has even been urged that as plants have no volition, natural selection is not applicable to them! In the literal sense of the word, no doubt, natural selection is a misnomer; but who ever objected to chemists speaking of the elective affinities of the various elements?—and yet an acid cannot strictly be said to elect the base with which it will in preference combine. It has been said that I speak of natural selection as an active power or Deity; but who objects to an author speaking of the attraction of gravity as ruling the movements of the planets? Everyone knows what is meant and is implied by such metaphorical expressions; and they are almost necessary for brevity. So again it is difficult to avoid personifying the word Nature; but I mean by Nature, only the aggregate action and product of many natural laws, and by laws the sequence of events as ascertained by us.[25]

I think Darwin was right. There is no God in his theory.

MORE CHRISTIAN CONTEXT

Natural selection and design are not the only places where Darwin is indebted to Christianity. As everyone knows, the force behind selection, the struggle for existence, is taken right out of the writings of the Reverend Thomas Robert Malthus. I like to describe Darwin as a great revolutionary but no rebel, and

[24] Charles Darwin, *On the Origin of Species by Means of Natural Selection; or, The Preservation of Favoured Races in the Struggle for Life* (London: John Murray, 1859), p. 84.

[25] Charles Darwin, *On the Origin of Species by Means of Natural Selection; or, The Preservation of Favoured Races in the Struggle for Life*, 3rd edn (London: John Murray, 1861), pp. 84–5.

the Malthus connection proves my point. Darwin was a child of the rich middle classes—his maternal grandfather was Josiah Wedgwood the potter—and he was not about to throw over this legacy, intellectually or emotionally. He took ideas given to him and then recast them in a whole new pattern—a bit like a kaleidoscope. Malthus was concerned about false hopes of progress and he argued that basically we cannot better our situation because population growth sets up pressures that preclude improvement. There will be inevitable 'struggles for existence' keeping things back.

The Darwin-Wedgwood family took this kind of thinking as a priori true—it was a good reason not to pay the people in their factories more money (they would only have more children)—but then Darwin transformed it by making it the motive force behind indefinite change. Not a rebel, but a revolutionary. The important thing, however, is that Malthus didn't think he was being anti-Christian in his thinking. The struggle was God's way of getting us up off our duffs and working. Without the spur, we would do nothing. Darwin internalized this completely, seeing the struggle as a good thing, not a terrible evil. Moreover, for Malthus the struggle had to bite or it would not do its job. Darwin felt the same. Later thinkers, for example the great naturalist John Muir, thought that the struggle had been overemphasized. 'I never saw one drop of blood, one red stain on all this wilderness. Even death is in harmony here.'[26] This was not the thinking of Parson Malthus or his follower Charles Darwin.

I could go on making similar points. As with Malthus, whether it be Christian or more likely deistic, Adam Smith's thinking about the 'invisible hand', that converts individual self-regard into harmony in the community, was nigh gospel for the Darwin-Wedgwood family. 'It is not from the benevolence of the butcher, the brewer, or the baker that we expect our dinner, but from their regard to their own interest.'[27] In taking advantage of the Industrial Revolution, the family was doing no more than God intended. (The Darwins were nominally Anglican, although belief lay lightly on the men. The Wedgwoods were Unitarian, with some Anglican connections. Deism was therefore in their life-blood.) Again, Darwin internalizes all of this but uses it for his own ends. Selection is never for the benefit of the group, but always for the individual (which could extend to the family). As in political economy, however, this is the basis of cooperation, not its nemesis.

Finally let me mention the tree of life, something with its roots (!) deeply embedded in Judaeo-Christian thinking. Here is Darwin again:

> The affinities of all the beings of the same class have sometimes been represented by a great tree. I believe this simile largely speaks the truth. The green and budding twigs may represent existing species; and those produced during each former year

[26] John Muir, *John of the Mountains: The Unpublished Journals of John Muir* (Madison, WI: University of Wisconsin Press, 1966), p. 93.

[27] Adam Smith, *An Inquiry into the Nature and Causes of the Wealth of Nations*, ed. R. H. Campbell, A. S. Skinner, and W. B. Todd, vol. 1 (Oxford: Clarendon Press, 1976), pp. 26–7.

> may represent the long succession of extinct species. At each period of growth all the growing twigs have tried to branch out on all sides, and to overtop and kill the surrounding twigs and branches, in the same manner as species and groups of species have tried to overmaster other species in the great battle for life.[28]

AFTER DARWIN

Not everyone was happy with Darwin's naturalistic agenda. His good old friend, the geologist Charles Lyell, went to his grave convinced that God had a hand in things. Alfred Russel Wallace turned his face firmly from the future, embraced spiritualism, and argued that unseen forces are at work, especially in human evolution. Above all, Asa Gray—a deeply committed evangelical Presbyterian—always hankered after guided variations. But the die was cast. There was no looking back. Thomas Henry Huxley may not have been very keen on natural selection, but promoting the naturalistic agenda was his lifelong mission. And so it has been through the twentieth century. This is not to say that no evolutionists have been religious. The greatest theoretician after Darwin was Ronald A. Fisher. He was a lifelong Anglican who used to preach in a chapel, and his religious beliefs influenced his thinking about science. The greatest promoter of evolution after Darwin was Theodosius Dobzhansky, a man with a huge number of students. He, too, was a deeply committed Christian and saw his religion and his science as in harmony. But neither they nor anyone else in the professional scientific community was about to bring God into the science story.

And yet! We come to the present and let us return for one final time to the most secular of them all, Richard Dawkins. He is not a great scientist, but he is a great science communicator, and his tremendous book *The Selfish Gene* caught the evolutionary mode of thinking as did no other. But just think about the central concept—the selfish gene. Dawkins used this metaphor to stress that natural selection works not for the good of the group but for the good of the individual. He is an ultra-Darwinian and hence, hardline naturalist though he may be, he reaches back (via Paley on design and Malthus on the struggle) to Adam Smith and the theology of the eighteenth century. I am far from the first to note these kinds of connections. The late Stephen Jay Gould was a strong critic of what he called pan-selectionism, where one sees natural selection as all-powerful in every case. He thought it wrong and he thought he knew why, although wrong, it is so tempting. As a secular Jew, brought up in a Marxist home, Gould could see that too many of today's Darwinians are caught in the mesh of their Christian training and fail to

[28] Darwin, *Origin of Species* (1859), p. 129.

realize that what they take to be truths of nature are in fact fictions of their faith or of the faith of their forefathers.[29]

Gould thought this a bad thing. Dawkins would think it a good thing. (In a condescending sort of way, in the *Blind Watchmaker* he is quite friendly towards Paley.) Trying to be a disinterested historian rather than an engaged philosopher, I will make no judgement; but, as I come to an end, I raise an item where I will break down and make a value claim. For all that I am somewhere to the right of Darwin on natural selection, I think that, on this matter, Gould may have a point that the Christian roots of contemporary evolutionary thinking mislead us. I refer to the place of humans in the evolutionary picture. We have seen that before Darwin, evolutionary thinking was explicitly progressionist, leading (as they used to say) from the monad to the man. Darwin was firmly against some kind of momentum, akin to embryological development, leading up to humans. But he was a committed progressionist and thought that we are a non-contingent outgrowth of the evolutionary process. His Christian training and his industrial family background combined to point inexorably to this conviction. More than this, he was prepared to put his conviction in the language of natural theology! Right at the end of the *Origin*, he wrote: 'Thus, from the war of nature, from famine and death, the most exalted object which we are capable of conceiving, namely, the production of the higher animals, directly follows.'[30] This is virtually identical to a sentiment he expressed in 1844, in the first full-length treatment of evolution. Then follows one of the most famous purple passages in the history of science, a passage that was virtually unchanged when, after years of waiting, Darwin went public.

> There is a simple *grandeur* in the *view* of *life* with its powers of growth, assimilation and reproduction, being *originally* breathed into matter under one or a few *forms*, and that whilst this our *planet* has gone circling on according to fixed laws, and land and water, in a *cycle* of *change*, have gone on replacing each other, and from so simple an *origin*, through the process of gradual selection of infinitesimal *changes*, endless *forms* most beautiful and most wonderful have been evolved.[31]

Note the words italicized and then compare with a passage written by the Scottish physicist David Brewster, in a review of a work by Comte that Darwin read in 1838, just after it had been published.

> In considering our own globe as having its origin in a gaseous zone, thrown off by the rapidity of the solar rotation, and as consolidated by cooling from the chaos of its elements, we confirm rather than oppose the Mosaic cosmogony, whether allegorically or literally interpreted....

[29] Stephen Jay Gould, *The Structure of Evolutionary Theory* (Cambridge, MA: Harvard University Press, 2002), p. 124.

[30] Gould, *Structure of Evolutionary Theory*, p. 490.

[31] Darwin, *Foundations*, p. 52. Italics added.

> In the grandeur and universality of these views, we forget the insignificant beings which occupy and disturb the planetary domains. Life in all its forms, in all its restlessness, and in all its pageantry, disappears in the magnitude and remoteness of the perspective. The excited mind sees only the gorgeous fabric of the universe, recognizes only its Divine architect, and ponders but on its cycle and desolation.[32]

Brewster loathed and detested evolution, which all goes to confirm what I said earlier about Darwin being a revolutionary not a rebel. He was not about to confirm the 'Mosaic cosmogony', but he had no qualms about drawing on the writings of its enthusiasts for his own purposes.

And so back to the present. Gould spotted the forces which led to thinking that evolution leads up to humankind, and he wanted nothing of them. Dawkins, however, buys right into the idea. Evolution is progressive and we won! Darwin's position is that there was a kind of evolutionary 'arms race' (not his language) and we came on top.

> If we look at the differentiation and specialization of the several organs of each being when adult (and this will include the advancement of the brain for intellectual purposes) as the best standard of highness of organization, natural selection clearly leads towards highness; for all physiologists admit that the specialization of organs, inasmuch as they perform in this state their functions better, is an advantage to each being; and hence the accumulation of variations tending towards specialization is within the scope of natural selection.[33]

This is precisely Dawkins's position. He is a bit of a computer geek and likes to put his ideas in that idiom. Real life arms races (in the military) have increasingly turned to electronic devices for support and it is significant to note analogously that humans have developed the biggest on-board computers. The fact that we are twenty-three times as smart as hippos is not definitive, but it surely tells us 'something'.

I do not want to make too much of all of this, but Gould is right. In the Darwinian world, whatever Darwin or Dawkins may think, progress is a bit of an iffy concept. Natural selection is no tautology, but it is relativistic. What makes for success and fitness in one context is not necessarily what makes for fitness in other contexts. There are no absolutes, not even brains. They may be good a lot of the time but they require masses of protein (in nature: chunks of dead animal) to keep them functioning. In the immortal words of the late Jack Sepkoski: 'I see intelligence as just one of a variety of adaptations among tetrapods for survival. Running fast in a herd while being as dumb as shit, I think, is a very good adaptation for survival.'[34]

[32] David Brewster, Review of *Cours de philosophie positive*, by Auguste Comte, *The Edinburgh Review* 67, no. 136 (1838): p. 301.

[33] Darwin, *Origin of Species* (1861), p. 134.

[34] Quoted in Michael Ruse, *Monad to Man: The Concept of Progress in Evolutionary Biology* (Cambridge, MA: Harvard University Press, 1996), p. 486.

Again, let me make my point clear. I am not against progress. I am not against biological progress. I favour humans over warthogs. It is just that I don't think it is a scientific concept or confirmed by Darwinian evolutionary biology. It is something we read into nature, rather than find in nature. And I am saying that we do not have to seek far for the origins of this belief for people like Darwin and Dawkins (and myself) who grew up in the culture that we did. And so, without further ado, I have my conclusion. In the centuries since the scientific revolution, we have seen the triumph of the methodological naturalistic approach in the life sciences. We do have Newtons of the blade of grass. But as good evolutionists, we should expect to find that the present is understandable only in terms of the past. This holds as much for evolutionary thinking itself as it does for the subjects of evolutionary thinking. This past was religious, Christian religious, and thus, as we should have expected, we find that today's biology, secular though it may be, echoes the thought systems of those with very different agendas from ours. Above all, it echoes the Christian agenda.

8

Christian Materialism and the Prospect of Immortality

Michelle Pfeffer

In 2015 the physicist, science writer, and religious sceptic Victor Stenger confidently announced the death of the immortal and immaterial soul and, along with it, the end of religion. The idea of the soul, in Stenger's judgement, was 'the most deeply personal and destructive superstition held by the bulk of humanity', and its demise at the hands of an advancing neuroscience would bring 'the final, fatal blow to religious belief'.[1] Stenger's convictions about the imminent obsolescence of religious belief may seem somewhat premature, given readily available sociological data about the scale of global religious belief and its likely trajectory. But the idea that religious commitment entails allegiance to the idea of an immortal soul is relatively uncontroversial. Belief in the afterlife, characteristic of a number of religions including Christianity, would seem to require something like an immaterial soul—a conception that locates the human person at the intersection of the material and spiritual, the natural and the supernatural. Opposed to this is a purely naturalistic conception of the person which, on Stenger's analysis, involves an alliance of materialism and atheism, supported by science.

The idea that materialist conceptions of the person are necessarily aligned with naturalism and atheism has a long history. The materialist philosophy of ancient atomists such as Democritus, Epicurus, and Lucretius led to their being maligned as atheists (even though they insisted on the existence of gods). In the seventeenth century atomistic theories played a key role in the emergence of

[1] Victor Stenger, 'Foreword', in Julien Musolino, *The Soul Fallacy* (New York: Prometheus Books, 2015), pp. 9–10. This statement recalls Sigmund Freud's infamous elucidation of the 'three blows' science has dealt to humankind: that the earth is not the centre of the universe, that humans do not hold a privileged place in the natural order and, finally, that human thought and consciousness are determined by unconscious drives. Sigmund Freud, 'Introductory Lectures to Psychoanalysis', in *The Standard Edition of the Complete Psychological Works of Sigmund Freud*, vol. 15, trans. James Strachey (London: Vintage Books, 1915), pp. 284–5.

the new science, but the traditional association of atomism with atheism continued to render forms of materialism religiously suspect. Its reputation was not enhanced when, in the eighteenth century, a number of radical French thinkers such as Julien Offray de La Mettrie and Paul-Henri Dietrich d'Holbach openly argued for both atheism and materialism, drawing upon contemporary developments in natural philosophy and medicine.[2] Looking back from the present, when ideas of the immaterial soul are under assault from contemporary neuroscience, history might seem to suggest a long-standing opposition between scientific materialism on the one hand and religious spiritualism on the other.

However, it is by no means clear that a materialist conception of the person necessarily rules out belief in post-mortem existence, or religious commitment more generally. In this chapter I will explore the work of a number of thinkers in the seventeenth and eighteenth centuries who explicitly argued for the compatibility of materialism and religion. At least thirty English writers in this period developed a materialist position that they argued was consistent with their Christian faith. They understood the soul to be a mortal and material substance, and held that the resurrection of the body, rather than the immortality of the soul, provided assurance of life after death. As for the idea of the immaterial and immortal soul—that was a heathen invention happily adopted by Roman Catholics to support belief in purgatory. Crucially, while these mortalist and materialist ideas were connected with developments in natural philosophy and medicine, they were not simply the result of a compromise forced upon the faithful by advances in the sciences but were argued to have a foundation in biblical anthropology.

In recent years historians have become more sensitive to the theological underpinnings of this form of mortalist-materialism.[3] While some scholars have claimed that mortalist-materialism was built upon scriptural exegesis and ideas of reform, these suggestions have not been adequately explored.[4] This chapter will begin to fill this gap. I argue that for mortalist-materialists, misinterpretation of scripture was the heart of the issue: advances in anatomy, natural philosophy, and medicine were less pressing than arriving at a genuinely biblical view of the person. These writers need to be understood

[2] La Mettrie is commonly thought to be the first person to describe himself as a 'materialist', but it was in fact Henry Layton, whose work is discussed below, who first took this step in *Observations upon a Sermon* [1692], p. 62.

[3] While the writers discussed in this chapter were materialists *and* mortalists, not all early modern materialists were mortalists and vice versa.

[4] The classic treatment of seventeenth- and eighteenth-century mortalism is Norman T. Burns, *Christian Mortalism from Tyndale to Milton* (Cambridge, MA: Harvard University Press, 1972). More recent works include Bryan W. Ball, *The Soul Sleepers: Christian Mortalism from Wycliffe to Priestley* (Cambridge: James Clarke & Co, 2008), and Ann Thomson, *Bodies of Thought: Science, Religion, and the Soul in the Early Enlightenment* (Oxford: Oxford University Press, 2008).

in the wider context of scholarly shifts towards philological and historical exegesis, as they creatively deployed existing exegetical methods that, at least in their mind, were aimed at defending Christianity by reforming its doctrine.

The chapter will begin with an overview of discussions of the soul in the period, before showing how mortalist-materialist writers interpreted four challenging biblical texts. Their historical-critical techniques were neither new nor unique, but their ultimate conclusions—that the Bible supports a mortalist and materialist understanding of the soul—were innovative and contentious. The third section will thus consider how mortalist-materialists justified their views by constructing a reform programme through two key moves. (1) They maintained that Protestants alone had the freedom to read the scriptures for themselves, and this was the reason (a) why the papacy had successfully sustained a false doctrine and (b) why Protestants alone could eradicate this doctrine. (2) By aligning themselves with earlier Protestant reformers, they justified their own radical work by suggesting that they too were campaigning against the errors of tradition.

THE SOUL AND BIBLICAL EXEGESIS

Seventeenth- and eighteenth-century England witnessed heated disputes over the doctrine of the immaterial and immortal soul. There was much at stake: the *imago dei*, the operations of human reason, the transmission of original sin, and the prospect of personal immortality were all said to hang upon the soul's immateriality. The brain, clammy and insipid, was typically considered incapable of producing something as refined as reason, and it seemed clear that corporeal substances were bound to decay. How could a brittle and breakable material body not only produce something as refined as human thought, but also subsist after death to experience the rewards of heaven or the punishments of hell? For many, the immaterial and immortal soul was the clear solution provided not only by medicine and natural philosophy but by scripture and the church fathers. To suggest anything else was seen not only as philosophically absurd, but as heretical, irreligious, and, crucially, anti-scriptural. Indeed, a cursory glance at seventeenth- and eighteenth-century England would lead many to think that Christianity was perpetually at war with materialism. For many early moderns, mortalism and materialism necessarily destroyed human spirituality, angels, God and, indeed, free-will, salvation, and eternal life.

In the face of these difficulties, mortalists and materialists sought to correct the idea that the denial of the immaterial and immortal soul was equivalent to denying revelation. The most prolific of these writers included the pamphleteer Richard Overton (*fl.*1640–64), the notorious Thomas Hobbes (1588–1679), the preacher George Hammon (*fl.*1655–60), the lawyer Henry Layton (1622–1705), the physician William Coward (1657–1725), the vice-chancellor of Cambridge Peter Peckard (1717–97), and the natural philosopher and theologian Joseph Priestley (1733–1804). John Milton (1608–74), two anonymous writers, Edmund Law (1703–87), and several other minor figures also held mortalist-materialist positions.[5] For these writers, while the Reformation had purged many false dogmas, popery could only be fully eradicated if the Church returned to the Hebrew conception of the human as material and naturally mortal. While their explanations of *how* matter could think varied, mortalist-materialists amassed a repertoire of arguments that remained remarkably stable throughout this period. While they harked from various professions and confessions, and most had no personal connections with each other, the consistency of their theological arguments meant that they laid the foundations for a tradition of mortalist-materialism that continues even today. They maintained that an omnipotent God could make matter think, that an immaterial soul could not pass on the effects of the Fall, and that the immortal soul rendered the Resurrection meaningless. They provided historical accounts of the heathen origins of the immortal soul, and contended that the Hebrew and Greek words for soul and spirit had been erroneously translated and misinterpreted.

While the Fifth Lateran Council in 1513 had called upon Christians to demonstrate the immortality of the soul through natural philosophy, the words of scripture remained of fundamental importance to early modern discussions of the soul and body. Ensuring that revelation and natural philosophy were in tune with each other was critical. In general the human soul of mainstream early modern English thought was an immaterial and immortal substance incapable of natural dissolution, and it was argued that the Books of Nature and Scripture unanimously supported this. For writers like Thomas Wadsworth (1630–76), in *Antipsychothanasia, Or, The Immortality of the Soul Explained and Proved by Scripture and Reason* (1670), the Genesis account revealed that the body and soul were created separately: one was made from dust, the other immediately created by God.[6] The soul 'actuated' the body 'by a vital union therewith' and thus while the soul could survive without the body, as James 2:26 states, 'the body without the Soul is dead'.[7] Reason and the scriptures attested that the soul was associated with ratiocination: how could the soul

[5] To this list could be added others who experimented with mortalism, materialism and other similar concepts: Margaret Cavendish (1623?–73), Charles Blount (1654–93), Thomas Tomkinson (1631–1710), John Locke (1632–1704), Isaac Newton (1641–1727), John Toland (1670–1722), Evan Lloyd (*fl.*1707), Anthony Collins (1676–1729), Samuel Strutt (*fl.*1730), Joseph Wimpey (*fl.*1741), Viscount Bolingbroke (1678–1751), John Jackson (1686–1763), and Francis Blackburne (1705–87). See also several 'fence-sitters' on the issue of the soul's materiality: David Abercromby (*d.*1701?), Francis Gregory (1623–1707), William Sherlock (1638/40–1707), Archibald Campbell (1691–1756), John Steffe (*fl.*1757–60), and William Kenrick (*c.* 1725–1779).

[6] Thomas Wadsworth, *Antipsychothanasia, Or, The Immortality of the Soul Explained and Proved by Scripture and Reason* (London: T. Milbourn, 1670), pp. 13–17.

[7] Wadsworth, *Antipsychothanasia*, pp. 15, 32.

think of France at one second and Egypt at the next if it was corporeal? Being immaterial, the soul was naturally incorruptible: the death of the 'terrestrial body' does 'not concern [the soul's] own essential life'.[8] Passages like Ecclesiastes 12:7 ('Then shall the dust return to the earth, as it was, and the spirit shall return unto God who gave it') only provided further support. It was argued that because scripture provided direct evidence of the soul's immateriality and immortality, to suggest anything else not only challenged revelation but would lead to theological absurdities. As Wadsworth explained, if the soul is made of atoms, God's wrath and Jesus's sacrifice was for nought but 'some irregular motions of a thin air...which is no more capable of offending God, than the falling of a Joynt-stool'.[9] In short, materialism not only turned its back on scripture but made a laughing stock of it.

MORTALIST-MATERIALIST EXEGESIS

In seventeenth- and eighteenth-century discussions of the soul, in a context in which truth and tradition were at stake, the antiquity of certain doctrines held profound weight. The recovery and restoration of Hebrew and early Christian doctrinal purity was critical. Scripture and sacred history needed to be examined closely: drawing on the humanist approaches of preceding centuries, early modern exegetes hoped to attend to the context of biblical writers, claiming that in order to fully appreciate the scriptures one had to understand the customs and languages of the Jews and Greeks. The aim was to read the scriptures as historical documents written by individuals living in a certain time and place. The funerary practices and surviving texts of the Hebrews, Chaldeans, Egyptians, and Greeks were habitually taken as direct proof that not only God's chosen people but all gentiles believed in an immortal soul. And, how else could every society in every age have believed in the soul if it was not a substantial reality? Philological as well as historical and anthropological expertise was central to exegesis, as was the close examination of scripture's grammar and syntax in its original languages.[10] The Hebrew words *nephesh* and *ruach* and the Greek *psyche* and *pneuma* were typically translated to 'soul' and 'spirit' respectively in the various English bibles available in the period. While exegetes noted the variety of possible meanings attached to these words, most concluded that their foremost meaning in the scriptures was an immortal substance distinct

[8] Wadsworth, *Antipsychothanasia*, p. 12. [9] Wadsworth, *Antipsychothanasia*, p. 130.

[10] Richard A. Muller, 'Biblical Interpretation in the Era of the Reformation: The View from the Middle Ages', in *Biblical Interpretation in the Era of the Reformation*, ed. Richard A. Muller and John L. Thompson (Grand Rapids, MI: Eerdmans, 1996), pp. 3–22.

from matter.[11] Immaterialists and immortalists were adamant that the scriptures confirmed their views.

Mortalists and materialists, however, claimed that their opponents' interpretations failed because they had not properly acquainted themselves with the contexts in which the word 'soul' appeared in the Bible, and the various contexts in which the biblical authors were writing. These writers proposed a different reading of history to their opponents: rather than being the universal, innate idea of all peoples, the immortal soul was actually the invention of heathen philosophers and magistrates who wished to enforce the concept of future rewards and punishments on the masses. Rather than being merely one soul-believing community among many, the Hebrews were described by mortalist-materialists as the possessors of the true understanding of humanness. Indeed, close and contextual reading of the Hebrew Bible revealed that God's chosen people had always been mortalists and materialists. On the other hand, the writers of the New Testament had occasionally allowed the assumptions of the Greek culture within which they were immersed to seep into their books. For mortalist-materialists, philological and historical-critical reading of the scriptures was thus vital to determining 'true' biblical anthropology.

Much mortalist-materialist exegesis rested upon an emphasis on the fluidity of the terms *nephesh*, *ruach*, *psyche*, and *pneuma*. For Richard Overton in the 1640s, the words were 'variously used upon various occasions' in the scriptures, representing the stomach (Proverbs 27:4), the eyes (Jeremiah 13:17), the heart (1 Samuel 17), the dead body (Psalm 16:10), the whole man (Leviticus 7:19), the breath (Acts 20:10), and life itself (Isaiah 53:17).[12] As such, Overton concluded that the Hebrew and Greek words could only refer to 'fleshy' things. Thomas Hobbes, for his part, insisted in *Leviathan* (1651) that Scripture never used the word 'soul' except to refer to life, and 'spirit' only to refer to 'thin, fluid, transparent, invisible body'.[13] Later materialists came to the same conclusions: as Edmund Law explained in his decisive appendix 'Containing the Use of the Words Soul or Spirit in Holy Scripture' (1755), no word in scripture ever 'stand[s] for a purely immaterial principle in man'.[14] On the contrary, *nephesh*, *ruach*, *psyche*, and *pneuma* convey ideas of 'materiality'. The knowledge of Hebrew possessed by many mortalist-materialists was limited; while the theologically trained Law and Joseph Priestley were capable Hebraists, the physician

[11] See, for example, Simon Harward, *A Discourse Concerning the Soule and Spirit of Man* (London: John Windet, 1604); John Brayne, *The Unknown Being of the Spirit, Soul, and Body, Anatomized* (London: Richard Moon, 1654); [Thomas Emes], *Vindiciae mentis* (London: H. Walwyn, 1702); William Salmon, *Ars Anatomica* (London: I. Dawks, 1714).

[12] R[ichard] O[verton], *Mans Mortalitie* (Amsterdam: John Canne, 1644), p. 18; R[ichard] O[verton], *Man Wholly Mortal* (London, 1655), pp. 62–3.

[13] Thomas Hobbes, *Leviathan, or, the Matter, Forme, & Power of a Common-Wealth, Ecclesiasticall and Civill* (London: for Andrew Crooke, 1651), p. 339.

[14] Edmund Law, 'Appendix: Containing the Use of the Words Soul or Spirit in Holy Scripture', in *Considerations on the Theory of Religion*, 3rd edn (Cambridge: J. Bentham, 1755), p. 381.

William Coward and the lawyer Henry Layton were not. The etymological and philological work of mortalist-materialists with little or no knowledge of Hebrew was largely derivative, and relied upon various dictionaries and lexicons. For example, Edward Leigh's *Critica Sacra* (1639), and more popular books like Wadsworth's *Antipsychothanasia* and Henry Hammond's *Paraphrase and Annotations upon the New Testament* (1653), did indeed suggest the fluidity of these Hebrew terms. However, in stark contrast to their sources, mortalist-materialists concluded that, in the words of Coward, not one of these words could 'by the most extorted Interpretation possible...be apply'd to denote a *Spiritual Substance* in Man'.[15] In what follows I will address how mortalist-materialists built upon these philological conclusions in their readings of Genesis 2:7, 1 Samuel 28, 1 Thessalonians 5:23, and Matthew 10:28.

Genesis 2:7 was discussed in most publications on the soul in this period. The Authorized Version renders it as follows: 'And the LORD God formed man of the dust of the ground, and breathed into his nostrils the breath of life; and man became a living soul'. Immaterialists generally followed Augustine in reading this as the moment when the immaterial soul entered the material body.[16] Mortalist-materialists undermined this interpretation by focusing on grammar. Overton, Hobbes, and George Hammon simply observed the implication of the passage: man did not *receive* a 'soul', but *became* a 'living soul' upon the reception of the breath.[17] Later in the century, Henry Layton added that while it was common to think that breath here equates to soul, this is

> plainly Confuted by the Grammatical Construction...: For the Words which would have suited to that intent, should have run thus: God breathed into his Nostrils the Breath of Life, and that, or it, *viz*, That Breath became a Living Soul, instead it says 'man became a living soul'.[18]

Coward also insisted that 'living soul' only signifies 'living person', and added that the word *nephesh* here is also used in Genesis 1:20 and 1:31 when the writer speaks of the many 'moving Creature[s] that [have] life'.[19] As many mortalist-materialists noted, if animals are called 'souls' in the scriptures (in Leviticus 22:11 and Revelation 16:3, to name two of the oft-quoted passages), the word must mean something material and mortal. The interpretation of the anonymous author of *The Materiality or Mortality of the Soul* (1729) was somewhat divergent: God did not breathe a soul into man, but man, before God breathed

[15] [William Coward], *Second Thoughts Concerning Human Soul* (London: for R. Basset, 1702), p. 200.

[16] Augustine, *The City of God*, trans. R. W. Dyson (Cambridge: Cambridge University Press, 1998), XIII 24.

[17] Overton, *Mans Mortalitie*, p. 1; Hobbes, *Leviathan*, p. 340; George Hammon, *A Discovery* (London: R. Ibbitson, 1655), p. 48; George Hammon, *Truth and Innocency* (London: Printed for the Author, 1660), p. 38.

[18] [Henry Layton], *Observations upon a Short Treatise* (London, 1697), p. 10.

[19] Coward, *Second Thoughts*, p. 264.

into him, was already a soul—albeit a lifeless one. Hence 'a *Dead Man* is called in the Scriptures a *Dead Soul*'.[20] For the author of *A Discourse Concerning the Soul* (1719), 'Scripture doth not say... God created a Soul or Spirit, and infused in him, but only breathed into his Nostrils the Breath of Life'. The author was disappointed in his immaterialist contemporaries: '[o]ne would think that these Divines have never read the Scripture'.[21]

A second key passage, 1 Samuel 28, told the story of how Saul, seeking guidance about an imminent Philistine attack, asked a witch to conjure the soul of Samuel. In the early modern period this passage was regularly cited as 'sufficient evidence' that the Hebrews believed that the soul was capable of existing separately from the body. Although not every commentator thought that the witch had successfully conjured Samuel's soul—Wadsworth argued that it would go against God's wisdom to allow a sorceress to control the soul of so good a man[22]—the fact that Saul *believed* that it was really Samuel's soul was evidence enough of a common Hebrew assumption of the soul's immateriality and immortality. As Richard Baxter explained, if the Jews did not believe that the soul was immortal, Saul would not have asked the witch to call up Samuel.[23] Coward's response was to argue that the whole scene was a sham. He pointed out that the witch was described by the Hebrew word '*Ob*'—'ventriloquist'—which provided a hint that she had used trickery to make it seem as if Samuel's soul was really there.[24] Layton's reaction was to point out that a single Hebrew believer in the soul's immateriality and immortality did not mean that all Hebrew people believed. 'This Practice of *Saul*'s', he concluded, 'proves no more, but that *a few bad Jews* of that time, may seem to have believed'.[25]

When it came to the historical context of the New Testament, mortalist-materialists emphasized the Hellenic atmosphere of the period. The Platonic idea of the soul's immateriality and immortality, they argued, was 'the prevailing Opinion in the World at that time'.[26] In their examination of New Testament writers, mortalists and materialists had to confront several problematic texts, particularly in Paul's epistles. Thus, a third key text was 1 Thessalonians 5:23, which embodied the recurrent biblical theme of a distinction between flesh and spirit. In this passage Paul prayed that the Thessalonians' 'whole spirit and soul and body [would] be preserved blameless'. While this verse was commonly taken as a list of the different aspects of a person, mortalist-materialists argued

[20] [Anonymous], *The Materiality or Mortality of the Soul* (London: John Noon, 1729), p. 2.

[21] [Anonymous], *A Discourse Concerning the Soul of Man* (Edinburgh: Printed for the Author, 1719), pp. 40–1.

[22] Wadsworth, *Antipsychothanasia*, pp. 55–6.

[23] Richard Baxter, *Richard Baxter's Dying Thoughts Upon Phil. I. 23.* (London, for B. Simmons, 1683), p. 42.

[24] Coward, *Second Thoughts*, p. 209.

[25] [Henry Layton], *Arguments and Replies in a Dispute Concerning the Nature of the Humane Soul* (London, 1703), pp. 107–8.

[26] [Henry Layton], *A Second Part of a Treatise Intituled a Search after Souls* (1706), p. 71.

that it was merely a metaphorical reference to the distinction between the 'carnal' and 'spiritual' aspects of humans. For Coward, the verse should be read as an idiom: 'Spirit, Soul and Body…being []joined signifies not much…because such Expressions…mean no more than to serve God frequently and zealously with…All or the Whole of you'.[27] Paul was simply adapting his expression to popular language and ideas, 'without giving any confirmation to the *truth* and *justness* of them'.[28] For Law, the text was 'purely figurative', and Hammon likewise suggested that 'soul' was used in scripture only 'for illustration's sake', sometimes meaning 'one thing, & sometimes another'.[29] Hobbes argued that these words were used in scripture 'metaphorically onely'.[30] Peter Peckard, for his part, suggested that Paul's discussion of spiritual and natural bodies was derived from his education: Paul was 'deeply versed in philosophic Wisdom'. In any case, there was a danger of 'adhering too closely to the Letter' of the text, for this 'kill[s] and reject[s] that free, popular Spirit, which animateth the Language'. Peckard emphasized 'the sense of the words', for he believed it was wrong to interpret 'Eastern Phrases in a literal Sense, and logical Way of reasoning'.[31]

A fourth challenging text was Matthew 10:28, which was particularly important because of its divergence from its counterpart Luke 12:4. Both passages recounted a story of Jesus instructing his disciples, and while Matthew has Jesus teaching them not to fear 'them which kill the body, but are not able to kill the soul', Luke has Jesus only cautioning them not to fear 'them that kill the body, and after that have no more that they can do'. While standard interpretations put the discrepancy down to a simple condensing of the story by Luke,[32] many mortalist-materialists took Luke's omission of the soul as an indication that while Matthew subscribed to the 'heathen' notion of the soul, Luke was a mortalist and materialist. Compared to Luke, who was a physician, Matthew was a mere publican and thus was not likely to have 'any great stock of Humane Learning'. Ergo, it is probable Matthew 'might easily be carried away by the common Opinion of that Time' within Greek culture.[33] It had been suggested in the seventeenth century that the early Christians had imitated pagan language and customs in order to facilitate proselytizing to pagans,[34] and mortalist-materialists deployed this historical-critical approach to explain how heathen ideas of the soul had illicitly found their way into one of the books of scripture. While Layton asserted that 'the Relations we find in St. Matthew are

[27] Coward, *Second Thoughts*, pp. 291–2. [28] Law, 'Appendix', p. 453.

[29] Law, 'Appendix', p. 414; Hammon, *Truth and Innocency*, pp. 35–6.

[30] Hobbes, *Leviathan*, p. 214.

[31] Peter Peckard, *Observations on the Doctrine of an Intermediate State between Death and Resurrection* (London: L. Davis and C. Reymers, 1756), pp. 17, 6–8.

[32] Wadsworth, *Immortality of the Soul*, pp. 63–5.

[33] Layton, *Observations upon a Short Treatise*, pp. 13–14.

[34] See, for example, Dmitri Levitin, 'John Spencer's *De legibus Hebraeorum* (1683–85) and "Enlightened" Sacred History: A New Interpretation', *Journal of the Warburg and Courtauld Institutes* 76, no. 1 (2013): pp. 69–70.

not Infallible Truths', Hammon and Coward voiced a less radical reading, describing Luke as 'the most exact Writer' and 'the best Expositor' respectively.[35] The suggestion was that Luke's judgement should be trusted: if Matthew had intended an immaterial being, Luke would have used the same expression.

REFORMATION AND ANTI-CATHOLICISM

Mortalist-materialists were concerned with deceptive interpretations of the scriptures: any reading that relied on the soul's immortality and immateriality, they claimed, was compromised by pagan and Catholic dogma. As I have suggested, while their basic exegetical techniques were far from novel, their ultimate conclusions were exceedingly provocative and called for justification. Reform rhetoric served to fill this lacuna. They made the case that all the 'evils' of popery—purgatory, prayers for the dead, veneration of saints—had their origins in the notion of the immortal soul, and thus Catholic priests had knowingly misread the scriptures to support these lucrative schemes. In the sixteenth, seventeenth, and eighteenth centuries, vitriolic polemic exposing various 'Catholic' sponsored plots and schemes left England, for the most part, robustly anti-Catholic, and mortalist-materialist reform rhetoric hinged on this context. In their emphasis on the recovery of primitive Hebraic ideas of the soul, mortalist-materialists presented their revisionist interpretations as part of the long reform of the Church.

Mortalist-materialists treated the scriptures as the highest authority, and maintained that their very ability to do so was a uniquely Protestant liberty. 'Let Scripture then alone, according to the Principles of our Reformers, be our Guide', Peckard advised.[36] Hobbes declared that he could not, 'without deserting the sense of Scripture', concede the soul's immateriality and immortality. The church should not be offended by his own 'doctrine', for while 'the church of Rome [pretends] to be above the Scripture', and 'forbiddeth' its members to read it, 'the church of England' did not.[37] This constructed opposition between the churches of England and Rome appears repeatedly in mortalist-materialist texts, and is best expressed by the author of *A Discourse*. Here Christians are advised 'not to live implicitly upon the Opinion of either Minister or Magistrate' but instead, 'like the *Bereans*'—those who in Acts 17:11 examined the scriptures to ensure Paul's teachings were trustworthy—Christians should 'compare the Doctrine with the Word.... which is a Privilege our Protestant Church alloweth

[35] Layton, *Observations upon a Short Treatise*, p. 17; Hammon, *Truth and Innocency*, p. 41; Coward, *Second Thoughts*, pp. 163, 281.

[36] Peckard, *Observations*, p. 44.

[37] Hobbes, *An Answer to Bishop Bramhall* (London: W. Crooke, 1682), p. 355.

to all her Members, and which the Popish Church denieth to theirs'.[38] While only Hammon would go so far as to intimate that 'the Church of *Rome* have also abused our *Greek* Copies.... to bear a face toward their opinion [of the soul]',[39] the church of England/church of Rome opposition remained a consistent mortalist-materialist tenet.

Mortalist-materialists insisted that if the soul were eliminated from Christian doctrine the Church would be wholly reformed. While the Protestant Church had already had its 'Doctrine, Worship, Discipline and Government... cleared, vindicated and asserted', the immaterial and immortal soul remained, which the author of *A Discourse* described as the 'most erroneous and ridiculous' doctrine of all.[40] Further reformation was required. 'When I am truly satisfied that the Church of England is so entirely pure, and Holy in all its Principles and Doctrines, that it needs no farther Reformation', Coward wrote, 'then I will, as bound in Conscience, silently and patiently acquiesce with full assent and consent to all it teaches'.[41] Even in the 1770s Joseph Priestley claimed that the soul's immateriality and immortality were 'in fact the remains of the same fabric of corruptions' that Protestant reformers had tried to clear away. The specious but still 'common' opinion of the soul, which had 'exceedingly altered and debased the true Christian system', continued to endure.[42] It was therefore expedient for Christianity to do away with the immaterial soul altogether, to ensure that the great edifice of 'popery' would be once and for all eradicated. Peckard wrote that because 'all the senseless Trumpery' of Catholicism was 'entirely founded' upon the immateriality and immortality of the soul, when the doctrine is given up 'the whole Business of Popery must absolutely very soon be demolished Root and Branch'.[43] In Priestley's words, '[t]he building itself has happily been thrown down; but I wish to dig up the very foundations, that they may never be built upon again'.[44] Mortalist-materialists constructed an opposition that placed Protestantism, materialism, and 'true Christianity' on one side and Catholicism, corruption, and the soul on the other.

By aligning themselves with past Protestant reformers, mortalist-materialists attempted to vindicate their rejection of tradition. Layton affirmed that if scripture or reason, 'grounded upon Nature and Experience', were to suggest something at odds with 'the Unanimous Judgement of all the Learned Divines in the World', he would always choose to believe the former authorities. By asserting as much, Layton wanted to demonstrate his allegiance to the 'Practices of our Reforming Divines, John Huus, Wicklif, Luther, Zuinglius, Calvin, Beza, and others, the first Reformers of our Church, who all rejected the Erroneous Doctrines and Practices of the Church of Rome'. Layton was not particularly

[38] [Anonymous], *A Discourse*, p. 3.

[39] Hammon, *Truth and Innocency*, p. 49.

[40] [Anonymous], *A Discourse*, p. 4.

[41] [William Coward], *The Just Scrutiny: Or, a Serious Enquiry into the Modern Notions of the Soul* (London: John Chantry, 1705), p. 10.

[42] Joseph Priestley, *Disquisitions Relating to Matter and Spirit* (London: J. Johnson, 1777), p. 232.

[43] Peckard, *Observations*, pp. 27–8.

[44] Priestley, *Disquisitions*, p. xxix.

interested in their views on the soul, but in their methods of reform: 'I think I follow [their] course in opposing the Doctrine of the Soul's separate subsistence'.[45] Priestley, meanwhile, referred to Luther and Tyndale precisely to show how their views on the soul were mortalist and materialist. Though Priestley knew that Calvin was 'violently opposed' to mortalist and materialist doctrine, the fact that Calvin bothered to publish texts against it at all suggested that 'thousands of the reformers were of a different opinion from him'. Drawing on the authority of 'Luther,... and the other reformers from popery', mortalist-materialists argued that '[i]f their conduct be justifiable,... why may not we of this age humbly presume to be reformers from popery also?'[46] To deny one's right to disagree with tradition, Coward powerfully observed, was to 'condemn[] the whole design of the Reformation, and throw[] Dirt upon the Reformers'.[47] In this sense the polemical aims of mortalists and materialists were completely typical of the period. They wanted to purge Christianity from paganism and popery.

Mortalist-materialists maintained that their endeavours to 'reform' the church were simply good Christian duty. In its simplest form, this sentiment was expressed in statements of authorial intent. Coward, for example, wrote of his book as follows: 'I do assure you, upon the Faith of a Christian, it is wrote with an honest Intention to convince the World...of the needless Error of feigning a Spiritual Substance united to the Body of Man'.[48] The author of *The Materiality and Mortality* was similarly unambiguous: finding the doctrines of immateriality and immortality nowhere in the scriptures, he 'thought it my Duty as a Christian, publicly to oppose it'.[49] As Peckard wrote, the denial of the immaterial, immortal soul 'establish[es] Christianity upon its true and proper Ground'.[50] Mortalist-materialists wanted to remove the wheat from the chaff and, in Priestley's words, reform the 'infidels' who hid under the name of Christianity into true 'rational Christians'.[51] For, after all, '[t]he doctrine of a separate soul most evidently embarrasses the true Christian system'.[52]

CONCLUSION

The key intervention of mortalist-materialists into seventeenth- and eighteenth-century debates over the soul was a philological and historical-critical reading of the Bible. They claimed that their approach undid the pagan and

[45] [Henry Layton], *Arguments and Replies* (London, 1703), pp. 16–17.
[46] Priestley, *Disquisitions*, pp. 231–2.
[47] Priestley, *Disquisitions* , p. 253. See also Coward, *The Just Scrutiny*, p. 214.
[48] Coward, *Second Thoughts*, unpaginated.
[49] [Anonymous], *The Materiality or Mortality of the Soul*, unpaginated.
[50] Peckard, *Farther Observations*, p. 4.
[51] Priestley, *Disquisitions*, p. xvii.
[52] Priestley, *Disquisitions*, p. 124.

Catholic assumptions of their opponents in order to uncover the underlying Hebraic truths of scripture. Despite the extent of English anti-Catholic sentiment, however, they had little success in rallying their fellow Protestants into action. Their opponents agreed that the elimination of the soul would undermine Catholicism, but they remained convinced that it would destroy Christianity *in toto*. In his critique of Coward's work, John Broughton gave Coward 'Acknowledgement for his Zeal in the Protestant Cause', agreeing that 'the Church of Rome' had drawn 'Ill Consequences' from the doctrine of the soul.[53] Nevertheless, Broughton concluded that this corruption was not enough to soil the fundamental truth of the doctrine. While Broughton supported critical and contextual exegesis more broadly, he declared that materialists 'turn and wrest the Scriptures' to such an extent that, if the church were to allow this method of interpretation, 'there is nothing so ridiculous, absurd or impious, but [that] the Scripture may be brought to Countenance'.[54]

Although their positions failed to gain significant traction at the time, mortalist-materialist arguments proved remarkably long-lived. In the later eighteenth and into the nineteenth century mortalist-materialism was taken up by various theologians, preachers, philosophers, historians, and physicians, and became closely allied with annihilationism (which denies the existence of hell) and conditionalism (in which immortality is granted only to the elect). Many facets of mortalist-materialist argument have endured into the present day, particularly in Seventh-day Adventist groups but also in some Christian responses to reductive neuroscientific theories, which attempt to harmonize materialist accounts of the person with 'true' Christian anthropology. In the twentieth century many early histories of the position, such as Le Roy Edwin Froom's *The Conditionalist Faith of Our Fathers* (1966), were written by Seventh-day Adventists, and openly aimed at 'the ultimate restoration of the Conditionalist faith'.[55]

Standard twentieth-century historical narratives assumed that while belief in the soul was expected in pre-scientific societies, early modern advances in science and medicine meant that the fall of the soul was inevitable and predictable. La Mettrie's anti-religious agenda cast a long shadow over the history of the material soul, suggesting an essential connection between naturalism, materialism, and irreligion. The history of psychology and the modern neurosciences

[53] John Broughton, *Psychologia* (London: for T. Bennet, 1703), sig.l3r.

[54] Broughton, *Psychologia*, sig.b3r.

[55] Le Roy Edwin Froom, *The Conditionalist Faith of Our Fathers* (Washington, DC: Review and Herald, 1965–6), p. 12. For an example of a Seventh-day Adventist manifesto that relies on the history of mortalist-materialism, see *Seventh-day Adventists Answer Questions on Doctrine* (Washington, DC: Review and Herald, 1957). A more recent example is Bryan W. Ball, 'The Immortality of the Soul: Could Christianity Survive Without It? Part I', *Ministry: International Journal for Pastors* 83, no.3 (2011): pp. 10–14. For mortalist-materialist responses to neuroscience see, e.g., Warren S. Brown, Nancey Murphy, and H. Newton Malony (eds), *Whatever Happened to the Soul? Scientific and Theological Portraits of Human Nature* (Minneapolis, MN: Fortress Press, 1998), and John W. Cooper, *Soul and Life Everlasting: Biblical Anthropology and the Monism-Dualism Debate* (Grand Rapids, MI: Eerdmans, 2000).

is often understood in these terms.[56] Historians of the seventeenth and eighteenth centuries have most commonly dealt with the soul under the rubric of the history of philosophy, particularly in the light of Descartes's mind-body scheme. While the Cartesian position might seem antithetical to materialism, it has been suggested that it actually opened the door to a set of radically naturalistic arguments. It was a small step, the story goes, from Descartes's notorious suggestion that animals were machines to the assertion that humans too were purely material. Thus, the *New Dictionary of the History of Ideas* (2005) describes eighteenth-century materialism simply as 'an extension of the seventeenth-century mechanical philosophy that was the hallmark of the scientific revolution.'[57] However, as Charles Wolfe has argued, materialism did not progress 'by simply adding further and further properties to Galilean or Cartesian extension like layers in a *millefeuille*'. Instead, ideas of thinking matter were the consequence of *criticism* of Cartesian concepts of inert matter, and notions of active organic matter were central to this development.[58]

As this chapter has argued, however, seventeenth- and eighteenth-century materialism was also a consequence of attempts to recover what was believed to be the 'Hebraic' notion of the soul. While the mortalist-materialists discussed in this chapter relied to a certain extent on medical and philosophical 'proofs' to support their ideas, we have seen that biblical exegesis and ideas of theological reform were powerful drivers. In recent years historians have become more familiar with the religious foundations of many mortalist and materialist ideas, but some continue to argue that '[t]he new science provided philosophers with a powerful motive to be materialists' and 'Christianity provided them with a powerful motive to resist materialism.'[59] While this is certainly the case for many seventeenth- and eighteenth-century English men and women—and the importance of a perceived nexus between materialism and atheism was in many ways a central assumption of the period and should not be overlooked—I suggest that Christianity also provided many with powerful motives *to be* materialists. While mortalists and materialists have been read as an outpouring of medical and natural philosophical advance, as radical heretics, or, by some confessional historians, as the fount of 'true' Christian anthropology, it is important that we situate them within the scholarly context in which they operated. We need to appreciate mortalists and materialists within the context of a historical-critical turn to the past and to the Christian scriptures.

[56] See, for example, J. R. Kantor, *The Scientific Evolution of Psychology*, vol. 2 (Chicago: Principia Press, 1969); Marc Jeannerod, *The Brain Machine*, trans. D. Urion (Cambridge, MA: Harvard University Press, 1985); J.-P Dupuy, *The Mechanization of Mind*, trans. M. B. DeBevoise, (Princeton, NJ: Princeton University Press, 2000).

[57] Peter Machamer and Francesca di Poppa, 'Materialism in Eighteenth-Century European Thought', in *New Dictionary of the History of Ideas*, vol. 4, ed. Maryanne Cline Horowitz (Detroit, MI: Charles Scribner's Sons, 2005), p. 1374.

[58] Charles T. Wolfe, *Materialism: A Historico-Philosophical Introduction* (Heidelberg: Springer, 2016), p. 11.

[59] Raymond Martin and John Barresi, *The Naturalization of the Soul: Self and Personal Identity in the Eighteenth Century* (London: Routledge, 2000), p. 131.

9

The Science of the Soul

Naturalizing the Mind in Great Britain and North America

Jon H. Roberts

Conventional narratives of the history of psychology often suggest that the scientific study of the mind began only in the late nineteenth century, when psychologists finally ceased viewing their work as a handmaiden of Christian theology. Such narratives are misleading. For one thing, they are predicated on the anachronistic notion that a scientific interpretation of psychology is dependent on a naturalistic and experimentally oriented approach to mental phenomena. No less importantly, they tend to ignore the important role that religious thinkers played in developing an approach to understanding the mind that was consistent with what thinkers in the late eighteenth and early nineteenth centuries characteristically envisioned as 'science'.

It is certainly true that, prior to the last quarter of the nineteenth century, theological considerations informed much of the discourse concerning the human mind in western Europe and North America alike. This is hardly surprising. The privileged status of mind in the overall scheme of things constitutes the intellectual foundation of a number of central Christian affirmations. For example, although believers have characteristically acknowledged that the nature of God defies precise description, they have long assumed that anthropomorphic categories provide the most adequate description of the Deity's creative, providential, and redemptive activities. Indeed, the very notion that God is personal has prompted Christians to employ vocabulary relating to mind—words such as *wisdom*, *benevolence*, and *volition*—in discussing the divine attributes.

Characteristics associated with mind have also played an especially important role in shaping Christian thinkers' understanding of both human nature and the encounter of humanity with God. Without denying that the human body participates in the goodness of the created order, they have nevertheless treated the mind as the seat of human identity, the source of the human capacity for moral activity, the basis for affirming that humanity had been created in God's image, and the primary grounds for belief in human immortality. In addition, many Protestants and Catholics alike have held that the very essence of piety is to be found in the inward dispositions of the psyche rather than in either ritual practices or institutional allegiances.

SOUL AND MIND

In early modern Europe and North America, the theological context within which much of the Christian discourse dealing with the human mind took place was discussion of the soul. During the sixteenth and early seventeenth centuries, most participants in that discussion endorsed an Aristotelian conception of the soul modified on occasion by their acceptance of an eclectic, not always entirely consistent, amalgam of ideas drawn from the Bible, Neoplatonism, the medical theories of Hippocrates and Galen, and sometimes even Hermetic and Paracelsian sources. The soul, they maintained, served as the principle of animation and the 'form' endowing each organism with its distinctive attributes.[1] From this perspective, Christian thinkers held that human beings were unique in possessing not only the powers of nutrition, growth, and reproduction associated with a 'vegetative' soul and the attributes of sensation, self-motion, appetite, imagination, and memory ascribed to an animal's 'sensitive' soul, but also a 'rational' soul—a divinely created spiritual substance comprising the faculties of understanding and will.[2] In keeping with this interpretation of the metaphysical status of the human soul, study of that subject was often discussed in the early modern period within the framework of 'pneumatology', generally regarded as that realm of investigation that dealt with incorporeal, spiritual substances.[3] And because they believed that it was the attributes associated with the rational soul that reflected, however dimly, the attributes of their Creator, believers typically located the 'divine power of the soule' in the '*Spirite of our mind*'.[4]

During the first half of the seventeenth century, the French philosopher René Descartes (1596–1650) abandoned the Aristotelian categories of matter

[1] Daniel Garber, 'Soul and Mind: Life and Thought in the Seventeenth Century', in *The Cambridge History of Seventeenth-Century Philosophy*, vol. 1, ed. Daniel Garber and Michael Ayers (Cambridge: Cambridge University Press, 1998), pp. 759–60.

[2] Garber, 'Soul and Mind', pp. 759–60 .

[3] Fernando Vidal, *The Sciences of the Soul: The Early Modern Origins of Psychology*, trans. Saskia Brown (Chicago: University of Chicago Press, 2011), p. 62.

[4] Richard Hooker, *Of the Laws of Ecclesiastical Polity: Preface and Books I to IV* (1593), ed. Georges Edelen, vol. 1 of *The Folger Library Edition of The Works of Richard Hooker*, ed. W. Speed Hill (Cambridge, MA: Belknap Press of Harvard University Press, 1977), p. 77.

and form, and ascribed both the vital principle and the organic functions previously assigned to the vegetative and sensitive elements of the soul to biological processes operating in accordance with mechanistic principles. In this way he constricted the category of *soul* to the human mind, with its self-awareness, its conscious experiences of reasoning, understanding, perceiving, sensing, willing, and imagining, and its ability to express thoughts in language.[5] More broadly, he divided the created order into two ontologically incommensurable kinds of substance: matter, which he characterized as extended, divisible, unconscious, and passive; and mind, which he described as spiritual, indivisible, self-active, and intelligent.[6] Yet, in spite of the fact that this dualism had the effect of exempting mind from the mechanistic processes that pervaded the material world, Descartes held that the mind operated in accordance with intelligible principles and that it was an appropriate object for systematic examination.

Although many of the views espoused by Descartes received harsh treatment at the hands of critics, most thinkers in the English-speaking world, the geographical area that serves as the focus of this essay, concurred with both his claim concerning the intimate association between the human soul and the mental faculties and his decision to exempt the human mind from the kind of naturalistic treatment accorded the realm of matter.[7] Even those Christians who continued to envision the soul as the form of the body and the principle accounting for animation in sentient beings, typically centred their discussions on the idea that the human soul served as the seat of a unified consciousness and the source of mental activity. In 1692, the clergyman Richard Bentley (1662–1742) articulated the conventional wisdom among believers when he identified the 'incorporeal inhabitant within us, which we call spirit and soul' as an immaterial, 'cogitative substance' that 'thinks and apprehends, and reflects and deliberates; that determines and doubts, consents and denies; that wills, and demurs, and resolves, and chooses, and rejects; that receives various sensations and impressions from external objects, and produces voluntary motions of several parts of our bodies'.[8] In effect, then, when discussing human nature,

[5] René Descartes, 'Objections and Replies', in *The Philosophical Writings of Descartes*, vol. 2, ed. and trans. John Cottingham, Robert Stoothoff, and Dugald Murdoch (New York: Cambridge University Press, 1984–91), pp. 243, 246; Descartes, 'Meditations on First Philosophy', in *Philosophical Writings*, vol. 2, pp. 19, 24.

[6] Descartes, 'Meditations on First Philosophy', vol. 2, pp. 17, 54, 59; Descartes, 'Principles of Philosophy', in *Philosophical Writings*, vol. 1, pp. 210, 232; Descartes, 'Rules for the Direction of the Mind', in *Philosophical Writings*, vol. 1, p. 42.

[7] Ann Thomson, *Bodies of Thought: Science, Religion, and the Soul in the Early Enlightenment* (Oxford: Oxford University Press, 2008), pp. 45–6.

[8] Richard Bentley, 'The Folly of Atheism, and (What is Now Called) Deism, Even with Respect to the Present Life' (Boyle Lectures, 1692), in *The Works of Richard Bentley, D. D.*, vol. 3, ed. Alexander Dyce (1838; repr., New York: AMS Press, 1966), pp. 34–5.

exponents of Christian thought made *soul* and *mind* essentially interchangeable concepts.[9] Both terms referred to an incorruptible, indivisible, incorporeal, self-active spiritual substance that enabled persons to act as purposive spiritual agents and served as the source of their moral and theological accountability.

Given the inclination of Christian thinkers to equate the human soul with mind, it is erroneous to assume, as have some historians, that the increasing tendency to substitute the term *mind* for *soul* in the early modern period signified a naturalization or a secularization of the inner life.[10] It may well be that thinkers who evinced little sympathy with the Christian worldview chose to abandon use of the term *soul* as a step toward dispensing with the notion that a special link existed between humanity and God. But proponents of Christian theology in Great Britain and North America also chose increasingly to employ the term *mind* rather than *soul*, not out of a desire to dismiss the soul's spiritual significance, but because they ascribed to the mind those attributes of selfhood most centrally involved in the divine–human encounter. That practice commonly led them to focus on cognitive, affective, and volitional attributes in discussing the interactions of human beings with God and with one another, while reserving use of the term *soul* to their discourse concerning human immortality.

Although exponents of Christian theology conferred upon the human mind an exalted place within the created order, many of them—especially Protestants, who dominated theological discussion in the English-speaking intellectual community during the early modern period—maintained that it was impossible to comprehend the status of humanity's mental faculties without an appreciation of the cataclysmic impact of Adam's rebellion against the dictates of God's will. That view, which led to efforts to assess the degree of degradation resulting from the Fall, stimulated interest in the nature of the human mind and helped to make the study of human nature a distinct area of inquiry.[11] The late-seventeenth and eighteenth centuries also witnessed a decided inclination on the part of Christian thinkers in Great Britain and North America to place greater emphasis on the ability of the human intellect to apprehend truth. Although most of those thinkers continued to affirm that Adam's sin had resulted in the corruption of the human will, their desire to relate the elements of their worldview to other intellectual pursuits and to defend their theology prompted them to emphasize that God had laid numerous important truths

[9] William J. Bouwsma, *The Waning of the Renaissance: 1550–1640* (New Haven, CT: Yale University Press, 2000), p. 165.

[10] Raymond Martin and John Barresi, *Naturalization of the Soul: Self and Personal Identity in the Eighteenth Century* (London: Routledge, 2000), pp. ix–x, 110–11.

[11] Peter Harrison, *The Fall of Man and the Foundations of Science* (Cambridge: Cambridge University Press, 2007), esp. pp. 3, 53–4, 64–5.

'within the reach of their [humanity's] natural Faculties'.[12] That emphasis, too, had the effect of generating greater interest in the nature and functions of mind.

MENTAL PHILOSOPHY AS DESCRIPTIVE SCIENCE

In keeping with the tendency among exponents of Christian theology to adopt a more optimistic conception of the capacities of the human mind, Thomas Reid (1710–96), a Scottish Presbyterian clergyman who went on to become Adam Smith's successor as professor of moral philosophy at Glasgow, produced an important series of works seeking to demonstrate that the human mind deserved respect as the 'noblest work of God'.[13] In developing that idea, Reid made a concerted effort to establish the trustworthiness of the testimony of consciousness, the instrument that provided the mind with immediate knowledge of its own operations and served as the medium for acquiring knowledge about external reality as well. Toward that end, he denounced the notion held by David Hume (1711–76) and other British empiricists that the mind was capable only of perceiving 'ideas'. That claim, he asserted, had served to bring 'death and destruction to all science and common sense' and had driven philosophers to scepticism.[14] Reid countered with a different approach. He acknowledged that the 'system of minds' was fundamentally different from the 'system of bodies', but insisted that God had provided the intellect with the ability to derive legitimate knowledge of both of those 'kingdoms' of the created order.[15] He thus maintained that the mind actually apprehended objects that inhabited the external world and existed independently of the mind.[16]

Reid coupled his affirmation of the reliability of the deliverances of perception with a robust view of the power and value of the human mental faculties. Dissenting from the notion that the mind did nothing more than passively receive impressions, he held instead that it played an active role in discerning the nature of reality. An examination of the data of consciousness, he maintained,

[12] John Locke, *An Essay Concerning Human Understanding*, ed. Peter H. Nidditch, 4th edn (Oxford: Clarendon Press, 1975), bk. 4, chap. 19, sec. 4, p. 698. See also Barbara J. Shapiro, *Probability and Certainty in Seventeenth-Century England: A Study of the Relationships Between Natural Science, Religion, History, Law, and Literature* (Princeton, NJ: Princeton University Press, 1983), pp. 78–9.

[13] Thomas Reid, *Practical Ethics: Being Lectures and Papers in Natural Religion, Self-Government, Natural Jurisprudence, and the Law of Nations*, ed. Knud Haakonssen (Princeton, NJ: Princeton University Press, 1990), p. 105.

[14] Thomas Reid, *An Inquiry into the Human Mind, On the Principles of Common Sense: A Critical Edition*, ed. Derek R. Brookes (University Park, PA: Pennsylvania State University Press, 1997), pp. 75–6; Reid, 'Essays on the Intellectual Powers of Man', in *The Works of Thomas Reid*, vols. 2–3, ed. Baruch A. Brody (1814–15; repr. Cambridge, MA: MIT Press, 1969), pp. 605–6.

[15] Reid, 'Intellectual Powers', pp. xxxiii, 118; Reid, *Inquiry into the Human Mind*, p. 170.

[16] Reid, *Inquiry into the Human Mind*, p. 168; Reid, 'Intellectual Powers', p. 627.

revealed that God had endowed the human mind with a set of intuitively derived 'first principles'—'principles of common sense'—that are prior to both the acquisition of experience and the process of reasoning.[17] Those principles, which 'are no sooner understood than they are believed', grounded the process of reasoning itself and augmented the 'simple apprehension' of the senses with a set of 'original and natural judgments'.[18] The list of principles that Reid provided encompassed necessary truths such as the axioms of mathematics and logic as well as a disparate set of other fundamental precepts, including the reality of the self, the existence of the external world, the consistent operation of laws of nature, and the ability of human beings to exercise 'some degree of power' over their actions.[19] Reid also maintained that the conscience, or moral sense, which he described as 'the faculty of distinguishing right conduct from wrong' and regarded as the source of the moral law that was written in each individual's heart, deserved inclusion as an 'original power of the mind'.[20] In time, he suggested, investigation of the mind would yield a knowledge of its 'powers and operations' that would be no less certain than the truths that natural philosophers had gleaned about the natural world.[21]

Reid's synthesis of intuitionism and empiricism and his belief in the viability of systematic investigation of mental phenomena became important components of the 'Scottish Enlightenment'. The tenets of common-sense realism, which were elaborated and in some cases modified as time went on, proved to be enormously resonant in the English-speaking world. This is hardly surprising. The notion that common sense should serve as the 'tribunal' in establishing first principles proved appealing to individuals who possessed no special learning or expertise, while preserving the notion that God had bestowed on human beings attributes of mind that set them apart from the rest of the created order.[22] In addition, Reid and his successors employed a philosophical vocabulary and a conceptual framework that seemed to validate many of the convictions that most people took for granted, such as the reality of the external world, the distinction between mind and matter, and, more broadly, the intelligibility of the created order.

Exponents of Christian theology were especially drawn to the philosophy of common-sense realism, because it seemed to confirm many metaphysical positions that they regarded as central to their worldview—beliefs such as the distinction between God and the created order, and the privileged status of mind.[23]

[17] Reid, *Inquiry into the Human Mind*, pp. 32–3, 71–2, 215; Reid, 'Intellectual Powers', p. 593.

[18] Reid, 'Intellectual Powers', pp. 593, 596; Reid, *Inquiry into the Human Mind*, p. 215.

[19] Reid, 'Intellectual Powers', pp. 30–40, 644–5, 340–3, 617–43.

[20] Reid, *Essays on the Active Powers of the Human Mind* [1788], in *The Works of Thomas Reid*, vols. 3–4; repr. Cambridge, MA: MIT Press, 1969, pp. 247, 365; p. 231.

[21] Reid, 'Intellectual Powers', p. xxxvii.

[22] Reid, 'Active Powers of the Human Mind', p. 360.

[23] E. Brooks Holifield, *Theology in America: Christian Thought from the Age of the Puritans to the Civil War* (New Haven, CT: Yale University Press, 2003), esp. pp. 175–80.

They especially valued the insistence of Scottish philosophers that all causal efficacy resided in mental activity.[24] Not only did this seem to affirm the reality of human agency, but it also seemed to provide support for the claim that 'the Infinite and Eternal Mind, the author of all power and wisdom, has given existence and motion to all things by that intrinsic power which mind possesses over matter, and over its own movements'.[25] More broadly, the conviction among proponents of common-sense realism that God had endowed the mind with intuitive powers enabling humanity to apprehend reality provided strong grounds for exalting the value of the human consciousness as a medium for disclosing religious truth.[26] At the same time, in espousing the virtues of empiricism while simultaneously emphasizing that the knowledge that could be obtained through that method was limited, intuitional realists appeared to lend credibility to the idea that God had supplemented the data of human consciousness with divine revelation.[27]

Thanks at least in part to the compatibility of intuitional realism with Christian theology, numerous works espousing the philosophy of 'Common Sense' began to appear in the libraries and on the lists of booksellers in North America as well as Great Britain even before 1800.[28] More significantly, for much of the nineteenth century, the claim that pre-rational intuitional principles were instrumental in allowing human beings to discern the nature of reality played a central role in shaping discourse concerning the nature of mental attributes and their activities in the English-speaking world. Although parish clergy contributed more than a little to that discourse, instructors in colleges and theological seminaries made the most concerted efforts to develop a 'science of the soul'. The Protestants who typically presided over courses devoted to *mental philosophy*, *moral philosophy*, and then, increasingly after about 1840, *psychology*, generally emphasized ideas associated with common-sense realism.[29] Beginning in the second quarter of the nineteenth century, however, textbooks addressing the nature and functions of mind became ever more eclectic,

[24] Thomas Reid to James Gregory, 14 June 1785, in *The Works of Thomas Reid, D. D. . . .*, vol. 1, ed. William Hamilton (1863; repr. Bristol: Thoemmes Press, 1994), pp. 65–7.

[25] Samuel Stanhope Smith, *The Lectures, Collected and Improved, Which Have Been Delivered for a Series of Years in the College of New Jersey; On the Subjects of Moral and Political Philosophy*, vol. 1 (Trenton, NJ: James J. Wilson, 1812), p. 290.

[26] Clement Long, 'Mental Philosophy', *American Quarterly Observer* 3 (1834): pp. 102–3; T. C. Upham, 'Immutability of Moral Distinctions', *Biblical Repository and Quarterly Observer* 6 (1835–36): p. 123.

[27] Holifield, *Theology in America*, pp. 177–8.

[28] David Lundberg and Henry F. May, 'The Enlightened Reader in America', *American Quarterly* 28 (1976): p. 270; Henry F. May, *The Enlightenment in America* (New York: Oxford University Press, 1976), pp. 346–7.

[29] Rand B. Evans, 'The Origins of American Academic Psychology', in *Explorations in the History of Psychology in the United States*, ed. Josef Brožek (Lewisburg, PA: Bucknell University Press, 1984), pp. 34, 39, 42–6; L. S. Hearnshaw, *A Short History of British Psychology 1840–1940* (1964; repr. Westport, CT: Greenwood Press, 1986), pp. 2–3; Rick Rylance, *Victorian Psychology and British Culture, 1850–1880* (Oxford: Oxford University Press, 2000), pp. 44–7.

drawing on a wide array of perspectives from Continental thought as well as Anglo-American sources.[30]

Proponents of mental philosophy put forward a number of different arguments to justify the study of the nature and activities of mind. Some, persuaded that 'mental philosophy is properly the basis of all science', argued that an understanding of that subject would 'facilitate our progress in the study of every department of truth'.[31] Others held that knowledge of mental philosophy played a key role in shaping theological perspectives. For example, Joseph Haven, a professor of mental and moral philosophy at Amherst, noted that conceptions of God, as well as views concerning the freedom of will and the nature of virtue, were intimately associated with convictions regarding mind.[32] Still others valued mental philosophy for the practical guidance it provided clergy in bringing human beings to God.[33]

Exponents of mental philosophy in Great Britain and North America acknowledged that human knowledge was limited.[34] They emphasized that because it was impossible to comprehend or even describe the actual substance of mind, it was necessary for mental philosophers to confine their investigations to 'the facts and laws of mental operation' as they were presented to consciousness.[35] Such investigations, they believed, would demonstrate that mental phenomena, like other aspects of the created order, could be described as a set of orderly and intelligible processes.[36] Christian mental philosophers adamantly denied that the phenomena associated with mind could be reduced to 'material entities or agents' or adequately described through the use of analogies based on the laws and processes associated with the physical universe; they insisted that introspective observation of the contents of consciousness was capable of yielding real knowledge about mental phenomena.[37]

In the period between about 1800 and 1870 most discussions of mental philosophy in the English-speaking world were predicated on two major claims.

[30] Jay Wharton Fay, *American Psychology Before William James* (New Brunswick, NJ: Rutgers University Press, 1939), pp. 50, 90–128.

[31] S. S. Schmucker, *Psychology; or, Elements of a New System of Mental Philosophy, on the Basis of Consciousness and Common Sense* (New York: Harper and Brothers, 1842), p. vii.

[32] Joseph Haven, *Mental Philosophy: Including the Intellect, Sensibilities, and Will* (Boston, MA: Gould and Lincoln, 1857), p. 23. See also A. B., 'Thoughts on the Relation of Mental Philosophy to Theology', *Christian Spectator* 7 (1825): p. 29.

[33] C. G. Finney, 'Sinners Bound to Change Their Own Hearts', in *Sermons on Various Subjects* (New York: S. W. Benedict, 1834), pp. 22, 27.

[34] [Thomas C. Upham], *Elements of Intellectual Philosophy* (Portland, ME: William Hyde, 1827), p. 12.

[35] Haven, *Mental Philosophy*, pp. 15–18. See also [Upham], *Elements of Intellectual Philosophy*, pp. 78–9.

[36] Leonard Woods, 'Philosophy of the Mind', *Literary and Theological Review* 2 (1835): pp. 578–9.

[37] Noah Porter, *The Human Intellect; With an Introduction Upon Psychology and the Soul* (New York: Charles Scribner, 1868), pp. 7, 52.

The first was that the mind comprises a number of innate powers, or 'faculties'.[38] Mental philosophers made it clear that they regarded faculties not as ontologically separate elements but simply as heuristic devices for describing the varied operations of the mind. They differed somewhat in the names that they gave to the faculties and in the ways in which they described their interaction, but most devoted the bulk of their attention to the intellect (or understanding), the will (or volition), and the affections (or feelings). Although they acknowledged that human beings shared some powers, such as mechanical reflexes, instincts, and certain emotional impulses, with other organisms, they insisted that the intellect, the will, and the moral judgement were limited to human beings (some said animals possessed rudiments of reason) and attested to their kinship with the Creator.[39] Often, moral philosophers employed expositions of faculty psychology by enjoining people to use their self-discipline to ensure that the powers of the mind would be used in accordance with the dictates of the divine moral law.[40]

The other claim that played a major role in discussions of the mind in the English-speaking world was that humans acquired knowledge by means of an association of 'simple ideas' that appeared regularly in conjunction with one another in time and place or, in some cases, successively.[41] This position, commonly known as associationism, was adumbrated in the late seventeenth century by John Locke (1632–1704) and then given more determinate form about a half century later by David Hartley (1705–57). Beginning in the late eighteenth century, especially in England, a few thinkers endorsed a radical version of the associationist position that ascribed all mental processes to sensations, ideas, and the effort to receive pleasure and avoid pain, and treated the mind as a kind of self-organizing mechanism. Perhaps the most well-known proponent of that position was the English philosopher John Stuart Mill (1806–73). In 1843, Mill presented a sustained defence of the idea that '*all we know* of objects is the sensations which they give us, and the order of the occurrence of those sensations'.[42] Rejecting the claims of the intuitionists, he asserted that all that could be known about the 'thinking principle' that apprehended the sensations

[38] Fay, *American Psychology*, esp. pp. 50–128.

[39] George I. Chace, 'Of the Moral Attributes of the Divine Being', *Bibliotheca Sacra and Theological Review* 7 (1850): p. 689.

[40] D. H. Meyer, *The Instructed Conscience: The Shaping of the American National Ethic* (Philadelphia, PA: University of Pennsylvania Press, 1972).

[41] Roger Smith, *The Norton History of the Human Sciences* (New York: W. W. Norton, 1997), pp. 250–8; Robert M. Young, *Mind, Brain and Adaptation in the Nineteenth Century: Cerebral Localization and its Biological Context from Gall to Ferrier* (1970; repr. New York: Oxford University Press, 1990), pp. 95–8.

[42] John Stuart Mill, *Collected Works of John Stuart Mill*, vols. 7–8: *A System of Logic Ratiocinative and Inductive; Being a Connected View of the Principles of Evidence and the Methods of Scientific Investigation: Books IV–VI and Appendices*, ed. J. F. Robson (Toronto: University of Toronto Press, 1974), p. 59.

was 'the series of its states of consciousness'.[43] Some influential students of the mind, such as the British psychologist Alexander Bain (1818–1903), joined Mill in making the association of ideas gleaned from sensory experience the foundation for understanding the intellect.[44] Most mid-century Anglo-American mental philosophers, however, continued to couple enthusiasm for the idea that knowledge was frequently obtained through the association of ideas acquired through the passive reception of sensations with the conviction that the mind possessed innate faculties and intuitive first principles enabling it to play an active role in imposing order and meaning on experience.[45]

By the middle of the nineteenth century, most exponents of mental and moral philosophy favoured an approach to the mind that was descriptive, taxonomic, and oriented toward the effort to discover empirical regularities in its operations. Notwithstanding the fact that their work remained closely intertwined with theological and ethical concerns, they took the position that their introspective approach to mental phenomena was grounded in 'experience, observation, and induction'. Such knowledge, they asserted, constituted a 'science of mind'.[46] Indeed, they maintained, 'psychology is as truly an inductive science as are the sciences of any other existences or classes of being'.[47]

MIND, THE LIFE SCIENCES, AND EXPERIMENTALISM

Well into the nineteenth century, mental and moral philosophers played the pre-eminent role in systematically describing mental phenomena. Those thinkers were not, however, the only individuals in English-speaking cultures who expressed fascination with those phenomena. Supporters of mesmerism, phrenology, and other varieties of 'folk psychology' also displayed a lively interest in the nature and activities of the mind. Although space constraints do not permit a sustained discussion of those movements, it should be emphasized that insofar as they seemed to countenance approaches to the mind that could be construed as an endorsement of materialism or mechanistic determinism, they encountered stiff opposition from partisans of mental philosophy, the vast majority of whom were committed to Christian theology. An illustration of this can be seen in the response of those thinkers to phrenology, a set of propositions about the nature of mind formulated during the late eighteenth and early nineteenth centuries by the Viennese-trained physician and neuroanatomist Franz Joseph Gall (1758–1828) and his student and eventual co-worker,

43 Ibid., p. 64.

44 Alexander Bain, *The Senses and the Intellect* (London: John W. Parker and Son, 1855), p. vi.

45 Rylance, *Victorian Psychology*, pp. 55–7.

46 Haven, *Mental Philosophy*, p. 16.

47 Porter, *Human Intellect*, p. 52.

Johann Gaspar Spurzheim (1776–1832). Phrenologists maintained that the brain is the organ of mind and that thought and other mental operations are simply functions of its physiological activity. Rejecting the notion that the faculties of mind are powers of an immaterial soul, they held instead that they consist of separate components within the brain and served the function of assisting individuals in adapting to their environments.[48]

During the 1830s and 1840s, many individuals in Great Britain and the United States alike embraced phrenology. Although most of those enthusiasts were primarily interested in using its insights to improve human character and behaviour, thinkers who championed approaches to mental philosophy that were strongly inflected with Christian theology typically denounced phrenology as an endorsement of the notion that 'the thinking principle is nothing more than organized matter' operating in accordance with fixed laws.[49] That notion, they charged, constituted both an endorsement of 'pure Materialism' and a denial of the ability of human beings to make the kind of real choices on which the very concept of moral responsibility was based.[50] Significantly, however, while their motivation in addressing phrenology was their concern about its theological implications, the tactic that they frequently used in assailing that doctrine was to dismiss it as 'mere *skull-guessing*' and hence antithetical to the scientific principles underlying mental philosophy.[51]

By 1850 ongoing research on the nervous system had done much to undermine the credibility of phrenology.[52] Much of that research also revealed, however, that neurological structures and processes played an important role in accounting for many mental phenomena. Of particular importance in that regard were investigations in the realm of sensory-motor physiology. During the half century or so after 1810, the work of experimentalists had demonstrated that the nervous system was the site of origin for both sensations and movement. Most of those scientists continued to regard the cerebral cortex as the 'organ' employed by a mind acting as actual agent for the higher mental functions of intelligence, perception, and will, while presiding over such lower functions as sensation and motion.[53] Still, their work clearly indicated that

[48] Useful overviews of phrenology and its critics include John van Wyhe, *Phrenology and the Origins of Victorian Scientific Naturalism* (Burlington, VT: Ashgate, 2004); John D. Davies, *Phrenology: Fad and Science: A Nineteenth-Century American Crusade* (New Haven, CT: Yale University Press, 1955).

[49] Nathan L. Rice, *Phrenology Examined, and Shown to be Inconsistent with the Principles of Phisiology [sic], Mental and Moral Science, and the Doctrines of Christianity; Also an Examination of the Claims of Mesmerism* (1848; repr. New York: Robert Carter and Brothers, 1849), pp. 96, 98, 103–6, 115–16.

[50] [Frederic Henry Hedge], 'Pretensions of Phrenology Examined', *Christian Examiner and General Review* n. s. 12 (1834–35): pp. 252–3.

[51] [Hedge], 'Pretensions of Phrenology', p. 263.

[52] Smith, *Norton History*, p. 411.

[53] Young, *Mind, Brain, and Adaptation*; Roger Smith, 'The Background of Physiological Psychology in Natural Philosophy', *History of Science* 11 (1973): pp. 75–123.

discussion of the attributes and activities of mind could not be detached from research associated with the experimental life sciences.

That insight was not lost on psychologists. During the period after 1850, a number of them emphasized that 'the time has now come when many of the striking discoveries of Physiologists relative to the nervous system should find a recognized place in the Science of Mind'.[54] In Great Britain, for example, during the 1850s and 1860s, Alexander Bain sought to integrate the idea that associationism constituted the means by which the mind interpreted experience with insights gleaned through research on the nervous system.[55] On a different front, William B. Carpenter (1813–85), Thomas Laycock (1812–76), and a number of other medically trained scientists made concerted efforts to show that the 'student of mental science, if he would be a successful student, must concentrate his researches upon the laws of action of the brain and nervous system, as they correlate the laws of thought and volition'.[56] Although proponents of that position did not typically maintain that mental phenomena could be altogether reduced to physical processes, they succeeded in making a strong case for using the results of neurophysiological research in accounting for human mental activities.[57]

Supplementing the increased knowledge of neurological structures and processes in altering psychologists' understanding of the human mind was the theory of organic evolution. The engineer-turned-philosopher Herbert Spencer (1820–1903) was the first to employ that theory in a sustained fashion in discussing psychology. In 1855 he placed associationism within an evolutionary context, ascribing the states and functions of the human mind and its adjustment to the environment to the collective experience of organisms throughout the history of life.[58] Initially, Spencer's evolutionary approach to psychology received little support.[59] It was not really until about 1870, when most natural historians had become converts to evolutionary theory, that students of human nature in the English-speaking world began to pay serious attention to its psychological implications. The work of Charles Darwin (1809–82), who made a determined effort to show that 'there is no fundamental difference between man and the higher mammals in their mental faculties', played an especially important role in convincing many Anglo-American psychologists of the legitimacy of Spencer's claim that 'mind can be understood only by observing how Mind is evolved'.[60]

[54] Bain, *The Senses and the Intellect*, p. v.

[55] Young, *Mind, Brain, and Adaptation*, pp. 101–33.

[56] Thomas Laycock, *Mind and Brain; Or, the Correlations of Consciousness and Organisation; Systematically Investigated and Applied to Philosophy, Mental Science and Practice*, vol. 1, 2nd edn (New York: D. Appleton, 1869), p. 4.

[57] Hearnshaw, *Short History*, pp. 19–22.

[58] Young, *Mind, Brain, and Adaptation*, pp. 150–96; Rylance, *Victorian Psychology*, pp. 203–50.

[59] Herbert Spencer, *An Autobiography*, vol. 1 (New York: D. Appleton, 1904), p. 546.

[60] Charles Darwin, *The Descent of Man, and Selection in Relation to Sex*, vol. 1 (1871; repr. Princeton, NJ: Princeton University Press, 1981), p. 35; Herbert Spencer, *The Principles of Psychology*, vol. 1, 2nd edn (London: Williams and Norgate, 1870–72), pp. 291–2.

A third set of developments also proved to be quite significant in shaping the way that psychologists in America and Great Britain thought about the mind: the experimental work of Wilhelm Wundt (1832–1920) and other German psychologists. Those psychologists were strongly committed to the use of meticulous measurements and controlled experiments within laboratory settings. They continued, however, to regard psychology as the study of consciousness and to value introspection as essential to their inquiries.[61] Wundt, for example, was a philosophical idealist who denounced efforts to reduce mind to neurophysiological processes as 'materialistic pseudo-science' and sought to promote the reality of mind as an active agent on experimental grounds.[62] The use of experimental controls and mathematical measurement to assess the results of introspection, he maintained, would enable psychologists to provide a more accurate description of the subjective elements of human experience.[63]

THE EMERGENCE OF THE 'NEW PSYCHOLOGY'

During the last quarter of the nineteenth century, the growing prominence of neurophysiology, evolutionary thought, and German experimentalism convinced a small but ever-growing number of academic psychologists in the United States and Great Britain that it was time to engage in a fundamental reassessment of the nature and functions of the mind. Those thinkers shared the conviction of William James (1842–1910) that the question of human nature had become 'a keen and urgent problem'.[64] In addition, like James, they believed that the time was ripe to construct a more rigorously scientific psychology, although they often expressed quite different views concerning what such a psychology would look like. In the United States the promotion of 'scientific psychology' was attended by the publication of new textbooks and journals, the creation of professorships, psychological laboratories, and professional societies, and efforts to establish autonomous departments of psychology in colleges and universities. By the beginning of the twentieth century, psychologists had made a good deal of headway in detaching their work from the discipline of philosophy.[65] In Great Britain, efforts to institutionalize the new psychology

[61] Kurt Danziger, *Constructing the Subject: Historical Origins of Psychological Research* (Cambridge: Cambridge University Press, 1990), pp. 17–18.

[62] Wilhelm Wundt, quoted in Theodore Mischel, 'Wundt and the Conceptual Foundations of Psychology', *Philosophy and Phenomenological Research* 31 (September 1970): p. 4.

[63] John M. O'Donnell, *The Origins of Behaviorism: American Psychology, 1870–1920* (New York: New York University Press, 1985), pp. 19–22.

[64] [William James], 'The Teaching of Philosophy in Our Colleges', *Nation* 23 (21 September 1876): p. 178.

[65] Thomas M. Camfield, 'The Professionalization of American Psychology, 1870-1917', *Journal of the History of the Behavioral Sciences* 9 (1973): pp. 66–75.

within higher education were largely unsuccessful until well into the twentieth century, but this did not prevent many students of human nature from committing themselves to the precepts of scientific psychology.[66]

Practitioners of the 'new psychology' were not the only individuals in the English-speaking world who sought to exploit scientific advances in the understanding of the mind. Advocates of scientific naturalism sought to show that the implications of the life sciences significantly undermined the credibility of the dichotomy between mind and matter that played a central role in traditional conceptions of Christian mental and moral philosophy. Thomas Huxley (1825–95), for example, asserted that the appropriate lesson to be drawn from psychological investigation of mental phenomena was that 'all states of consciousness . . . are immediately caused by molecular changes of the brain-substance'.[67] Similarly, the American polymath Lester Frank Ward (1841–1913), who expressed hostility to supernaturalism in all its forms, wrote scathingly of 'the vague and ethereal conceptions to which the terms soul, mind, spirit, etc., give rise' and insisted that neurophysiology constituted the real 'basis of all psychical phenomena'.[68]

Not surprisingly, Christian thinkers, who still played an important role within the Anglo-American intellectual community, strenuously resisted these claims. More surprising, perhaps, most proponents of scientific psychology in the late nineteenth century joined them in resisting efforts to reduce mind to the activities of the nervous system and in continuing to regard consciousness as the focus of psychological investigation. At a time when many of those individuals were trying to extricate themselves from departments of philosophy and establish psychology as an autonomous discipline, few of them relished the idea of allowing their field to become a sub-discipline of biology. Those students of the mind were not immune to the allure of neurophysiological research. Some even highlighted 'the relation of psychology to physiology and of the line of demarkation [*sic*] between them' as one of the central questions facing their discipline.[69] Most of them made it clear, however, that they were unwilling to give neurological investigation the last word on the nature of mind and its activities. For example, George Trumbull Ladd (1842–1921), a Yale psychologist who authored one of the first American texts in physiological psychology, defined the subject matter of scientific psychology as 'states of consciousness' and insisted that 'if cerebral psychology is the only scientific psychology, then

[66] Mitchell G. Ash, 'Psychology', in *The Cambridge History of Science*, vol. 7: *The Modern Social Sciences*, ed. Theodore M. Porter and Dorothy Ross (Cambridge: Cambridge University Press, 2003), p. 254.

[67] Thomas Huxley, 'On the Hypothesis That Animals Are Automata, and Its History', in *Method and Results: Collected Essays by T. H. Huxley*, vol. 1 (1917; New York: Greenwood Press, 1968), pp. 242–4 (quotation on p. 244).

[68] Lester F. Ward, *Dynamic Sociology; Or, Applied Social Science as Based Upon Statical Sociology and the Less Complex Sciences*, vol. 1 (1883; repr., New York: Greenwood Press, 1968), p. 124.

[69] George Stuart Fullerton, 'Psychology and Physiology', *Psychological Review* 3 (1896): p. 2.

there is no science of psychology'.[70] The Yale experimental psychologist E. W. Scripture (1864–1945) similarly repudiated 'the unjustified subordination of psychology to physiology'. 'Mind and brain', he declared, 'are not the same'.[71] In 1884, while surveying 'The New Psychology', John Dewey (1859–1952) maintained that 'no direct conclusions regarding the nature of mental activities or their causes can be drawn from the character of nervous structure or function'. Rather, 'explanations of psychical events, in order to *explain*, must themselves be psychical and not physiological'.[72] In 1878 William James opined that nothing was 'more ludicrously false than the assertions so loudly made by some authors that the only sound psychological science is that founded on physiology'.[73] By the time his *The Principles of Psychology* appeared in 1890, James had become receptive to the idea of attempting to ground psychology on 'the empirical correlation of the various sorts of thought or feeling with definite conditions of the brain', but he remained convinced that psychology must address the realm of the subjective and continued to regard the nature of the actual connection that existed between thought and brain as the 'ultimate of ultimate problems'.[74] Psychology could aspire to becoming truly scientific, he asserted, only when it chose to 'renounce' efforts to arrive at 'ultimate solutions' concerning the ontological status of mind.[75]

In Great Britain, where psychology was considerably less well established institutionally, influential proponents of the systematic study of the mind also resisted the reduction of studies of the mind to neurophysiology. Thomas Hill Green (1836–82), for example, adamantly denied that the physiology of the nervous system was capable of generating the products of consciousness.[76] Similarly, James Ward (1843–1925) held that while an intimate relationship could sometimes be discerned between neurophysiological activity and mental phenomena, efforts to describe thought and volition in terms of 'the monotonous interplay of molecules in the cavity of a skull' were unlikely to pay rich dividends.[77]

[70] George Trumbull Ladd, 'Psychology as So-called "Natural Science"', *Philosophical Review* 1 (1892): pp. 34, 50.

[71] E. W. Scripture, 'The Problem of Psychology', *Mind* 16 (1891): pp. 308, 309.

[72] John Dewey, 'The New Psychology', in *The Early Works of John Dewey, 1882–1898*, vol. 1, ed. Jo Ann Boydston (Carbondale, IL: Southern Illinois University Press, 1969), pp. 52, 55.

[73] William James, quoted in Ralph Barton Perry, *The Thought and Character of William James*, vol. 2 (Boston, MA: Little, Brown, 1935), p. 28.

[74] William James, *The Principles of Psychology*, vol. 1 (New York: Henry Holt, 1890), pp. vi, 177–82; Gerald E. Myers, *William James: His Life and Thought* (New Haven, CT: Yale University Press, 1986), p. 9.

[75] William James, 'A Plea for Psychology as a "Natural Science"', *Philosophical Review* 1 (1892): pp. 147–50.

[76] Rylance, *Victorian Psychology*, pp. 315–16.

[77] James Ward, *Naturalism and Agnosticism: The Gifford Lectures Delivered Before the University of Aberdeen in the Years 1896–1898*, 4th edn (London: A & C. Black, 1915), pp. 303–4; James Ward, 'Psychological Principles', *Mind* 8 (1883): p. 169. See also Roger Smith, *Free Will and the Human Sciences in Britain, 1870–1910* (London: Pickering and Chatto, 2013), pp. 12–13, 45–55, 69–72.

For their part, few neurophysiologists made strong efforts to bring psychology within the purview of their discipline. Nor were most of those scientists intent on taking a dogmatic stand in favour of neurophysiological reductionism. Some shared the view of the English neurophysiologist David Ferrier (1843–1928), who maintained that 'no purely physiological investigation can explain the phenomena of consciousness'.[78] Others embraced the stance of psychophysical parallelism, which allowed them to focus on the nervous system while ceding the discussion of mental phenomena to psychology, metaphysics, and theology.[79]

The desire of psychologists to preserve disciplinary autonomy and the indifference of neurophysiologists to expanding their reach were not the only factors accounting for the new psychologists' unwillingness to obliterate the distinction between the mind and the nervous system. Another important consideration was at work. Psychologists would have had to have been almost unimaginably obtuse not to recognize that if their efforts to establish their credentials as a separate 'scientific' discipline within institutions of higher education were to bear fruit, it would be highly impolitic to cast that discipline in the role of ally to the village atheist. During the last quarter of the nineteenth century, it became apparent that some college officials were already suspicious of the religious implications of the new psychology. In 1877, for example, the University Senate at Cambridge University denounced an attempt to make experimental psychology a distinct discipline as an effort to 'insult religion by putting the human soul in a pair of scales'.[80] Given such suspicions, academic psychologists understandably concluded that in dealing with an area as highly charged as the human mind, their best hope for ingratiating themselves with the officials in colleges and universities lay in emphasizing the compatibility of their discipline with Christian theology. This was especially the case because many of those institutions designated the exposition of the nature and activities of the mind as the introduction to study of philosophy.

Many college presidents and trustees made it clear that they looked favourably on efforts that sought to show the compatibility of psychology and religion. For example, in its report of 1875–6, Harvard's Board of Overseers approved the development of a course dealing with the relations of psychology and physiology as a way of countering materialistic views concerning the mind.[81] In 1885 G. Stanley Hall (1844–1924), who was teaching courses in experimental psychology at Johns Hopkins, made a point of emphasizing the 'utter incommensurability' of

[78] David Ferrier, *The Functions of the Brain* (New York: G. P. Putnam's Sons, 1876), p. 255.

[79] Smith, *Norton History*, pp. 645, 824–5.

[80] Quoted in L. S. Hearnshaw, *The Shaping of Modern Psychology: An Historical Introduction* (London: Routledge and Kegan Paul, 1987), p. 125.

[81] Sheldon M. Stern, 'William James and the New Psychology', in *Social Sciences at Harvard 1860-1920: From Inculcation to the Open Mind*, ed. Paul Buck (Cambridge, MA: Harvard University Press, 1965), p. 184.

the 'cells and fibers' of the nervous system with the revelations of consciousness. The 'new psychology', he declared, should be regarded as 'Christian to its root and center'.[82] Similarly, Dewey insisted that the new psychology 'finds no insuperable problems in the relations of faith and reason'.[83] Ladd, too, denied that physiological psychology had done anything to undermine the status of 'man as a spiritual and rational being'.[84]

Practitioners of the new psychology also made it clear, however, that there were limits to their efforts to emphasize the congruence of psychology with religious interests. In particular, they commonly complained that 'theological formulae' and 'insidious orthodoxies' had sometimes impeded careful investigation of the 'great open questions' of their discipline.[85] More important, perhaps, in spite of the fact that a sizable number of psychologists remained committed to religious beliefs and values in their personal lives, they recognized that the norms of scientific discourse and explanation no longer permitted psychologists to invoke the supernatural or to employ peculiarly theological concepts such as the soul. That recognition prompted them to eliminate God-talk and other references to spiritual matters from their descriptions and analyses, and to limit the focus of their discussions to natural entities and processes.[86] William James thus promised readers of *The Principles of Psychology* (1890) that he would remain 'positivistic and non-metaphysical' in dealing with the mind.[87] Although he did not always keep that promise, his work clearly indicated that he associated the development of a scientific psychology with a 'fair and square and explicit *abandonment* of such questions as that of the soul, the transcendental ego, the fusion of ideas or particles of mind stuff'.[88] Even Ladd, who insisted on 'the explanatory value of the metaphysical postulate of a mind, or soul', acknowledged that 'the postulate should not be intruded into the science of psychology'.[89] Hall, notwithstanding his insistence that psychology could be used as a weapon against materialism, simply ignored the issue of religious truth in developing his research agenda.[90]

One of the clearest manifestations of the impatience that most scientific psychologists had with realms of experience that seemed to transcend their understanding of the natural order was their hostile response to spiritualism and psychical research. To be sure, a few psychologists, most notably William

[82] G. Stanley Hall, 'The New Psychology', *Andover Review* 3 (1885): pp. 127, 247.
[83] Dewey, 'The New Psychology', vol. 1, p. 60.
[84] George T. Ladd, *Elements of Physiological Psychology: A Treatise of the Activities and Nature of the Mind* (New York: Charles Scribner's Sons, 1887), p. iii.
[85] G. Stanley Hall, 'Philosophy in the United States', *Mind* 4 (1879): pp. 89–90.
[86] Jon H. Roberts and James Turner, *The Sacred and the Secular University* (Princeton, NJ: Princeton University Press, 2000), esp. pp. 19–60.
[87] James, *Principles of Psychology*, vol. 1, pp. vi, 182.
[88] James, 'Plea for Psychology', pp. 147–50 (quotation on p. 149).
[89] Ladd, 'Psychology as So-called "Natural Science"', p. 52.
[90] O'Donnell, *Origins of Behaviorism*, pp. 119–20.

James, made efforts to bring paranormal research within the purview of scientific inquiry. Most of them, however, shared the view of James's Harvard colleague, Hugo Münsterberg (1863–1916), who held that if spiritualists and other defenders of psychic phenomena were correct in believing that 'supernatural connections' existed between the 'physical and psychical worlds', then the claims of scientific psychologists were false and the world was 'controlled by inanity and trickery'.[91]

Few Christian thinkers in Great Britain and America expressed great enthusiasm for the new scientific psychology. Persuaded as they were that mind—divine and human alike—possessed a privileged status within reality, they were understandably leery of tendencies to bring it and the nervous system into closer convergence and of efforts to narrow the distance between the human mind and that of other species. Conservative Christians, confronted by a mechanistic conception of the physical universe that appeared to exclude the supernatural from consideration and by a biological theory of evolution that seemed to undermine belief in the divine inspiration of the scriptures, tended to dismiss or disparage the proclamations of the new psychologists as ominous indications of the irreligious tendencies of modern science. But even Christians of more liberal theological inclinations supported late-nineteenth-century scientific psychology rather half-heartedly. To be sure, those believers tended to accept the pronouncements of the new psychology at face value. Moreover, some liberals exploited insights drawn from a new subspecialty within psychology that arose at the turn of the twentieth century—the psychology of religion—in revising their approach to such important concerns as conversion and apologetics.[92] Nevertheless, those Christians also made it clear that their theological commitments imposed decided limitations on the kind of claims about the nature of mind that they were willing to accept. For example, although they were not in principle opposed to physiological psychology, they insisted on describing the nervous system as 'the servant of personality for the gathering of sensations, and its organ for the expression of its life and action'.[93] They also coupled their acceptance of the theory that humans had originated by means of an evolutionary process with an affirmation of the idea that at some point during the course of that process, the mental attributes of human beings became different in kind from those of other animals.[94] In turn, that conviction

[91] Hugo Münsterberg, 'Psychology and Mysticism', *Atlantic Monthly* 83 (1899): pp. 67–8, 85. See also Deborah J. Coon, 'Testing the Limits of Sense and Science: American Experimental Psychologists Combat Spiritualism, 1880–1920', *American Psychologist* 47 (1992): pp. 143–51.

[92] Christopher G. White, *Unsettled Minds: Psychology and the American Search for Spiritual Assurance, 1830–1940* (Berkeley, CA: University of California Press, 2009), esp. pp. 134–57.

[93] William Newton Clarke, *An Outline of Christian Theology* (1898; New York: Charles Scribner's Sons, 1922), p. 187.

[94] Jon H. Roberts, *Darwinism and the Divine in America: Protestant Intellectuals and Organic Evolution, 1859–1900* (Madison, WI: University of Wisconsin Press, 1988), pp. 175–80; Smith, *Free Will*, esp. p. 8.

provided them with a springboard for arguing for the privileged nature of mind within the created order and for emphasizing the need to begin with mind rather than matter in the quest to understand reality.

Just as significantly, liberal Christians also indicated—on some occasions implicitly, on others explicitly—that the preoccupations of scientific psychology only rarely converged with their theologically oriented concerns about the status of mind within reality. For example, while they shared the view of the new psychologists that the human consciousness was nicely adapted to respond effectively to its environment and even to alter that environment, their commitment to the exalted status of mind prompted them to elevate the human personality to the pinnacle of the created order. The elements of personality, they declared, constituted the Creator's crowning work and 'the organ for the deepest and truest revelations of God'.[95] No less importantly, they remained committed to the idea that human interactions included not only horizontal relationships with natural surroundings, but also a vertical dimension—a dimension characterized by a relationship with God. That commitment prompted them to regard the 'environment' in spiritual as well as naturalistic terms. In fact, they maintained, 'the relation of man as a spirit to God as a Spirit above him' constituted the very essence of religion itself.[96]

In the period after 1900 Anglo-American exponents of Christian theology found additional reasons for expressing reservations about, and often even opposition to, many of the pronouncements made in the name of a scientific approach to mental phenomena. During the early twentieth century, a period marked by the increasing visibility and influence of both psychoanalysis and behaviourism, psychologists characteristically became increasingly more aggressive and outspoken in their insistence on detaching their discussions of human nature from God-talk and other rhetoric associated with religion. Sigmund Freud (1856–1939), for example, famously dismissed religious belief as an 'illusion' and lauded a rigorously naturalistic approach to science as the most promising instrument for understanding the human condition. The behaviourists sought to make psychology 'an undisputed natural science' by interpreting the behaviour of human beings, like that of other animals, as largely the result of the formation of habits developed through the association of stimulus and response.[97] John B. Watson (1878–1958), the individual most closely associated with behaviourism in its early years, made no secret of his contempt for religion. In 1925, for example, he ascribed the origin of religion to the work of indolent 'medicine men' who used fear of the unknown to force others in their tribe to share with them the fruits of their labours. 'If the fear

[95] Samuel Harris, *The Philosophical Basis of Theism* (New York: Charles Scribner's Sons, 1883), p. 528.

[96] Clarke, *Outline*, pp. 1, 147.

[97] John B. Watson, 'Psychology as the Behaviorist Views It', *Psychological Review* 20 (1913): pp. 158–77 (quotation on p. 163).

element dropped out of any religion', he asserted, 'that religion could not long survive'.[98] Although many behaviourists found some of Watson's programmatic statements to be overblown, they shared his unwillingness to invoke religious concepts in describing human behaviour. For their part, proponents of the Christian worldview frequently found themselves denouncing psychoanalysis, behaviourism, and other naturalistic approaches to the mind as inimical to many of their most cherished ideals and values.[99] Indeed, cultural commentators on the prowl for evidence of tension between science and religion can hardly do better than examine the always complicated, often tempestuous relationship that has existed between Christian thought and scientific psychology.

In recent years proponents of the Christian worldview have not been alone in voicing reservations concerning the merits of the prevailing scientific approaches to psychology. They have been joined by a number of other thinkers such as Raymond Tallis and Thomas Nagel who explicitly reject theological interpretations of mind, but insist that reductionist and materialist approaches to psychological phenomena are incapable of accounting satisfactorily for phenomena associated with the human consciousness.[100] Although their claims have been quite controversial, it is not necessary to accept the legitimacy of their views to recognize that no clear consensus has yet emerged concerning the ontological status of mind.

ACKNOWLEDGEMENTS

I am indebted to Christine Axen for her research assistance. I also wish to thank Peter Harrison, Ron Numbers, the Boston Historians of American Religion, and participants in the Conference on 'Science Without God: Religion, Naturalism, and the Sciences', for their insightful comments on an earlier version of this paper. Thanks, too, to Patrick Walsh for his skilful proofreading and conscientious editing. Finally, I continue to cherish the support I receive from the members of my family: Sharon (ILYS) and Jeff.

[98] John B. Watson, *Behaviorism* (New York: W. W. Norton, 1925), pp. 3–4 (quotations on p. 4).

[99] See, for example, Jon H. Roberts, 'Psychoanalysis and American Religion, 1900–45', in *When Science and Christianity Meet: From Augustine to Intelligent Design*, ed. David C. Lindberg and Ronald L. Numbers (Chicago: University of Chicago Press, 2003), pp. 225–44; and John C. Burnham, 'The Encounter of Christian Theology with Deterministic Psychology and Psychoanalysis', *Bulletin of the Menninger Clinic* 49 (1985), pp. 321–52.

[100] Raymond Tallis, *Aping Mankind: Neuromania, Darwinitis and the Misrepresentation of Humanity* (Bristol, CT: Acumen, 2011); Thomas Nagel, *Mind and Cosmos: Why the Materialist Neo-Darwinian Conception of Nature is Almost Certainly False* (New York: Oxford University Press, 2012).

10

Down to Earth

Untangling the Secular from the Sacred in Late-Modern Geology

Nicolaas Rupke

What does it mean to conduct 'geology without God'? The present chapter considers this question, describing the historical process of the disappearance from the geological literature of references to the Bible and God. From the late eighteenth century onwards, such references no longer were to be found in the main text of key publications, not even in forewords or in concluding chapters and epilogues. Perhaps more generally, this change was indicated by a morphing of the meaning of the word 'creation' as it changed from 'God's handiwork' to 'the natural world', exemplified by the title of the influential treatise on evolution, *Natürliche Schöpfungsgeschichte* [Natural history of creation] (1868) by Darwin's 'German bulldog', the Jena biologist Ernst Haeckel (1834–1919).

In early modern times, the scientific study of the earth had been part of an 'origines sacrae' tradition that treated both the Bible and nature as equal sources of reliable data in reconstructing the history of the universe and, in particular, of the earth, which was assumed to be coeval with the history of mankind. A classic of the genre was Edward Stillingfleet's *Origines Sacrae: or a Rational Account of the Grounds of Natural and Revealed Religion* (1662). Perhaps better remembered today in the literature on the history of geology is Thomas Burnet's *Telluris Theoria Sacra, or the Sacred Theory of the Earth* (Latin 1681–9; English 1684–90), which sketched the past, present, and future of the earth in a context of biblical history. Similar works of natural history continued to be published throughout the eighteenth century, famous among them Johann Jakob Scheuchzer's *Lithographica Helvetica* (1726), in which he misidentified the fossil skeleton of a giant salamander as the remnants of a human being who had drowned in Noah's deluge, the '*Homo diluvii testis*' ('*das betrübte Beingerüst eines alten Sünders*').

Towards the end of the eighteenth century and through the first half of the nineteenth, Bible- and God-talk disappeared from the geological literature—or rather from much of the literature that emanated from a majority of the members of leading academic institutions. The study of rocks and fossils opened up a new and vast perspective of earth history, indicating that the history of the earth had not covered the same stretch of time as the history of mankind, but extended back immeasurably before the appearance of humans. Influential in this context was Georges-Louis Leclerq, Comte de Buffon (1707–88), whose *Les époques de la nature* (1778) gave scientific credibility to the emerging belief in a long prehistory. Buffon was censured by the Sorbonne's Faculty of Theology and retracted some of his heterodox views. Still, it became possible to talk about rocks and fossils—about 'geology', as Jean André Deluc (1727–1817), a Genevan Calvinist, who moved to London to become Reader to the Queen, called this branch of the study of nature—without reference to biblical human history. Among the coryphaei of the new geology, especially in the German-speaking world, was Johann Friedrich Blumenbach (1752–1840), who in his *Beyträge zur Naturgeschichte* (Contributions to natural history) (1790), in dealing with prehistory, now omitted references to the Bible and God, facilitated by the fact that the new University of Göttingen, where he was based, had freed the non-theological faculties from the censorship right of the faculty of theology. As a result, the utilitarian advantages of geology—stratigraphy in mining, paleontology in comparative anatomy—became the main justification for dealing with the subject, rather than religious relevance.

By contrast, in the Anglo-American realm, Bible- and God-talk survived in the form of the argument of design and, more generally, of natural theology. This happened at approximately the same time that the new geology was introduced at the Universities of Oxford and Cambridge. Here the subject was made to serve an ecclesiastical, Anglican purpose by being stripped of its utilitarian connections and placed on the epistemological footing of Cuvierian functionalism, which meshed well with Paleyan natural theology and its design argument. One of Britain's leading representatives of the new geology was Oxford's William Buckland (1784–1856), whose comprehensive treatment of historical geology appeared as a Bridgewater Treatise on 'the Power Wisdom and Goodness of God as Manifested in Creation', under the title *Geology and Mineralogy Considered with Reference to Natural Theology* (1836).

The catalyst for the disinvestment of religion in English geology was the establishment of several scientific institutions in London—the Royal Institution, the Hunterian Museum, followed later by the Natural History Museum, the University of London, both University College and King's College, the Museum of Practical Geology which became the School of Mines—and the concomitant emergence of a metropolitan culture in which science did not serve the function it had at Oxbridge. The omission of 'Bible and God' was part of a metropolitan drive for independence from Oxbridge, which was achieved not only by cultivating science for its socio-economic utility but by discarding the functionalist approach in favour of a transcendentalist epistemology borrowed from Continental sources and introduced, in part, via Edinburgh. Transcendentalism stripped geology from religious language because, although the approach did not necessarily carry anti-religious implications, it severed the link of geology with the Paleyan design argument and thus made it non-ecclesiastical in its applicability. Moreover, by being from the Continent and Scotland, it represented science cultivated for its own sake.[1]

Let us explore in some detail how and why this happened by looking at the book that, at the time, was perhaps the most popular trendsetting treatise on the study of the earth, the planets, and the stars: *Kosmos*, written by Blumenbach's student Alexander von Humboldt (1769–1859).

HUMBOLDT AND *KOSMOS*

Humboldt was a major trendsetter of early- and mid-nineteenth-century science. He significantly contributed to the removal from scientific discourse of references to God, human immortality, and Christian doctrine by means of his influential *Kosmos* (1845–62).[2] This widely admired and holistic treatment of celestial and terrestrial phenomena avoided references to God and to divine design. Reactions across the world of Western scientific culture varied, depending on both geographical and ideological space. The many, and often lengthy, reviews of the successively published *Kosmos* volumes took clear and opposing stands. Both Catholics (e.g. in Austria) and Anglicans (in Great Britain) objected. 'Free-thinkers' in Germany and France applauded. By and large, the liberals drew comfort from Humboldt's mostly (but not entirely) implicit insistence that the realm of nature offers us knowledge that is of a different kind from that of 'higher speculative views'.[3]

Humboldt's *Kosmos* played a leading role in determining the place of scientific knowledge in mid-nineteenth-century European society, including its location in relation to religion. Humboldt produced *Kosmos* towards the

[1] Rupke, Nicolaas, 'The Secularization of Geology in Nineteenth-Century England', in *The Secularization of Science*, ed. David C. Lindberg and Margaret Schabas (symposium papers, University of Wisconsin, Madison, 21–2 September 1990), pp. 73–108.

[2] *Kosmos* appeared in five volumes under the title *Kosmos: Entwurf einer physischen Weltbeschreibung*: vol. 1 (1845), vol. 2 (1847), vol. 3 (1850–1), vol. 4 (1858), vol. 5 (1862), the last published posthumously.

[3] Humboldt to Bunsen, January 4, 1846, in *Briefe von Alexander von Humboldt an Christian Carl Josias Freiherr von Bunsen (1816–1856)* (Leipzig: Brockhaus, 1869), p. 72. For further details on Humboldt and *Kosmos*, see Nicolaas Rupke, introduction to *Cosmos: A Sketch of the Physical Description of the Universe*, vol. 1, edited by Alexander von Humboldt, trans. E. C. Otté (Baltimore, MD: Johns Hopkins University Press, 1997), pp. vii–xlii.

end of his long life, when he was in his late seventies and eighties, and is best known for this book, although he was internationally celebrated long before its first volume appeared. *Kosmos* constituted both the summary of Humboldt's lifelong interests and a holistic digest of the scientific study of celestial and terrestrial phenomena.

By the middle of the nineteenth century, Humboldt was thought of not merely as a German scientist, but as the consummate 'Mr European scientist', who formed the hub of an international network of correspondence, knew a variety of languages, had lived in both Paris and Berlin, wrote with equal facility in French and German, possessed a genial temperament, commanded an encyclopaedic knowledge, was a nobleman, a statesman, and a patron of the sciences—in short, the person most qualified to sum up and weigh the natural sciences at the mid-century. As one contemporary exulted:

> Were the republic of letters to alter its constitution, and choose a sovereign, the intellectual sceptre would be offered to Alexander von Humboldt. The New World would send deputies across the Atlantic to assist at his installation, and the princes and philosophers of every clime,—the autocrats of the East and the democrats of the West,—would hail the enterprising traveller who trod the mountain crests of Europe, ascended the American Cordilleras, and explored the auriferous beds of the Uralian chain.[4]

Kosmos was an immensely popular book, a great success, both for its author and the publishers. It enjoyed five nineteenth-century authorized German editions, including a German-American one, and a string of translations. International interest in *Kosmos* was so strong, and the commercial potential of translations so considerable, that renditions into other languages began appearing as soon as volume 1 was completed. By the time the fourth volume had come out, *Kosmos* had been translated into no fewer than eleven different languages, and in some languages more than one translation had been produced. Into English, for example, there were three. Considering just the original edition, published by J. G. Cotta in Stuttgart and Tübingen, twenty-two thousand copies of volume 1, twenty thousand of volume 2, and fifteen thousand each of volumes 3, 4, and 5 were printed. In terms of sales, *Kosmos* made its author the most successful of his generation. Moreover, a burgeoning *Kosmos* spin-off industry developed, an example of which was the 5-volume *Briefe über Alexander von Humboldt's Kosmos* (1848–60), a commentary for 'educated laymen', to which the Freiberg geologist Carl Bernhard von Cotta (1808–79) made a major contribution, and which appeared nearly in tandem with the successive *Kosmos* volumes.[5]

[4] [David Brewster], review of *Kosmos: A General Survey of the Physical Phenomena of the Universe*, by Humboldt, *North British Review* 4 (1845): p. 202.

[5] For bibliographical details see Horst Fiedler and Ulrike Leitner, *Alexander von Humboldts Schriften: Bibliographie der selbständig erschienenen Werke* (Berlin: Akademie Verlag, 2000), pp. 382–434.

Humboldt's name became synonymous with the title of his last book, and came to denote the totality of the sciences as well as its popular treatment. The *Monatsschrift für die gesamten Naturwissenschaften* (monthly for all the sciences), edited by the physics lecturer Georg Krebs (1833–1907) who intended to bring to the educated public the results of modern science in a popular form, was simply called *Humboldt*. Several different Humboldtian scientific periodicals were founded that carried the title *Kosmos*.

Kosmos went through a long gestation. By Humboldt's own account, he began the project in 1819, in France and in French, entitling it 'Essai sur la Physique du Monde'. The concept began to take shape with a series of salon lectures by Humboldt, delivered during the years 1825–28, first in Paris and subsequently in Berlin. Then in 1834, Humboldt chose the title *Kosmos* for his planned book, in order to emphasize that he was not writing a conventional physical geography/geology ('physische Erdbeschreibung'), but that both heaven and earth were an integral part of his conception. Also, this title, in addition to indicating the vast scope of his book, gave expression to Humboldt's aesthetic-holistic epistemology, because in Homeric times, the word 'cosmos' had meant 'ornament' and 'elegance', and later had come to denote the order or harmonious arrangement of the world. Yet there was more to the choice of the title, in that *Kosmos* made a break with the contemporary terminology for the whole of the physical world, at the time referred to as either 'nature' or 'creation'. Lorenz Oken's *Allgemeine Naturgeschichte* (1833–42) or Hermann Burmeister's *Geschichte der Schöpfung* (1851) are two of many possible examples to be cited.

Humboldt argued that it was a far superior accomplishment to perceive connections than to study isolated facts. To help readers grasp his holistic concept of the universe, Humboldt spoke of 'Naturgemälde' (painting of nature; inadequately translated by Otté as 'delineation of nature'),[6] adding the element of an aesthetic appreciation of nature to its scientific study. Use of the fine-arts metaphor and of a poetic presentation formed the vehicle for bringing the results of specialized scientific research to the public at large, integrating the study of nature with its human, societal context, giving science a spiritual dimension. It also constituted a truncated discourse of nature, without reference to religious language. Nature was sublime, but not divine, or if it were so, this was not stated with any emphasis. God was not mentioned, and neither was Christianity, at least not in the usual sense as the supreme religion—the victorious, ultimate outcome of the growth and development of civilization. Such a mention of Christian creed was common towards the end of general treatises on 'natural history'. Burmeister, for one, concluded his *Geschichte der Schöpfung* by drawing attention to the 'Verheißung' (religious promise) of Christianity as the coming world religion.[7] By contrast, Humboldt relativized the beliefs of 'Christian Fathers' and 'Hebrew Writers' by intercalating them in

[6] Humboldt, *Cosmos*.

[7] Hermann Burmeister, *Geschichte der Schöpfung*, 3rd edn (Leipzig: Otto Wiegand, 1848), p. 578.

a series of descriptions of nature 'at different Epochs and among different Races of Man' that contained views of nature by the Greeks, the Romans, the Indians, the Persians and, concluding the list, by Goethe.

In considering the role played by Humboldt's last major book, we need to look at the reactions to it, how it was read, and how it was used. If we want to appreciate *Kosmos* for the contribution it made to the dynamics of historical change, it is less important to know what Humboldt actually wrote and what precisely he meant than to explore the impact on its readership. A crucial question is: What did *Kosmos* mean to Humboldt's contemporaries and what did they read in it? One approach to answering this question is a survey of the reviews that were written of Humboldt's last book.

REVIEWS

Across Europe and in the United States substantial reviews in leading periodicals appeared, written by some of the best-known scientists or literati. Reactions to *Kosmos* differed from country to country. In Germany, almost without exception, superlatives of praise were added to expressions of national pride. The *Heidelberger Jahrbücher für Literatur* (1845), for example, in the person of the aged Heidelberg physicist Georg Wilhelm Muncke (1772–1847), raised *Kosmos* on the high pedestal of being beyond the critique of a single learned reviewer: the grand task of providing a description of the universe in the coherence of its parts had previously been attempted in part by the physicists/astronomers Johann Heinrich Lambert (1728–77), Joseph Johann von Littrow (1781–1840), and Pierre-Simon, Marquis de Laplace (1749–1827), yet its completion had been left to Humboldt.

In France, too, Humboldt and his book were strongly praised, although not as unexampled. It was gently pointed out that *Kosmos* contained no new or penetrating insights, nor novel contributions to any scientific specialty, but was a condensation of principal discoveries and theories. In a forty-page essay for the *Revue des deux mondes* (1846), the Parisian zoologist and later holder of the chair of the natural history of man at the Muséum d'Histoire Naturelle, Jean Louis Armand de Quatrefages de Bréau (1810–92), commented that in the individual branches of science on which Humboldt had worked (chemistry, botany, geology, and zoology), he had his superiors in Antoine-Laurent de Lavoisier (1743–94), Antoine-Laurent de Jussieu (1748–1836), Leopold von Buch (1774–1853), and Georges Cuvier (1769–1832), but that as 'physicien du globe' (physicist of the globe) he deserved to be ranked alongside these 'rois d'intelligence' (kings of the mind).

By far the longest reviews, and in the largest numbers, were written in Britain. One reason for this was that reviewing periodicals, in which major essays appeared, in some cases with substantial extracts from the discussed texts, were more numerous in the British Isles than on the Continent. The Scottish physicist and writer David Brewster (1781–1868), writing in the *North British Review*, and his fellow Scot the physicist-cum-geologist and Edinburgh professor of natural philosophy James David Forbes (1809–68), in the *Quarterly Review*, used the occasion to write a brief biography of Humboldt. Brewster's biography of Humboldt was particularly good, and his essay was translated into French, attracting praise also in the German *Magazin für die Literatur des Auslandes*.

Yet not all was praise. In fact, *Kosmos* was criticized severely for Humboldt's omission of God. As the gentleman-scholar John Crosse (1810–80), writing in the Benthamite *Westminster Review*, remarked: 'a sketch of the universe in which the word "God" appears nowhere, but the spirit of God is supposed everywhere, will perhaps be regarded as dangerously Atheistical by the stickler for the word'.[8] To most of the British reviewers, Humboldt's demonstrations of the harmonies and beauty of the physical world required the mention of a Supreme Harmonizer. The absence of 'proofs of divine design' was noted with dismay. Whereas in France the positivistically inclined orientalist-theologian Ernest Renan (1823–92), in his thirteen-page *Kosmos* review for *La liberté de penser*, explicitly praised Humboldt for having avoided the language of natural theology 'as it is understood in England', the British reviewers sorely missed references to 'the power, wisdom and goodness of God as manifested in the creation', as the multi-author *Kosmos*-of-a-sort, the Bridgewater Treatises, had put it.[9] Forbes rebuked:

> We conceive it to be impossible for any well-constituted mind to contemplate the sum and totality of creation, to generalize its principles, to mark the curious relations of its parts, and especially the subtle chain of connexion and unity between beings and events apparently the most remote in space, time, and constitution, without referring more or less to the doctrine of final causes, and to the design of a superintending Providence. We call it the highest pedantry of intellect to put to silence suggestions which arise spontaneously in every mind, whether cultivated or not, when engaged in such contemplations; and we are sorry to observe in the work before us a silence on such topics so pointed as must attract the attention of at least every English reader.[10]

The clergyman-astronomer Thomas John Hussey (1797–1866/7?) in his thirty-eight-page essay for the Congregational *British Quarterly Review* went further, chiding Humboldt not only for the omission of proper references to

[8] [John Crosse], review of *Kosmos: A Sketch of a Physical Description of the World…*, *Westminster Review* 44 (1845): p. 154. This essay combined a review of *Kosmos* and the anonymously published *Vestiges of the Natural History of Creation*.

[9] Ernest Renan, review of *Cosmos*, by Humboldt, *La liberté de penser: Revue philosophique et littéraire* 2 (1848): p. 571.

[10] [J. D. Forbes], review of *Kosmos: Entwurf einer physichen Weltbeschreibung*, by Humboldt, *Quarterly Review* 77 (1845): pp. 163–4.

God but for surreptitiously introducing Hegelian pantheism. This reproach echoed a wider concern that Humboldt might be making common cause with Berlin's Hegelian radicals. Humboldt had shown 'the very height of affectation, or something worse', by completing a treatise on the harmony of the natural world 'without one reference to Him whom Faith recognizes as the Source and the Life of all things'. Hussey did not feel that his review was the place for a conventional natural theology, but, he continued:

> if Baron Humboldt is at liberty to refer us on almost every page to the eternal order and the eternal laws of Nature, it were surely hard that we should be denied all right of reference to Him, the alone eternal, without whose preordination we assert that this order had never been, and without whose co-ordination, these laws had been powerless as the infant's whisper, to direct or control the worlds which hang upon them.[11]

To many of his British critics, it was inconceivable that a popular exposition of science should be without the stated aim that the study of nature leads up to nature's God, and a noticeable feature of the British reviews was that they added the argument of design to *Kosmos*. Thus Humboldt's book was 'domesticated' for an English readership by making it consonant with natural theology.

The British press stood not alone when rebuking Humboldt for failing to raise the banner of orthodox Christianity. Religious conservatives in Prussia were in fact deeply disappointed that the author of *Kosmos* did not join the victorious battle for the Christian revelation as foundation of German unity in church and state. Also on the Continent, especially in Catholic-conservative circles, accusations were made of Hegelian radicalism or heterodoxy with respect to the biblical creation story, and the question was disquietly posed: 'Does the author of *Cosmos* ever talk about God?'. The *Journal historique et littéraire* (1858) answered in the affirmative, yet the passages cited in proof from *Kosmos* demonstrated that Humboldt in fact avoided using the name God, merely referring to 'a first impulse'.[12]

Humboldt read the British reviews with great interest, but pointed out that the accusations of atheism were unfounded. Already in January 1845, when personally translating for the French edition of *Kosmos* the introductory 'Considérations' (reflections), he added a self-defence in the form of a reference to 'the very Christian Immanuel Kant' and in the English version stated:

> The ultimate object of the experimental sciences is ... to discover laws, and to trace their progressive generalization. All that exceeds this goes beyond the province of the physical description of the universe, and appertains to a range of higher speculative views. Emanuel Kant, one of the few philosophers who have escaped the

[11] [T. Hussey], review of *Kosmos: A General Survey of the Physical Phenomena of the Universe*, by Humboldt, *British Quarterly Review* 6 (1846): p. 354.

[12] 'L'Auteur du Kosmos ne parle-t-il jamais de dieu?', *Journal historique et litteraire* 24 (1857–8): pp. 493–7.

imputation of impiety, has defined with rare sagacity the limits of physical explanations, in his celebrated essay On the Theory and Structure of the Heavens, published at Königsberg in 1755.[13]

For Humboldt to keep natural theology out of *Kosmos* was unlikely to have been a matter of unbelief or atheistic sympathies, but reflected a conviction that an expository scientific treatise was not the proper place for the argument of design.

The language of natural theology played a mediating role in bringing people of different Christian creeds together in a latitudinarian pursuit of science, but this applied to the English-speaking world, not to continental Europe. In a German-speaking world of 'cuius regio, eius religio', religion was a force of particularism and, with that, of autocratic rule. If science was to act as a force of national unification and freedom, religion had to be kept out of the picture. The mediating mission of *Kosmos* was first and foremost at the level of socio-political divisions, nationally and internationally. An American commentator, James Davenport Whelpley (1817–72), in an essay on *Kosmos* in the Whig *American Review*, commented that those who studied the world and contemplated the connections of its multifarious phenomena were knitted together in a cosmopolitan network, 'making common cause against ignorance and prejudice. If the world is ever to be harmonized, it must be through a community of knowledge, for there is no other universal or non-exclusive principle in the nature of man'.[14] London's *Athenaeum*, which had previously criticized the Bridgewater Treatises, used similar language in a review of volume 2 of *Kosmos*:

> Never since the world began has there been any epoch so marked as the present by the wonderful application of the powers of nature to the want of man. We hold the key by which we may lock in one common brotherhood all the nations of Europe—and finally the world; making peace the universal desire and the interchange of thought the universal instinct of every people.[15]

People who were scientifically literate and knowledgeable about the orderly arrangement of the world would rise above party division. *Kosmos* could thus become an argument from nature in the promotion of liberal, social, and political causes. Those who disapproved of revolutionary chaos on the one hand or reactionary despotism on the other, in the sequence of events from the French Revolution to the revolution of 1848, could take up the *Kosmos* banner. As Humboldt stated at the beginning of volume 1, nature is the realm of freedom. By implication, the proper study of nature would lead to liberty, away from religious and political absolutism and oppression.

[13] Humboldt, *Cosmos*, vol. 1, p. 50.

[14] James Welpley, 'Humboldt's *Cosmos*', *American Review* 3 (1848): p. 603.

[15] Review of *Cosmos: Sketch of a Physical Description of the Universe*, by Humboldt, trans. E. Sabine, *Athenaeum* (12 February 1848), p. 162.

DISCOURSES OF 'GEOLOGY AND GOD'

Let us map out in a general, systematic manner Humboldt's omission of God from the study of the cosmos and the various reactions to this. One way of doing so is to narrate the history of the relationship between 'Bible' and 'geology', including 'geology without God', as a story of multiple discourses that have existed alongside each other for much of the period from 1750 to today. 'Discourse' is used here in a social theory sense, denoting a coherent set of contentions that generates its own regime of validity inside a particular constituency with distinct socio-political values. Some five discourses can be recognized, each identifiable by a particular hermeneutic strategy in dealing with the Bible and geology, four of them taking shape at the time geology emerged as a scientific discipline.[16]

Humboldt's *Kosmos* approach, which owed much to Immanuel Kant (1724–1804), became the leading discourse. It considered the relationship between the Bible and science as unproblematic for the simple reason that it is non-existent—there is no meaningful connection between the two. Bible and geology do not clash, because they share no common ground, but have separate spheres of validity, the one of moral conduct, the other of physical reality. 'Hebrew scripture' has no bearing on the modern study of the physical world or the other way around. Bible and science each belong to wholly separate domains of reality. The Bible pertains to the sphere of morality and spirituality; science, by contrast, rules when it comes to physical things. Biblical descriptions of the natural world may have a poetical quality, but the belief that these passages should contain divine revelations about the natural world in accord with modern scientific discoveries and theories is absurd. If people wish to see in these poetical passages of the Old and New Testaments some elements of divinity, inspiration, and revelation, that is admissible; but inerrancy and literal truth are out of the question, as the Bible bears many imperfections that mark its origins as an historical document.

Little if any reference to the Bible occurs in Humboldt's voluminous scientific oeuvre, including *Kosmos*, as we have seen above, except that he discussed the Bible, especially the Old Testament, for its poetic descriptions of nature. Damning with faint praise, Humboldt downplayed the importance of the Bible by extolling its qualities as 'Hebrew poetry'.[17]

German Jews appreciated Humboldt's respect for the literary quality of the Old Testament, but orthodox Christians raised the alarm: the great, trendsetting scientist was removing God and religion from the scientific study of nature.

[16] See Nicolaas Rupke, 'Five Discourses of Bible and Science 1750–2000', in *A Master of Science History: Essays in Honor of Charles Gillispie*, ed. Jed Z. Buchwald (Dordrecht: Springer, 2012), pp. 179–96.

[17] Humboldt, *Cosmos*, vol. 2, p. 57.

As also mentioned above, Humboldt defended himself by appealing to the example set by 'the very Christian Immanuel Kant',[18] who, in his *Allgemeine Naturgeschichte und Theorie des Himmels* (1755), had separated science from scripture, making no reference in his cosmogonical treatise to the Mosaic cosmogony and relegating God-talk to the sphere of metaphysics. Kant's authority and example were appealed to by other scientific Christians who, while continuing to hold the Bible in respectful esteem, dissociated their scientific work from scripture.

In Great Britain, a leading representative of the 'separate spheres' view was Baden Powell (1796–1860), the Anglican clergyman and Savilian Professor of Geometry at Oxford. Powell addressed the issue of the Bible and science in a series of publications, from his *Revelation and Science* (1833) to *The Unity of Worlds and of Nature* (2nd edn, 1856). The Bible is a source of moral and spiritual intimations 'which are, in their essential nature, alien from physical consideration', Powell insisted.[19]

This discourse merged with higher criticism of the Bible. Already around 1800, Lutheran theologians in Germany had criticized the Mosaic geologists for taking the Genesis account of creation and flood literally. *Moses und David keine Geologen* (Moses and David [were] no geologists) (1799), declared the title of a book by the Helmstedt and later Göttingen theologian David Julius Potter (1760–1838). Genesis 1 was to be bracketed with Psalm 104 and represented a 'Schöpfungshymnus', a creation hymn, the main purpose of which had been the ordination of the sabbath week. In a later critique, too, the original purpose of the creation story was not to give an account of how the world came about, but to assert the monotheistic view and fight polytheism.[20] More radical than the hermeneutic revisions that were forced on many Bible believers by science, were those from this critical tradition within theology, leaders of which ranged from Potter's older Göttingen colleague Johann Gottfried Eichhorn (1752–1827) to the latter's later successor Julius Wellhausen (1844–1918), both orientalist-theologians. Whereas geology appeared to remove a literal meaning from Genesis, the documentary hypothesis of the historical school in biblical studies went further, reducing the entire Pentateuch from a unitary record of divine revelation to a product of historical change, cobbled together from a variety of pre-existent sources and repeatedly altered in a process of editorial changes. The Creation and Deluge stories, for example, were traced back from the Bible to Babel. Scripture appeared diminished from

[18] Humboldt, *Cosmos*, vol. 1, p. xxv.

[19] Baden Powell, *The Unity of Worlds and of Nature: Three Essays on the Spirit of the Inductive Philosophy; the Plurality of Worlds; and the Philosophy of Creation* (London: Longman, Brown, Green, Longmans, and Roberts, 1856), p. 300.

[20] Eckart Otto, 'Auf welche Fragen antwortet eine antike Schöpfungstheologie im alten Orient und in der Bibel? Die Falle des Kreationismus', in *Evolutionstheorie und Kreationismus—ein Gegensatz*, ed. Otto Kraus (Stuttgart: Franz Steiner Verlag, 2009), pp. 17–26.

a divinely revealed, inerrant account of the grand scheme of the world—past, present, and future—to a fallible product of human contrivance, in part plagiarized from pagan sources.

Higher criticism did not become a topic of major public debate in Britain before the appearance of the *Essays and Reviews* (1860),[21] questioning the Mosaic authorship and with that, the authenticity of the Pentateuch as a divinely inspired account of history. Powell was one of the seven members of the Established Church who authored *Essays and Reviews*, one of the 'Septem contra Christum', as a troubled Cantabrigian called the team of authors. In his contribution, 'On the Study of the Evidences of Christianity', Powell reiterated that the 'region of spiritual things' and the 'domain of physical causation' were two unrelated spheres.[22] Biblical miracles, because they go against the laws of physics, are not believable, and Powell pleaded for a Christianity without miracles. Biblical stories that are claimed to be historical but contradict scientific fact must be transformed into truths taught by parables. A fellow contributor, the lawyer and Egyptologist Charles W. Goodwin (1817–78), insisted that neither the literal interpretation of Genesis nor the figurative one does justice to the nature of the 'Hebrew cosmogony'. It is not a divine 'but a human utterance, which it has pleased Providence to use in a special way for the education of mankind'.[23]

This discourse, by its very nature, was more practised than written about, and Powell was somewhat of an exception publicly to express his convictions about the separateness of Bible and science. Others did in their work what Powell put in writing. Among them were such Christian giants of Victorian science as John Herschel (1792–1871), Charles Lyell (1797–1875), and Richard Owen (1804–92). Tellingly, none of these religious men of astronomy and geology consented to having their name put to the so-called Scientists' Declaration, which in response to *Essays and Reviews* affirmed the essential harmony between the Holy Writ and physical science. Owen, for one, when asked for his endorsement of the Declaration, declined by retorting that modern science, not scripture, was the means by which God revealed natural truth.[24]

The Humboldtian discourse, although mainstream, was by no means the only one. More crudely dismissive of God was the atheistical and materialist approach. Its representatives asserted that the Bible is a ragbag of antiquated stories and in part harmful notions from which geology helps set us free. Already some eighteenth-century deist and atheist adversaries of physico-theology

[21] Important, too, was the bishop of Natal, John William Colenso's, *The Pentateuch and the Book of Joshua Critically Examined* (London: Longman, 1862).

[22] Baden Powell, 'On the Study of the Evidences of Christianity', in *Essays and Reviews*, ed. John William Parker, 10th edn (London: Longman, Green, Longman, and Roberts, 1862), p. 152.

[23] Charles W. Goodwin, 'On the Mosaic Cosmogony', in *Essays and Reviews*, p. 305.

[24] Nicolaas Rupke, *Richard Owen: Biology Without Darwin* (Chicago: University of Chicago Press, 2009), p. 245.

wrote with contempt about the Bible and, by implication, God. Of later nineteenth-century authors, writing disparagingly about scripture, a few examples may suffice. Carl Vogt (1817–95), the geologist, zoologist, materialist, and revolutionary politician, known for his participation in the Revolution of 1848, systematically denounced and ridiculed the Mosaic hexaemeron, the story of Noah's Ark, and especially, too, the unity of humankind in Adam and Eve.[25]

An example from Britain was Joseph McCabe (1867–1955) who, having left the Catholic priesthood, became a freethinker and atheist who not only rejected the literalist and the harmonist interpretations of the Bible (see below), but higher criticism as well. In one of his many publications, *The Bankruptcy of Religion* (1917), he argued that the approach to the Bible by Eichhorn and his followers was merely a ploy by liberal theologians to save what could be saved from the sinking ship of scripture. Higher criticism had been a strategy to preserve religion and theology from destruction by the unstoppable march of truth, giving Christianity an opportunity to hold on to the Bible by dealing with it in a more rational, acceptable manner. The Old Testament and its early books were childlike and demonstrably wrong, with 'numerous palpable blunders and inconsistencies'. The churches and the clergy 'imposed upon ignorant Europe a colossal delusion' of the Bible as a supernatural document. Moses not being the inspired author of the Pentateuch, the position of Christ is directly affected, being reduced to that of 'a human and fallible person'.[26]

It would be a mistake to assume that these two discourses, although shared by many of the leading scientists of the late-modern period, led to the complete untangling of the secular from the sacred in late-modern geology. Religious scientists, among them reputable geologists, stood their ground by persisting with Bible-and-science discourses that connected geological history to God. The older of these discourses has treated the Bible as a guidebook for geology, continuing the early-modern 'origines sacrae' tradition that was never fully abandoned but has survived until today in the form of young-earth creationism. The Bible is divinely inspired, literally true, and the textual passages that deal with the natural world are imperatively valid for science. Most important to the literalist view have been creation (Gen. 1–2), the flood (Gen. 6–8), and the age of the world as calculated on the basis of the genealogies of the ante- and post-diluvial patriarchs (Gen. 9–11). On the validity of these signposts to the early history of the world depends, they believed and believe, the entire scheme of a divinely guided, eschatological history and Christian soteriology. As the English clergyman-naturalist and convert to Methodism Joseph Townsend (1739–1816) commented: 'The science of geology becomes of infinite importance, when we consider it as connected with our immortal

[25] Carl Vogt, *Köhlerglauben und Wissenschaft: eine Streitschrift gegen Hofrath Rudolph Wagner in Göttingen* (Giessen: Ricker, 1855).

[26] Joseph McCabe, *The Bankruptcy of Religion* (London: Watts & Co., 1917), pp. 137, 141–2, 145.

hopes. These depend on the truth of revelation, and the whole system of revealed religion is ultimately connected with the veracity of Moses.'[27]

Another, less literalist discourse has argued that scripture is adjustable. Apparent discrepancies between the Bible and geology and science disappear if we interpret certain biblical texts the right way, not in a literal but a figurative sense. During the two or three decades following the French Revolution, some of the Protestant leaders of the new historical geology created space for their science by reinterpreting the creation and deluge stories in a non-literal sense. Already during the second half of the eighteenth century, the Swiss Calvinist and geological naturalist Deluc helped shape a different discourse from the literalist—one that accepted modern science, yet also stuck to a belief in the Bible as God's word.[28] Apparent discrepancies between scripture and science can be resolved by taking relevant biblical expressions and stories in the right sense, in many instances metaphorically. The Bible is scientifically accurate, if only we interpret the texts correctly. Scientific Christians and theologians alike, concerned that the book of nature not be at variance with scripture, put forward a variety of harmonization schemata.[29]

A genre of literature developed that dealt with the congruence of the Bible and science. Not uncommonly, geological textbooks would include a chapter on how to reconcile the new earth history with the biblical accounts of creation and deluge. The Yale University scientist Benjamin Silliman (1779–1864) added a lengthy supplement on 'Consistency of geology with Sacred History' to the American edition of Robert Bakewell's *An Introduction to Geology* (1833). Edward Hitchcock (1793–1864), president of Amherst College and professor of natural theology and geology, included in his frequently reprinted *Elementary Geology* (1840) a chapter on the 'Connection between geology and natural and revealed religion'. James D. Dana (1813–95), Silliman's student and successor at Yale as professor of natural history and geology, appended to all four editions of his authoritative *Manual of Geology* (1863; 1874; 1880; 1895) a Genesis and geology harmonization scheme.

The disappearance of Bible and God from the study of rocks and fossils has for many decades been discussed by historians of science and religion under the heading provided by the title of Charles Gillispie's classic study, *Genesis and Geology*, which depicted this eclipse as a form of progressive secularization.

[27] Joseph Townsend, *The Character of Moses Established for Veracity as an Historian, Recording Events from the Creation to the Deluge* (London: Longman, Hurst, Rees, Orme and Brown, 1813), p. 430.

[28] Marita Hübner, *Jean André Deluc (1727–1817): Protestantische Kultur und moderne Naturforschung* (Göttingen: Vandenhoeck & Ruprecht, 2010).

[29] Nicolaas Rupke, 'Christianity and the Sciences, 1815–1914', in *The Cambridge History of Christianity*, vol. 8: *World Christianities, c. 1815–c. 1914*, ed. Sheridan Gilley and Brian Stanley (Cambridge: Cambridge University Press, 2006), pp. 164–80.

Secularization theory states that society increasingly has been turning its back on religion and that science is the single most powerful force in bringing such change about. Keith Thomas, Owen Chadwick, and various others long ago reaffirmed the belief in progressive secularization, which they defined in terms of a decline of belief in miracles. More recently, this view has been much criticized, and secularization theory revised. Ron Numbers, for one, states: 'I refer to the virtual disappearance of religious language from professional literature as privatization'.[30]

Through the nineteenth century, the traditional mention of Bible and God largely, but not wholly, disappeared from the language of professional geology. This was not simply due to the secularizing influence of scientific studies of earth and planetary science; rather, an 'Ent-theologisierung' (removal of theology) took place, when 'the book of nature' and 'the book of scripture' each were assigned their own, professional domain: one scientific, the other theological. But while adopting methodological naturalism in their publications, many geologists privately held on to their religious beliefs.

[30] Ronald L. Numbers, 'Science, Secularization, and Privatization', in *Eminent Lives in Twentieth-Century Science and Religion*, ed. Nicolaas Rupke, 2nd edn (Frankfurt: Peter Lang, 2009), pp. 349–62.

11

Naturalizing the Bible

The Shifting Role of the Biblical Account of Nature

Scott Gerard Prinster

In 1859, nearing the end of his life, the New England Unitarian minister and abolitionist Theodore Parker reminisced about the many years he had spent studying German innovations in biblical scholarship, marvelling that 'Germany [was] the only land where theology was... studied as a science, and developed with scientific freedom'.[1] Worrying that American theologians were failing to engage with the increasingly scientific spirit of their age, Parker had promoted these radical approaches among his New England colleagues in the late 1830s and 1840s, delighting in the detached academic rigour of the Germans, who seemed unencumbered by traditional concerns about leaving doctrine and everyday practice undisturbed.[2]

It is tempting for historians of science to disregard religious activity in the same way that we once dismissed alchemy and astrology: religious concerns strike us as being not only non-science, but anti-science. Beginning in the 1860s and 1870s, popular authors and speakers began to characterize the relationship between scientific knowledge and religious knowledge as one of inescapable conflict. However, as twentieth-century historians of science came to recognize religious scholarship as an intellectual activity overlapping scientific questioning, rather than only as its nemesis, we came to acknowledge how

[1] Theodore Parker, 'The Letter from Santa Cruz, Called "Theodore Parker's Experience as a Minister"', in *Life and Correspondence of Theodore Parker, Minister of the 28th Congregational Society, Boston*, vol. 2, ed. John Weiss (London: Longman, Green, Longman, Roberts, and Green, 1864), p. 462.

[2] Dean Grodzins, *American Heretic: Theodore Parker and Transcendentalism* (Chapel Hill, NC: University of North Carolina Press, 2002), p. 70.

valuable it might be to clarify the role of religious thought in the historical development of science. Theodore Parker and his promotion of 'scientific' biblical scholarship is one particularly vivid example of an influential religious leader who did not view the shifting American intellectual landscape in terms of conflict, but as an opportunity to recognize scriptural knowledge as essentially compatible with scientific assumptions and methods.

This chapter will examine the field of *biblical criticism*, which draws upon a variety of interpretive methods to determine the authenticity, authorship and editing, and relevance of the Old Testament (Hebrew scriptures) and New Testament (Christian scriptures). Biblical criticism is not a modern approach, but became an apparent necessity even as the leaders of the early Christian Church were first constructing the official canon of the Bible; for example, the Assyrian theologian Tatian (*c*.120–180 CE) tried to reconcile the very different narratives of the four Gospels into a single story about the life and teachings of Jesus.[3] I will focus here on approaches to biblical interpretation characterized by scholars and laypeople as 'scientific', that is, appropriating scientific methods, assumptions, and vocabulary, or otherwise claiming 'scientific' results.

Exploring the appearances of naturalism in the history of biblical criticism helps us to highlight intellectual activity that is often poorly represented in the history of the canonical sciences. Contrary to the popular triumphalist view, in which science advances at the expense of religious claims and authority, a considerable number of scholars and laypeople alike have engaged in scientific criticism of the Bible texts and interpretation of the natural world from a religious viewpoint. When we chronicle this activity throughout the nineteenth century, we see a number of emerging expressions of naturalism, that is, tentative explanations of phenomena in terms of natural factors alone. Within religious circles in particular, we also see antithetical uses of the language of naturalism among proponents and opponents of scientific biblical criticism. When conservative opponents reviled these critical approaches as 'naturalism' they were criticizing them as *metaphysical naturalism*, the complete denial of supernatural entities. However, actual metaphysical naturalists did not typically participate in the scientific interpretation of the Bible, since they considered its usefulness negligible. Rather, the scientific biblical criticism circulating among sympathetic Protestant scholars and laypeople produced a range of subtle and often implicit expressions of *methodological naturalism*, the description of natural phenomena solely in terms of natural forces.[4]

[3] Peter Byrne and Leslie Houlden (eds), *Companion Encyclopedia of Theology* (London: Routledge, 1995), p. 111.

[4] On metaphysical and methodological naturalism, see Paul Draper, 'God, Science, and Naturalism', in *The Oxford Handbook of Philosophy of Religion*, ed. William J. Wainwright (Oxford: Oxford University Press, 2005), p. 279.

SCRIPTURE AS SELF-EVIDENCING KNOWLEDGE

By the eighteenth century, scholars had generally settled on a number of basic principles about the proper interpretation of biblical knowledge. First, although a few radical critics had suggested that the Bible's various books and passages differed in authenticity and value, scholars in general considered scripture a divinely inspired *unity*, such that one text could be clarified or interpreted by referring to other texts. Second, although the Old Testament and the New Testament were put into written form hundreds of years apart, Christian doctrine reconciled their differing styles and emphasis by linking them as prophecy and fulfilment. Third, scholars could generally avail themselves of three accepted methods of interpreting the Bible: an allegorical approach, in which readers searched the text for symbolic messages; a moral approach, in which the texts communicated lessons for individual or social conduct; and a doctrinal approach, in which the narrative was interpreted to justify the Church's teachings. It is important for modern readers to recognize that the claimed truth of the canon as a whole was not identical to the claims we are accustomed to in today's debates over biblical authority; the philosophically rigorous assertion that each passage in isolation is 'literally true' did not become common until the late nineteenth century.[5]

These principles give us a preliminary understanding of how scholars traditionally considered the Bible texts a special form of knowledge and approached their interpretation differently than that of literary and historical documents, and why the application of scientific principles to biblical criticism has been a controversial endeavour. The dominant interpretive principle, developed in the early modern period, asserted that scripture is *self-evidencing*. Bible commentaries as early as the Protestant Reformation declared that scripture could be read sufficiently by its own light. Although orthodox scholars might also be willing to consider external evidence such as geological discoveries that reassured them of God's rational and benevolent nature, it was understood that the most crucial source of evidence for the truth and authority of the Bible was the Bible itself. Protestant leader John Calvin called the biblical accounts *autopiston* (αὐτόπιστον), or self-authenticating: 'Scripture, carrying its own evidence along with it, deigns not to submit to proofs and arguments, but owes the full conviction with which we ought to receive it to the testimony of the Spirit.'[6] In addition to *sola scriptura*, the doctrine that all other forms of knowledge were subordinate to the Bible, it was this principle of self-authentication that placed Protestantism in an especially complicated relationship with

[5] Hans W. Frei, *The Eclipse of Biblical Narrative: A Study in Eighteenth and Nineteenth Century Hermeneutics* (New Haven, CT: Yale University Press, 1974), pp. 1–3.

[6] John Calvin, *Institutes of the Christian Religion* (London: Bonham Norton, 1599), bk. 1, chap. 7, section 5.

science. Both Catholicism and Judaism acknowledge the authority of other forms of religious knowledge, such as Church tradition and rabbinic interpretation respectively, and have typically resolved their apparent conflicts with scientific claims more easily than Protestantism. This essay will focus on the particularly interesting debates over naturalistic biblical criticism that arose in the interactions between American Protestantism and natural knowledge in the nineteenth and early twentieth centuries.

EARLY NATURALISM AND EARLY BIBLICAL CRITICISM

American scholars of nature first began to exclude supernatural interpretations from their studies in the late eighteenth and early nineteenth centuries, albeit gradually and inconsistently at first. Particularly in the cases of meteorological and medical phenomena—lightning, earthquakes, and sickness, for example—observers were by this time commonly omitting supernatural factors from their explanations without comment.[7] One common way of introducing a naturalistic interpretation was simply to sidestep its potentially controversial nature; authors often described puzzling phenomena in terms of natural causes with no discussion of alternative supernatural explanations. When their natural writings invoked the Bible at all, it was increasingly as a source of historical information rather than as testimony of supernatural influences.[8]

British and American biblical scholars in the early nineteenth century widely acknowledged that the most innovative and significant research was being done in universities in Germany, which some celebrated as the 'Athenaeum of modern times'.[9] A number of social and political developments in the German Confederation helped to transform their provincial approaches to higher education, producing a system of modern universities conducive to advances in biblical research, and attracting influential scholars from abroad. Prussian reformers opened schools under the control and support of the state, rather than of the church. German scholars were also not required to abide by doctrinal statements, as their American and British counterparts were.[10] Innovative

[7] Ronald L. Numbers, 'Science without God: Natural Laws and Christian Beliefs', in *When Science and Christianity Meet*, ed. David C. Lindberg and Ronald L. Numbers (Chicago: University of Chicago Press, 2003), pp. 269–72.

[8] For example, see *The Medical Repository*, the first American journal of proto-science, founded in 1804.

[9] 'Account of the Unitarian Churches in Transylvania', *Christian Examiner and Theological Review* 2 (1825): p. 262. On this point, see also D. Young, 'Essays and Dissertations in Biblical Literature', *Princeton Review* 2 (July 1830): pp. 324–5.

[10] Victor Shea and William Whitla, *'Essays and Reviews': The 1860 Text and its Reading* (Charlottesville, VA: University Press of Virginia, 2000), p. 55.

biblical scholarship could thus be conducted in Germany with less concern for the everyday practice of faith than in other Western nations.[11]

The first signs of naturalism in biblical interpretation appeared in the early rationalist and historicist approaches of liberal German scholars. These innovators enthusiastically read and absorbed rationalist Enlightenment works, notably the British philosopher John Locke's *Paraphrase and Notes on the Epistles of Paul* (1707). A number of theologians, including Johann Gottfried Eichhorn (1753–1827), Johann Gottfried von Herder (1744–1803), Johann David Michaelis (1717–91), and Johann Jakob Griesbach (1745–1812), incorporated this rationalist spirit into their own work around the turn of the nineteenth century and began to question the traditional interpretation of the Bible texts.[12] Another prominent development was the introduction of historical criticism, most notably by a mid-nineteenth-century network of scholars known as the Tübingen School.[13] This circle of specialists in biblical scholarship, who had largely either studied or taught at the University of Tübingen, began to coalesce around theologian Ferdinand Christian Baur (1792–1860) in the early 1830s. Baur adopted the philosophy of history of Georg Hegel, who depicted the flow of time as a struggle between opposing forces, inexorably generating the spirit of reason. Challenging the supernaturalist tradition then dominating the Tübingen faculty, Baur and his colleagues promoted a rational and historical (*historisch-kritische*) approach to the Bible, interpreting each text separately as a product of its particular time and place, and as reflecting the often uneasy compromise among dissenting factions in the early Christian Church.[14]

Histories of biblical criticism generally classify German biblical interpreters as either supernaturalists or rationalists, based on how the scholars reconciled the Bible and natural evidence. The supernaturalists insisted on the literal truth of scriptural claims but could no longer maintain this argument merely on the traditional principle of self-evidence—the rationalist and historicist turn in German biblical studies compelled even conservative theologians to satisfy scholarly standards for philosophical rigour and historical factuality.[15] The supernaturalist camp was exemplified by theologian Ernst Hengstenberg (1802–69), whose traditionalist position dominated German theological activity before Baur began to promote the historical-critical movement. Hengstenberg led the orthodox faculty at Berlin and edited the anti-rationalist theological

[11] Andreas W. Daum, '*Wissenschaft* and Knowledge', in *Germany, 1800–1870*, ed. Jonathan Sperber (Oxford: Oxford University Press, 2004), pp. 137–61.

[12] Jerry Wayne Brown, *The Rise of Biblical Criticism in America, 1800–1870: The New England Scholars* (Middletown, CT: Wesleyan University Press, 1969), pp. 15–23.

[13] August Tholuck, 'Theological Encyclopaedia and Methodology', trans. Edwards A. Park, *Bibliotheca Sacra* 1 (1844): p. 356.

[14] Henning Graf Reventlow, *History of Biblical Interpretation: From the Enlightenment to the Twentieth Century*, trans. Leo G. Perdue (Atlanta, GA: Society of Biblical Literature, 2010), pp. 247, 277.

[15] Reventlow, *Biblical Interpretation*, pp. 286–7.

journal *Evangelische Kirchen-Zeitung* from the late 1820s. In addition to having significant influence through this periodical, Hengstenberg's work also attracted considerable attention overseas, especially among conservative Reformed scholars in the United States. Hengstenberg's successful influence can be attributed partly to the credible explanations that he gave to apparent contradictions in the Pentateuch. He also distinguished himself as the first biblical scholar to make use of the considerable body of new archaeological knowledge emerging from Napoleon's military campaigns in Egypt.[16] The rationalists, on the other hand, demanded that even biblical narratives appear reasonable and be consistent with one another and with nature. The most conspicuous representative of this group was Heidelberg biblical scholar Heinrich Paulus (1761–1851), who interpreted miraculous claims in the Bible by explaining them in purely natural terms such as meteorological phenomena.[17] In general, however, rationalist scholars were not ready to advocate fully naturalistic interpretations, but promoted instead more moderate positions.

Both of these camps found it difficult to maintain their influence amid the rapid developments in German scholarship. The supernaturalists risked irrelevancy by refusing to accommodate new evidence from the natural sciences, and rationalist arguments were generally unable to withstand Immanuel Kant's influential critique that unaided reason was not a defensible intellectual position.

The most influential representatives of this German theological ferment were the historicist innovator Wilhelm Martin Leberecht de Wette (1780–1849) and Baur's student David Friedrich Strauss (1808–74). De Wette's work—both in German and in English translation—found particular welcome among New England liberals, who eagerly adopted his reconciliation of Christianity and reason. Rather than embrace the Enlightenment tendency toward secularism, he strove to harmonize rationalism and supernaturalism in a way that incorporated elements of both.[18] Historian Robert Pfeiffer singled out De Wette's *Beiträge zur Einleitung in das Alte Testament* (Contributions on the introduction to the Old Testament, 1806–7) as the most important biblical criticism in the first half of the century, successfully disseminating a moderate stance between traditional interpretation and naturalism.[19] Strauss, on the other hand, rejected both supernaturalism and rationalism as ultimately two expressions of the same failing: the assumption that rational consistency was an essential condition of modern knowledge. He did not consider the higher truth of the Bible subject to the standards of rational interpretation, but rather encouraged scholars to

[16] Reventlow, *Biblical Interpretation*, pp. 297–8. Also, Alexander J. Schem, 'Hengstenberg and his Influence on German Protestantism', *The Methodist Review* 44 (January 1862): pp. 108–28.

[17] Reventlow, *Biblical Interpretation*, pp. 204–10, 286–98.

[18] Siegfried B. Puknat, 'De Wette in New England', *Proceedings of the American Philosophical Society* 102, no. 4 (August 1958): pp. 376–95.

[19] Robert H. Pfeiffer, *Introduction to the Old Testament* (New York: Harper, 1948), p. 47, quoted in Brown, *Rise of Biblical Criticism*, p. 164.

tolerate the tensions between scriptural and scientific knowledge. Strauss's solution to the problem of biblical miracles was to interpret them as the mythology of more primitive cultures, a position that he made especially prominent in his 1835 volume *Das Leben Jesu* (The Life of Jesus). Even more influential abroad than the Tübingen School's historical-critical approach, this work analysed Christianity's enduring message in terms of historical change and natural evidence.[20]

GERMAN BIBLICAL SCHOLARSHIP ARRIVES IN NEW ENGLAND

Although many orthodox American scholars in the early nineteenth century ultimately found German biblical criticism too radical in its approach and too dangerous in its conclusions, the boldness and philosophical rigour of this scholarship attracted liberal and conservative students alike. One prominent example was Moses Stuart (1780–1852), Andover Professor of Sacred Literature, widely considered the father of American biblical criticism and, for a conservative scholar, unusually knowledgeable about German critical literature. In fact, Stuart's mastery of the German language raised suspicions about his own orthodoxy: 'It was whispered, that I was not only secretly gone over to the Germans, but was leading the Seminary over with me, and bringing up, or at least encouraging, our young men to the study of deistical Rationalism.'[21] Stuart and a small group of like-minded traditionalists nevertheless insisted on the value of German biblical scholarship, studied cautiously. Baptist theologian Barnas Sears, who edited the intellectually ambitious quarterly *The Christian Review*, concurred: 'He who will not study a German book cannot be a critical student of the original Scriptures.'[22]

Perhaps the most influential importer of German scholarship to the United States was the young Unitarian minister Joseph Stevens Buckminster (1784–1812), whose interest in the rational criticism of the Bible led him to assemble one of the finest theological libraries in New England. He arranged for the publication of an American edition of Griesbach's New Testament criticism in 1809—the first scholarly treatment of the Bible in the United States—which was promptly adopted as a textbook at Harvard University. Buckminster's scholarship was promoted in both the magazine *Monthly Anthology* and the theological journal *General Repository and Review* such that,

[20] Reventlow, *Biblical Interpretation*, pp. 245–62.

[21] Moses Stuart, letter to the editor, *The Christian Review* 6, no. 23 (September 1841): p. 455.

[22] Barnas Sears, 'German Literature: Its Religious Character and Influence', *The Christian Review* 6, no. 22 (June 1841): p. 272.

when the clergyman died in 1812, his collection of German texts was already well known, and many scholars and students gathered to bid for a share of the three thousand volumes.[23]

This growing enthusiasm for German scholarship did not pass unnoticed by conservative critics, who expressed concerns about the repercussions of such radical ideas. Based on his thorough knowledge of German language and scholarship, Stuart warned in 1819 that the greatest extant source of heretical interpretation was Germany, which 'in half a century, has produced more works on criticism and sacred literature than the world contains besides'.[24] Princeton theologian Samuel Miller (1769–1850), an ardent participant in the internal controversies of the Presbyterian Church, implicated the newly formed Unitarian denomination as the primary culprit in importing these destructive German writings into the United States. Miller published in 1821 a collection of essays harshly critical of Unitarianism, accusing adherents, among other things, of 'standing on German ground' in their disparagement of 'the Divinity of Christ, and of the Trinity in Unity... as among the most pestiferous of errors' and their 'revolt at the suggestion of anything mysterious in religion'.[25] The Unitarian response to Miller's accusations was perhaps not as convincing as it might have been:

> We should as little think of attributing the *naturalism* (as it is called) of the German critics and theologians to their rejection of the doctrine of the Trinity, as of attributing the infidelity of the philosophers and many of the dignified clergy in France before the revolution, to their rejection of the doctrine of transubstantiation. We do not... think their theories and arguments so plausible and seductive, that much is to be feared from a more general knowledge of their writings.[26]

Whatever merit Miller's allegations and the Unitarians' denial had, American scholars in the 1820s were already giving increasing attention to the possibility of naturalistic interpretations of the Bible.

The Unitarians' unruffled response to Miller's denouncement was not representative of American theological journals, which were beginning to give continuous attention to German biblical scholarship by about 1825. Although a few scholars in the Presbyterian journals *Biblical Repertory* and *Bibliotheca Sacra* reported these new interpretive approaches with dispassionate interest, others began to express alarm. 'This mode of interpretation is finally dangerous and pernicious to Divine Truth', warned Lutheran pastor Charles Christian Tittman, 'the advocates of the historical mode of interpretation, are also the

[23] Brown, *Rise of Biblical Criticism*, pp. 23–4, 27–9.

[24] Moses Stuart, 'Letters to Dr. Channing on the Doctrine of the Trinity', in *Miscellanies* (Andover, MA: Allen, Morrill and Wardwell, 1846), p. 177.

[25] Samuel Miller, *Letters on Unitarianism; Addressed to the Members of the First Presbyterian Church, in the City of Baltimore* (Trenton, NJ: George Sherman, 1821), pp. 55, 71.

[26] William Ellery Channing, quoted in 'Strictures on Extracts from Dr. Miller's Letters', *Christian Register* 1, no. 50 (26 July 1822): p. 198.

earnest asserters of the system Rationalism'.[27] Still others approvingly cited influential German supernaturalists such as Hengstenberg and Tholuck in defence of orthodoxy, further increasing the visibility of the struggles over biblical naturalism among their colleagues.[28]

In the wake of Buckminster's untimely death, his Unitarian colleague Theodore Parker (1810–60) emerged as the American scholar most enthusiastic about the possibility of scientific biblical criticism. His first serious engagement with biblical scholarship came while he was a divinity student at Harvard. As one of the editors of the monthly journal *Scriptural Interpreter*, Parker was instrumental in exposing his New England colleagues to the work of Eichhorn, Michaelis, Paulus, Strauss, De Wette, and other significant German scholars. His fascination with the 'scientific' biblical scholarship of these figures led him to amass more than thirteen thousand volumes, most in foreign languages, and which Thomas Wentworth Higginson called 'the richest public library in Boston'.[29] Parker praised Strauss's mythical approach with reservations, concluding that the Bible can be considered neither completely historical nor completely fictional, and he credited Strauss for helping Christians move past their dilemma between 'the frozen realm of stiff supernaturalism, and lifeless rationalism' to the expanses of free religious thought.[30] However, it was the 'feeling' rationalism of De Wette, so conducive to Transcendentalism, and the 'deep vein of piety' in his work that most attracted Parker's attention. De Wette also affirmed Parker's conviction that the Old Testament miracles could not be admitted as historical facts, but rather deserved reverence as mythology. Parker translated and annotated De Wette's *Introduction to the Old Testament* in 1843—the first complete work of De Wette's made available in English—and this translation attracted considerable attention among liberal New England scholars.[31]

Parker's scholarly activity was not representative of his contemporaries but constituted the most radical edge of naturalistic biblical criticism occurring in the United States in his time. Although his efforts to establish a scientific tradition of biblical criticism in America were not ultimately successful, his publications provoked his opponents into revealing explicitly their intellectual loyalties. Throughout his career he concluded from his extensive studies that

[27] Charles Christian Tittmann, 'Translation of Tittmann on Historical Interpretation', *Biblical Repertory: A Collection of Tracts in Biblical Literature* 1, no. 1 (1825): pp. 143–8.

[28] Karl Friedrich Stäudlin, 'History of Theological Knowledge and Literature, from the Beginning of the Eighteenth Century to the Present Time', *Biblical Repertory* 1 (1825): pp. 201–34; Ernst Hengstenberg, 'Causes of the Denial of the Mosaic Origin of the Pentateuch', trans. E. Ballantine, *The American Biblical Repository* 11, no. 30 (April 1838): pp. 416–48.

[29] *31st Annual report of the Boston Public Library* (30 April 1883), quoted in Puknat, 'De Wette in New England', p. 386.

[30] Theodore Parker, 'Strauss's Life of Jesus', in *The Critical and Miscellaneous Writings of Theodore Parker* (Boston, MA: J. Munroe, 1843), p. 304.

[31] Grodzins, *American Heretic*, p. 70; Puknat, 'De Wette in New England', p. 378.

the world of matter and the world of human spirit were simply analogous lawful expressions of the same divine Mind, and their kinship led him to reject both the biblical miracles and the miraculous authority of the Bible.[32] Mindful of the controversial nature of his position, he articulated these views carefully, and often only implicitly. His rejection of the traditional supernaturalism of the Old Testament was also motivated by moral as well as logical reasoning. For example, he chafed at conservatives' willingness to accept at face value the Bible's attribution of evil and unjust acts to God. However, other religious scholars refused to join him in criticizing the Bible's special supernatural authority. Even other Unitarians were so scandalized at his efforts to reduce the miraculous to natural explanations that they attempted to force him out of the denomination.[33]

Parker's career illustrates a pattern common during the nineteenth century: the early proponents of scientific biblical criticism seldom used the explicit language of naturalism positively to describe their conclusions. However, critics—even liberal ones—were not reluctant to censure bold interpretations as 'naturalism', or worse. For example, after Parker concluded a series of lectures on religion in 1841 to packed houses at Boston's Masonic Temple, his Transcendentalist colleague, Orestes Brownson, lamented:

> We had both placed the origin and ground of religion in a religious sentiment natural to man; but while I made that sentiment the point of departure for proving that religion is in accordance with nature and reason, he made it his starting-point for reducing all religion to mere naturalism, another name for downright Pantheism, or rather, Atheism.[34]

As Brownson's words suggest, the defenders of traditional biblical interpretation used 'naturalism' to mean metaphysical naturalism rather than methodological naturalism. Throughout his career, Parker insisted that his scholarship was naturalistic only in method, defending the reality of the supernatural and condemning the growing popularity of philosophical atheism, even as he denied the legitimacy of the Bible's miraculous claims.[35]

One important factor in understanding and evaluating Parker's claims of a scientific biblical criticism was that the nineteenth-century categories of 'science', 'scientific', and 'scientist' were neither clearly defined nor stable. What we call science today was, in the eighteenth and early nineteenth centuries,

[32] 'As in nature [God's] influence was modified only by the capacities of material things, so [in spirit] must it be modified by the capabilities of spiritual things.... This conclusion follows undeniably from the analogy of God's presence and activity in matter.' Parker, *A Discourse of Matters Pertaining to Religion* (Boston, MA: Charles C. Little and James Brown, 1842), p. 181.

[33] Grodzins, *American Heretic*, pp. 153–4, 446ff.

[34] Orestes Brownson, *The Convert; or, Leaves from My Experience* (New York: Edward Dunigan and Brother, 1857), p. 345.

[35] Dean Grodzins, 'Theodore Parker', *Dictionary of Unitarian and Universalist Biography*, http://uudb.org/articles/theodoreparker.html (posted 9 September 2002).

typically described as either *natural philosophy* or *natural history*. Scholars used the word 'science', consistent with its historical roots in the Latin *scientia*, to mean systematized knowledge of any subject, and this category included areas of study surprising to us today, such as economics, linguistics, and theology. Around the mid-nineteenth century, as practitioners of natural philosophy excluded supernatural explanations from their research, their uses of the word 'science' took on an increasingly narrow naturalistic meaning.[36] Parker was surprisingly knowledgeable about these developments, reading and regularly citing the latest work of natural philosophers and experimentalists. For example, he was already referring to Auguste Comte in 1837, even before the radical French philosopher was well known in Europe; Parker found his work too hostile to religious thought. Far more congenial was the work of English polymath William Whewell, whose treatment of natural knowledge left room for religious conclusions. In other words, where Comte's outlook amounted to metaphysical naturalism, Whewell advocated only a methodological naturalism.[37]

Ultimately, Parker's efforts to establish a scientific tradition of American biblical criticism were unsuccessful, largely because his careful distinction between methodological and metaphysical naturalism was lost on his audiences. Reviewers in conservative channels, such as the Baptist periodical *Christian Review*, associated him with the notorious infidels Thomas Paine and Voltaire. Even Parker's liberal allies disapproved of his methods of interpretation; for example, Transcendentalist and educator Elizabeth Peabody preferred that he lead religion toward supernaturalism liberated from 'superstition and metaphysical absurdities', rather than in the direction of naturalism.[38]

In addition to the widespread resistance to Parker's naturalism, his goal of grafting German philosophical concerns onto American biblical thought could not overcome the incompatibility between the two varieties of intellectual stock. Unlike in Germany, biblical studies in the United States was deliberately segregated from secular university activity and was taught with practical pastoral concerns in mind, rather than as a detached philosophical exercise. American clergy also generally occupied themselves with social concerns such as abolitionism, revivalism, and missionary and Bible societies, and gave less attention to what they considered esoteric intellectual pursuits. Parker's Unitarian colleague Ezra Stiles Gannett warned that Parker had 'resorted to the aid of a philosophy which irreverently invades the provinces of both religion and

[36] On the historical meanings of 'science', see Ronald L. Numbers, 'Aggressors, Victims, and Peacemakers: Historical Actors in the Drama of Science and Religion', in *The Religion and Science Debate: Why Does It Continue?*, ed. Harold W. Attridge (New Haven, CT: Yale University Press, 2009), p. 16ff.

[37] Grodzins, *American Heretic*, pp. 272–3.

[38] 'Mr. Parker's Discourse', *The Christian Review* 7, no. 26 (June 1842): pp. 161–81; Grodzins, *American Heretic*, p. 270.

theology, and the inculcation of which from the pulpit would be injurious alike to the Christian faith and temper'.[39] Furthermore, the most influential professional religious scholars participating in debates over modern criticism—both proponents and opponents—died in the 1850s and 1860s, leaving no significant successors. Finally, the advent of the Civil War and its concomitant moral concerns presented a significant additional distraction from the scholarly exploration of the Bible. Although early scholarship had involved relatively bold considerations of naturalistic biblical interpretation, much of the later research arose separately, from the efforts of religiously observant scholars of nature rather than from scholars of religion.[40]

AMERICAN NATURALISTS AND NATURALISTIC CRITICISM

As cautious as American biblical authorities in the early nineteenth century had been in offering even subtly naturalistic interpretations, scholars of nature showed less reluctance to exclude supernatural explanations. Whatever their personal religious beliefs, by the turn of the nineteenth century natural philosophers increasingly limited their explanations to natural forces. However, there was also a significant number of observant Christians among these scholars, some of them even Protestant clergy, who did not conform to the widespread expectation of methodological naturalism. The most visible efforts at mid-century to reconcile the biblical account with natural evidence came not from professional theologians and biblical experts, but from these pious men of science. For example, many Christian students of earth history harmonized the Mosaic account of creation with geological evidence of an ancient earth by interpreting the biblical 'days' as geological epochs. Critics, however, responded to these attempts at reconciliation with strong opposition.[41]

The most prominent American promoter of the 'day-age' theory, interpreting the days of creation as geological eras, was the Yale professor Benjamin Silliman, both an evangelical Protestant and a geologist and chemist. Already well-known as the founder and editor of the *American Journal of Science and Arts*, the nation's longest-running scientific journal, Silliman generated a considerable controversy when he published in 1829 an American edition of Robert Bakewell's celebrated British volume *Introduction to Geology*, adding a lengthy appendix reconciling geological discoveries with the biblical account of

[39] Ezra Stiles Gannett, review of *A Discourse on the Transient and Permanent in Christianity*..., by Theodore Parker, *The Monthly Miscellany of Religion and Letters* 5, no. 1 (July 1841): pp. 45–7.

[40] Brown, *Rise of Biblical Criticism*, pp. 180–2.

[41] Ronald L. Numbers, *Science and Christianity in Pulpit and Pew* (New York: Oxford University Press, 2007), pp. 46, 101–3.

creation.[42] Because this textbook quickly became the standard for geology education in the United States, Silliman's efforts at mediation became widely known and provide a vivid example of the widespread resistance faced by harmonizers.

One group of critics, secular figures who resented the intrusion of religion into science, can be exemplified by the colourful infidel Thomas Cooper, who lambasted Silliman for the 'unconditional surrender of his common sense to a clerical orthodoxy' and 'a most injudicious attempt to intermingle what the peace of mankind requires to be kept separate'.[43] Although Thomas Jefferson had praised him as 'one of the ablest men in America', Cooper had been repeatedly thwarted in his careers as a judge and a geology professor, based on accusations that he was an 'infidel, atheist, Deist, Unitarian or almost anything heterodox', labels that at least one biographer considered plausible.[44] Cooper concluded from these attacks that religious leaders 'look with a jealous eye at every scientific discussion', and forbid 'all investigations that do not harmonize with their own theological creed.... No printer, no editor of a scientific journal, dare insert an article in favor of any opinion which the clergy have pronounced heterodox.' He was particularly sensitive to any indication that religious institutions were pursuing greater legal status, either in government or in education.[45] Cooper had long corresponded with Silliman about scientific questions, such as the medical applications of chemistry, but he regarded the use of religious explanations in geological research as an unacceptable intrusion. Cooper's attack on Silliman's harmonization efforts, beginning in 1829, provided America's first exposure to an explicit criticism of the biblical account of the geological record.[46] Cooper published his increasingly vitriolic censure through a variety of public channels, including Silliman's own *American Journal*, but their exchanges came to an end after Cooper concluded in a private letter: 'You must take your own course... I am persuaded that the Bible is in many respects a detestable, and in all respects an unauthenticated book: that religion is the great enemy of science'.[47] Cooper's position continued to spread through

[42] Benjamin Silliman, 'Outline of the Course of Geological Lectures, given in Yale College', appended to *An Introduction to Geology: Comprising the Elements of the Science in its Present Advanced State, and All the Recent Discoveries; with an Outline of the Geology of England and Wales*, by Robert Bakewell, 1st American edn (New Haven, CT: Hezekiah Rowe, 1829).

[43] Thomas Cooper, *On the Connection between Geology and the Pentateuch* (Boston, MA: Abner Kneeland, 1837), pp. 68, 72.

[44] Thomas Jefferson to Joseph C. Cabell, 27 June 1810, in *Early History of the University of Virginia*, edited by Nathaniel Francis Cabell (Richmond, VA: J.W. Randolph, 1856), pp. 1–2; Charles F. Himes, *Life and Times of Judge Thomas Cooper, Jurist, Scientist, Educator, Author, Publicist* (Carlisle, PA: Dickinson School of Law, 1918), p. 31.

[45] Thomas Cooper, 'Appendix On the Clergy', *Boston Investigator*, 27 July 1832, p. 1.

[46] Rodney Lee Stiling, 'The Diminishing Deluge: Noah's Flood in Nineteenth-Century American Thought', PhD dissertation, University of Wisconsin, 1991, p. 106ff.

[47] Thomas Cooper to Benjamin Silliman, 17 December 1833, in *The Papers of Joseph Henry*, vol. 2, ed. Nathan Reingold (Washington, DC: Smithsonian Institution Press, 1975), pp. 134–7.

liberal networks such as the freethinker newspaper *Boston Investigator*, but scientific institutions were slower to communicate an explicit methodological naturalism. For example, it was not until 1864 that *The Geological Magazine* was willing to inform a reader that '"Scriptural Geology" is not within the scope of the Geological Magazine. We shall be glad, however, to receive any notices of any geological facts he may have to contribute.'[48]

The second, larger group of critics comprised scriptural literalists who resented the intrusion of naturalistic science into biblical faith. This position was articulated by the Presbyterian minister Gardiner Spring, an old college friend of Silliman's, who scolded that the subject of the Genesis narrative was ever beyond the reach of geological knowledge. Despite the growing authority of the sciences, Spring and his orthodox contemporaries were unwilling to allow physical evidence to trump or even contaminate the established biblical accounts. Spring had met Silliman socially in 1854 aboard a steamship at a public celebration, and the two struck up a conversation, then a correspondence, about the proper relationship between geological facts and the Mosaic narrative. Silliman had sent the clergyman his appendix to Bakewell's textbook, as well as some other writings, and Spring's response was firm: 'a science which is thus in its infancy may not diminish my confidence in the literal narrative of the creation as given in the first chapter of Genesis'.[49] Like his religiously orthodox colleagues, Spring's resistance was not to science but to science 'falsely so-called', including any nascent field such as geology that appeared too speculative to compete with established wisdom. Spring's critique was an extensive catalogue of the past and present disagreements among geologists, but he also clearly believed that even a mature naturalistic discipline would never be able to overturn the philological evidence for the Mosaic account: 'The work of creation was a great *miracle*. It is not by geology that we shall ever understand it'.[50]

A third source of criticism, less obviously related to concerns over encroaching naturalism, were the specialists in biblical studies who resented men of science straying beyond their proper professional domain to voice opinions on religious matters. The most prominent representative of this concern was Andover Professor of Biblical Exegesis and Literature Moses Stuart, often called the father of biblical science in America, who protested, 'I am unable to see how the discoveries of modern science can determine the meaning of Moses' words'.[51] Unlike his other religiously orthodox colleagues, Stuart was proficient in German and conversant with the latest developments in naturalistic biblical criticism, so this was more than conservative rhetoric; he thoroughly

[48] Gardiner Spring, quoted in 'Correspondence', *The Geological Magazine* 1 (1864): p. 95.

[49] Francis C. Haber, *The Age of the World: Moses to Darwin* (Baltimore, MD: The Johns Hopkins Press, 1959), p. 261.

[50] Haber, *Age of the World*, p. 263, also pp. 258–64.

[51] Moses Stuart, 'Critical Examination of Some Passages in Gen. 1; with Remarks on Difficulties that Attend Some of the Present Modes of Geological Reasoning', *Biblical Repository* 7, no. 21 (January 1836): p. 49.

understood the claims of liberal theologians and naturalist harmonizers. A friend of Silliman's from when they were both tutors at Yale in 1804, Stuart had long disapproved of the geologist's reinterpretation of the Mosaic accounts to accommodate geological evidence. In fact, it was in response to their correspondence that Silliman had compiled his 1829 appendix to Bakewell's textbook. Their debate reached a wider audience and invited additional opinions through the pages of the Andover theological journal, *Biblical Repository*, and Silliman's *American Journal of Science*. Although their correspondence ended in 1836, Silliman publicly acknowledged the theologian's death in 1852, lamenting that his harmonizing efforts had apparently failed to persuade Stuart of the legitimate participation of scientists in biblical scholarship.[52]

Clearly, it had already become more difficult by mid-century to defend the idea that the biblical view of nature and the scientific view of nature had anything to say to each other. Silliman had presciently concluded in the second edition of his textbook that 'it will be no new case, should a mediator between hostile armies fail to conciliate either party, and only provoke the artillery of both; nor would it create either surprise or displeasure, should the writer of these remarks be regarded as too geological for the theologians, as he is certainly too theological for some of the geologists'.[53] Although other harmonizers would also try to reconcile naturalistic explanations and the biblical worldview, the criticism of these efforts only increased—not only from the growing domain of science, but also from religious professionals.

BRITISH CONTROVERSIES AND AMERICAN POPULARIZATION

Although British involvement in the naturalization of biblical scholarship was negligible before 1860, the spread of German critical methods thereafter among theologians and clergy triggered a series of prominent controversies and attracted popular attention there and in the United States.[54] Especially significant among these developments was the 1860 publication of *Essays and*

[52] Stuart, *A Hebrew Chrestomathy, Designed as the First Volume of a Course of Hebrew Study* (Andover, MA: Flagg and Gould, 1829), pp. 117–18; Silliman, 'Supplement by the Editor', in *An Introduction to Geology: Intended to Convey a Practical Knowledge of the Science, and Comprising the Most Important Recent Discoveries; With an Explanation of the Facts and Phenomena which Serve to Confirm or Invalidate Various Geological Theories*, by Robert Bakewell, 2nd American edn (New Haven, CT: Hezekiah Howe, 1833), pp. 389–466; Stuart, 'Some Passages in Gen. 1', pp. 46–106; James L. Kingsley, 'Remarks on a "Critical Examination of Some Passages in Gen. 1; with Remarks on Difficulties that Attend Some of the Present Modes of Geological Reasoning. By M. Stuart, Prof. Sacred Lit. Theol. Sem. Andover"', *American Journal of Science* 30 (July 1836): pp. 114–30; George P. Fisher, *Life of Benjamin Silliman, M.D., LL.D., Late Professor of Chemistry, Mineralogy, and Geology in Yale College*, vol. 2 (New York: Charles Scribner, 1866), pp. 115–16.

[53] Silliman, 'Supplement by the Editor', p. 466.

[54] Brown, *Rise of Biblical Criticism*, p. 116 n. 13.

Reviews, a collection of treatises by seven progressive English churchmen praising the prominence of naturalism in modern scriptural scholarship. The shared message of the essays was summed up by the final contributor, Benjamin Jowett, who concluded bluntly: 'Interpret the Scripture like any other book'.[55] This collection far outsold *The Origin of Species* and generated considerable attention among theologians and leaders in the Church of England, producing what one historian has called the 'greatest religious crisis of the Victorian age'.[56] A year later the well-known Anglican bishop John William Colenso, serving in the South African colony of Natal, published his own naturalistic criticism of selected Bible narratives, which secular and religious periodicals alike criticized harshly.[57] A third controversy was the heresy trial of Scottish biblical scholar William Robertson Smith for his articles in the ninth edition of the *Encyclopaedia Britannica* questioning the authorship of the book of Deuteronomy and the truth of biblical history.[58] These three events involved prominent leaders of British Protestantism, and the ensuing publicity served to attract considerable and sustained attention in both Britain and America, first through popular religious newspapers then through the secular press. Although the controversies originated largely independently, they gave the appearance of an organized movement of naturalistic infidelity, a coordinated attack with German weapons upon the divinely conferred special status of the Bible.[59]

One especially important consequence of the widespread popular reporting of the British controversies, especially the response to *Essays and Reviews*, was the growing participation of non-scholars in American debates over the naturalistic interpretation of the Bible. Although even liberal biblical scholars were generally cautious about attacking the supernatural origin and authority of scripture, those outside scholarly circles did not always exercise such restraint. For example, Henry Preserved Smith, whose conclusions were so liberal that he would be charged with heresy and leave the Presbyterian ministry for Congregationalism, reassured his colleagues in 1882 that 'one rule of ordinary historical criticism cannot be applied to the scriptures. It is generally assumed that accounts of miracles or of phenomena which can only be accounted for by

[55] Shea and Whitla, *'Essays and Reviews'*, p. 33.

[56] Ieuan Ellis, *Seven Against Christ: A Study of 'Essays and Reviews'* (Leiden: Brill Publishing, 1980), p. ix.

[57] Timothy Larsen, 'Bishop Colenso and His Critics: The Strange Emergence of Biblical Criticism in Victorian Britain', in *The Eye of the Storm: Bishop John William Colenso and the Crisis of Biblical Interpretation*, ed. Jonathan A. Draper (London: Continuum International, 2003), pp. 42–63; John W. Rogerson, 'Colenso in the World of Nineteenth-Century Intellectual Ferment', in Draper, *Eye of the Storm*, pp. 127–35.

[58] Gordon Kempt Booth, 'William Robertson Smith: The Scientific, Literary, and Cultural Context from 1866 to 1881', PhD thesis, University of Aberdeen, 1999.

[59] Shea and Whitla, *'Essays and Reviews'*, pp. 4, 6, 42.

the supposition of supernatural agents, must be rejected at once as legendary.'[60] In contrast, popular periodicals associated with liberal Protestantism readily spread the news of naturalistic advances in scholarship. The editors of the Unitarian magazine *Christian Examiner* even chastised themselves for lacking the boldness of Bishop Colenso, and the radical newspaper *Unity* regularly included naturalistic essays and reviewed books such as John White Chadwick's *The Bible of To-day*, in which the author gave an unapologetically naturalistic interpretation of the Bible accounts.[61]

Conservative responses to the popularization of modern biblical criticism generally remained confined within denominations until leaders realized how extensively these methods had spread, corrupting even the work of influential figures such as seminary educators and Sunday School teachers. Although the Presbyterian minister Charles A. Briggs was removed from ministry in 1892 for his dangerously liberal ideas, the conservative response to this event was surprisingly muted. A commentator in *The Presbyterian and Reformed Review*, for example, found him sincere but misguided and blamed his errors on the Broad Church movement.[62] This explanation of heterodox interpretation as merely aping intellectual fashion, rather than as an indication of a legitimate scholarly movement, was not an uncommon response from conservative scholars.[63] The biblical scholar William Henry Green noted prematurely, 'May we not say of the latest critical attempt to roll the Pentateuch off its old foundation, that it has not achieved success?'[64]

While not expressing significant concern about the existence of naturalistic biblical criticism among scholars, Protestant leaders did become quite anxious at the thought of heterodox interpretations spreading through their churches. The Congregationalist minister Washington Gladden, for example, recalled a ministers' meeting at which all agreed on the apocryphal nature of the Scripture passage 1 John 5:7, but none present would willingly discuss this conclusion with members of their congregations.[65] Genuine expressions of concern from religious leaders began to appear regularly in American secular newspapers around 1884, as it became clear that suggestible laypeople were also keenly

[60] Henry Preserved Smith, 'The Critical Theories of Julius Wellhausen', *The Presbyterian Review* 3, no. 10 (April 1882): pp. 357–88.

[61] 'Review of Current Literature', *Christian Examiner* 74, no. 1 (January 1863): pp. 133–5; review of *The Bible of To-day: A Course of Lectures*, by John Chadwick White (New York: G. P. Putnam's Sons, 1879), *Unity* 2 (1878): pp. 204–5.

[62] Talbot W. Chambers, 'The Inaugural Address of Professor Briggs', *The Presbyterian and Reformed Review* 2, no. 7 (July 1891): p. 493.

[63] Mark A. Noll, *Between Faith and Criticism: Evangelicals, Scholarship, and the Bible in America* (Vancouver, BC: Regent College Publishing, 1991), p. 20.

[64] William Henry Green, 'Prof. Robertson Smith on the Pentateuch', *Presbyterian Review* 3, no. 9 (January 1882): pp. 155–6.

[65] Washington Gladden, *Recollections* (Boston, MA: Houghton Mifflin, 1909), pp. 260–1.

interested in the possibility of employing methodological naturalism in the realm of biblical interpretation. The Union College professor Tayler Lewis worried that 'the impression is produced, especially on the minds of the young, that there has somehow arisen some new and terrible form of doubt, some most formidable difficulty unknown to other times, and unassailable by any former arguments'.[66]

NATURALISTIC CRITICISM AND THE POLARIZATION OF PROTESTANTISM

By 1910 it was clear that the widely publicized cases of heterodox biblical scholarship, such as the heresy trial of Charles Briggs, were not isolated examples but signs of a larger movement that demanded an organized militant response. The conservative Minneapolis Baptist minister William Bell Riley identified the University of Chicago Divinity School as one particularly dangerous source of liberal biblical interpretation. The Chicago Divinity professor George Burman Foster's 1906 work, *The Finality of the Christian Religion*, had been one of the first clear signs that the school was disseminating heterodoxy to student ministers, religious educators, and interested laity.[67] Riley helped to organize an impressive grassroots response, including the development of regular Bible Conferences and Prophetic Conferences, in which beleaguered conservatives could take stock of their numbers and organize around a number of credal points, in particular the literal truth and infallibility of the Bible texts. The Christian Fundamentalist movement arose out of this network, explicitly in response to concerns over modern biblical criticism, which were later refashioned into the prominent campaign against evolutionary teachings.[68]

The continued pursuit of biblical scholarship among conservative American Protestants explicitly excluded naturalistic premises or conclusions, which they found incompatible with orthodox interpretive tradition. However, many liberal religious scholars were willing to explore naturalistic interpretation as an intellectual exercise, separate from the pastoral consequences of a Bible reduced to natural phenomena. Their early forays into scientific biblical criticism involved subtle, often implicit use of methodological naturalism, because imprudent statements of such an approach were typically silenced with accusations of infidelity or atheism. As long as naturalistic speculations about the

[66] Tayler Lewis, 'The Old Way the True Way', *The Congregationalist* 37, no. 18 (April 1885): p. 2.

[67] Ferenc M. Szasz, 'Three Fundamentalist Leaders: The Roles of William Bell Riley, John Roach Straton, and William Jennings Bryan in the Fundamentalist-Modernist Controversy', PhD dissertation, University of Rochester, 1969, pp. 38–9.

[68] William Vance Trollinger, *God's Empire: William Bell Riley and Midwestern Fundamentalism* (Madison, WI: University of Wisconsin Press, 1990), pp. 30–1.

Bible accounts remained confined to scholarly circles, criticism was generally minor or confined to denominational networks, but the spread of this heterodoxy to pastors, Sunday School teachers, and laypeople elicited a furious and far-reaching backlash from conservative Protestants. The traditions of naturalistic biblical criticism have continued to thrive in liberal Protestant circles, generating a variety of interpretive positions, from ambivalent naturalistic explanations to methodological naturalism to even full metaphysical naturalism. Contrary to the popular claim that scientific progress simply rendered religious thought irrelevant, the scholarly and popular activity of reconciling the biblical accounts of nature with scientific findings generated new varieties of biblical Protestantism and new networks of observant Christians wrestling faithfully with the repercussions of natural evidence.

12

Anthropology and Original Sin

Naturalizing Religion, Theorizing the Primitive

Constance Clark

The striking jacket of William King Gregory's *Our Face from Fish to Man* (1929) illustrates a lasting conundrum for anthropology. Gregory was a leading American palaeontologist and evolutionary thinker, and the visual introduction to this popular work depicted an evolutionary sequence, arranged in ladder-like fashion, ascending from a fish at the bottom, to 'man' at the top, represented by the face of 'a Roman athlete'. Just beneath the Roman athlete, and above a chimpanzee, dwelt a face labelled 'a Tasmanian' (Figure 12.1).[1] This was a very particular Tasmanian: Trucanini, famous as 'the last Tasmanian'. The picture in the diagram came from a photograph of Trucanini taken by Charles Woolley for Melbourne's 1866 International Exhibition and widely reprinted in books of the late nineteenth century. The face in the original photograph looks tragic and haunted. Trucanini died in 1876, the last of a small remnant population of her people, a group wiped out by war, forced displacement, and disease after the settlement of the British in Tasmania in 1803.[2]

[1] William King Gregory, *Our Face from Fish to Man: A Portrait Gallery of Our Ancient Ancestors and Kinsfolk Together with a Concise History of Our Best Features* (New York: G. P. Putnam's Sons, 1929).

[2] Vivienne Rae-Ellis, 'Representing Trucanini', in *Anthropology and Photography, 1860–1920*, ed. Elizabeth Edwards (New Haven, CT: Yale University Press, 1992), pp. 230–3; James Ryan, 'Images and Impressions: Printing, Reproduction and Photography', in *The Victorian Vision: Inventing New Britain*, ed. John M. MacKenzie (London: V&A Publications, 2001), pp. 214–39. On the diagram, see Constance Areson Clark, *God—or Gorilla: Images of Evolution in the Jazz Age* (Baltimore, MD: Johns Hopkins University Press, 2008), pp. 150–2. For brief but moving accounts of the fate of the last population of Tasmanians, see also George W. Stocking, Jr., *Victorian Anthropology* (New York: Free Press, 1987), pp. 274–83; Henrika Kuklick, *The Savage Within: The Social History of British Anthropology, 1885–1945* (Cambridge: Cambridge University Press, 1991), p. 252. For an important nineteenth-century anthropological discussion of Tasmanians as representatives of the primitive condition, see Edward Burnett Tylor, 'On the Tasmanians as Representatives of Palaeolithic Man', *Journal of the Anthropological Institute of Great Britain and Ireland* 23 (1894): pp. 141–52.

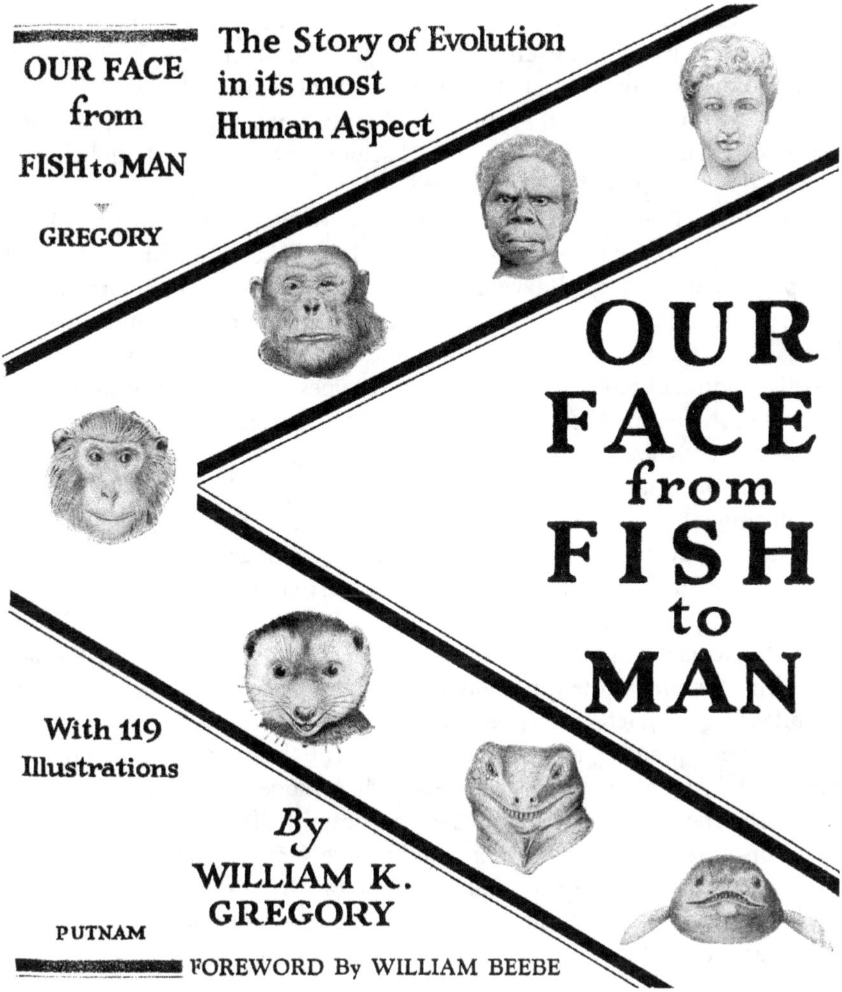

Figure 12.1. Jacket of William K. Gregory, *Our Face from Fish to Man* (New York: G. P. Putnam's Sons, 1929).

Like ghosts of the anthropological theory that made the Tasmanians so fascinating to Victorians, Trucanini's image lasted long beyond the end of the scientific consensus that assigned her to a 'missing link' rung on the evolutionary ladder. It might seem odd that a woman who died in 1876 should occupy the evolutionary space below that of a 'Roman athlete'. But this was the point of the theory. The Tasmanians represented, according to this theory, the evolutionary beginnings of the ascent of human societies from the most primitive early stages to the achievement of modern civilization. Primitive peoples, including the Tasmanians, were relics of the Stone Age past, living fossils, and contemporary survivals of Palaeolithic societies.

Professionalized, late nineteenth-century Anglo-American anthropology characterized itself, quite emphatically, as naturalistic. Anglo-American anthropologists defined their discipline as a profession and their subject of study as a branch of natural history in the context of the Victorian push for a naturalistic science. The founders of the discipline, especially Sir John Lubbock (1834–1913) and Edward Burnett Tylor (1832–1917), belonged to the cohort of scientists associated with Thomas Henry Huxley (discussed in Chapter 13 and elsewhere by Bernard Lightman),[3] committed to a belief in a natural order that was predictable, orderly, and deterministic, and they insisted that both human history and human nature should be studied as part of the natural world. Human societies that seemed foreign could be categorized just as naturalists classified animals and plants.[4] Frequent analogies to other biological species illustrated the point. The evolutionary model of a trajectory from savage, to barbarian, to civilized society, grew out of older traditions, but in the context of the late nineteenth century the old ideas of a Great Chain of Being and of Enlightenment notions of progress and of stages of civilization took on essentially new implications. These ideas about the stages of development of human societies, and even implicitly religious narratives of human origins, all became naturalized. They became the subjects of scientific study.

The Victorian era founders of Anglo-American anthropology in the late nineteenth century wrote with meticulous attention to methodology and conscientious scrutiny of their sources, and avoided God-talk. They were, nevertheless, extremely preoccupied with religion, and with the historical 'development' of morality. Among the essential tenets of their model of the development of human societies was the postulate of the 'psychic unity' of the species. In addition, their concerns grew out of earlier traditions very much shaped by religious concerns. Their notion of the human trajectory from primitive, or savage, cultures to 'civilization' was not just the story of the progression of technology, learning, and intellect; it was also an assertion of the evolution of morality.

[3] Bernard Lightman, 'Science and the Public', *Wrestling with Nature: From Omens to Science*, ed. Peter Harrison, Ronald L. Numbers, and Michael H. Shank (Chicago: University of Chicago Press, 2011), pp. 337–75; see also Ronald L. Numbers, 'Science without God: Natural Laws and Christian Beliefs', in *When Science and Christianity Meet*, ed. David C. Lindberg and Ronald L. Numbers (Chicago: University of Chicago Press, 2003), pp. 265–85; Daniel P. Thurs and Ronald L. Numbers, 'Science, Pseudoscience and Science Falsely So-Called', *Wrestling with Nature*, pp. 281–306; Jon H. Roberts and James Turner, *The Sacred and the Secular University* (Princeton, NJ: Princeton University Press, 2000), pp. 28–41; Evelleen Richards, *Darwin and the Making of Sexual Selection* (Chicago: University of Chicago Press, 2017), pp. 379–81.

[4] Idus Murphree, 'The Evolutionary Anthropologists: the Progress of Mankind, the Concepts of Progress and Culture in the Thought of John Lubbock, Edward B. Tylor, and Lewis Henry Morgan', *Proceedings of the American Philosophical Society* 105, no. 3 (June 1961): pp. 265–300; Stocking, *Victorian Anthropology*; George W. Stocking, Jr., *Race, Culture, and Evolution: Essays in the History of Anthropology* (Chicago: University of Chicago Press, 1968); J. W. Burrow, *Evolution and Society: A Study in Victorian Social Theory* (Cambridge: Cambridge University Press, 1970); Frank Miller Turner, *Between Science and Religion: The Reaction to Scientific Naturalism in Late Victorian England* (New Haven, CT: Yale University Press, 1974).

The trajectory was deterministic, guided by inexorable natural laws, and yet still informed by an implicitly Judeo-Christian view of 'human nature'. Although these anthropologists had by and large adopted, by the 1870s, a naturalistic methodology which excluded religious explanations from scientific theories, underlying concerns shaped in large part by religious traditions and debates persisted.

There were significant continuities in the thought of the evolutionary anthropologists, including Tylor's, with earlier work in ethnology that was not explicitly naturalistic and that embraced biblical themes. Tylor considered the Quaker physician and ethnologist James Cowles Prichard (1786–1848) to be one of the founders of the discipline. Indeed, Tylor had himself been raised a Quaker, although he resigned his membership in 1864. Prichard and Henry Christy (1810–65), also a Quaker, contributed ethnological principles and techniques and raised questions that continued to shape the field. They were associated with the Aborigines Protection Society, founded in 1837 with the motto *ab uno sanguine*—'of one blood'. Many of its members were Quakers, evangelical Christians, and dissenters, and some of their coterie, including Prichard, were among the organizers of the Ethnological Society of London, a group with a scientific agenda that included debates on the question of whether all humans shared a single ancestry. The question of the 'unity of man' at a time of the expansion of empire, of increasing acquaintances with people and societies very different from their own, had both religious and political implications. Their allegiance to monogenism was consistent with their commitment to religious tolerance, and offered scientific support for arguments against the exploitation of—and in favour of the civilizing mission towards—non-European peoples.[5] Prichard, in particular, contributed methodology and a focus on philology. He was also absolutely committed to the view of revealed religion as a direct communication from God.[6] The ethnologists of the early nineteenth century shaped later concerns of anthropology in fundamental ways, both methodologically in developing linguistic analyses of the connections among groups of people, methods adopted from natural history, anatomy, and linguistics, and by their emphases on the questions of monogenism versus polygenism—whether the various types of humans were descended from a single strain or were essentially separate species.[7]

[5] Efram Sera-Shriar, 'Race', in *Historicism in the Human Sciences in Victorian England*, ed. Mark Bevir (Cambridge: Cambridge University Press, 2017), pp. 56–7.

[6] George W. Stocking, Jr., 'From Chronology to Ethnology: James Cowles Prichard and British Anthropology, 1800–1850', in *Researches into the Physical History of Man*, ed. James Cowles Prichard (Chicago: University of Chicago Press, 1973), pp. ix–cx.

[7] For more on this see David N. Livingstone, *Adam's Ancestors: Race, Religion and the Politics of Human Origins* (Baltimore, MD: Johns Hopkins University Press, 2008); G. Blair Nelson, '"Men Before Adam!": American Debates over the Unity and Antiquity of Humanity', in *Science and Christianity Meet*, ed. David C. Lindberg and Ronald L. Numbers (Chicago: University of Chicago Press, 2003), pp. 161–81.

The Victorian anthropologists most responsible for shaping the discipline also insisted that the development of religion could be understood naturalistically: E. B. Tylor devoted much of his magnum opus, *Primitive Culture*, to a discussion of the evolution of religion through the stages of evolution from savage to civilized society. By the late nineteenth century when anthropologists shaped their discipline, God was not an available explanatory hypothesis. Yet, for Tylor, the development of religion was a central part of human history that should be studied as natural history. This was a fundamental tenet of Tylor's naturalistic philosophy. At the same time, the terms in which such studies occurred were very much inflected by earlier concerns that belonged in part to Christian tradition.[8] The effacement of God as a hypothesis did not necessarily remove questions once in the province of religion as matters of concern.

SCIENTIFIC NATURALISM AMONG THE VICTORIANS

Archaeological and palaeontological discoveries had, by the nineteenth century, begun to reveal an expanse of time that made it possible and necessary to think beyond many of the old verities about human history; but although fossil and archaeological discoveries uncovered apparent sequences in technology, the character of prehistoric human societies was harder to trace through fossil evidence alone. And as European travellers came home bearing tales of people whose societies seemed less technologically 'advanced' than Euro-American society, questions about the unfamiliar customs of these people also arose.

Victorian evolutionary anthropologists identified 'primitive' societies as remnant populations of people arrested at early stages of evolutionary development, representatives of Europeans' own ancestry. Nineteenth-century evolutionary anthropology postulated a trajectory from the primitive to civilization: extant primitive societies could be arranged in a natural sequence along a scale that led from 'primitive' or 'savage' to 'barbarian', to 'civilized'. Most anthropologists agreed that Trucanini's people belonged to the lowest—or earliest—of these stages. The school of evolutionary anthropology saw this evolution from the primitive to the civilized as a natural process: human nature was unitary and determined by natural law. The trajectory existed in nature.

This was a teleology, and although it was evolutionary, it was not Darwinian. Whereas the evolution of animal species was, in Darwin's view, contingent and open-ended, evolutionary anthropology set out a fixed and deterministic path of human society. The model drew on a metaphor drawn from what at the time was regarded as a scientific principle—recapitulation. Developed

[8] Kuklick, *The Savage Within*, pp. 80–1.

from embryology, this was the idea that during its development *in utero*, the individual organism recapitulated the steps of the evolutionary history of its species. The metaphor applied to human societies went from the reassuring assertion that just as evolution itself was 'just like' the development of the individual from infancy to adulthood, so human civilization had evolved from the childlike primitive through stages—from savage to barbarian to civilization.[9] Groups called 'savage' represented the childhood of the race, and the children of 'advanced' societies resembled savages. Thus Robert Louis Stevenson could write that children 'dwell in a mythological epoch, and are not the contemporaries of their parents.'[10] The theory of recapitulation, used in this way, implied that living people who fit into the category of the primitive were also not, in an important sense, contemporaries of the civilized.[11]

Tylor and the other evolutionary anthropologists were not the first to refer to non-European peoples as primitive, savage, barbaric, or childlike. People have defined themselves against other societies, ethnicities, races and cultures for a long time. The term 'barbarian', for example, comes from ancient Greek descriptions of outsiders.[12] The idea that human civilization developed by a series of stages would have been familiar to Enlightenment readers of Condorcet, Turgot, and Adam Ferguson. Many of the terms of Victorian anthropology can be traced to Enlightenment thinkers who saw human history as marked by stages of progress toward civilization. The term 'civilization' did not begin to take on its familiar meaning until the middle of the eighteenth century, and with it came a new sense of the word 'progress'. The newly defined 'civilization' was necessarily singular, and remained so even as anthropologists began, gradually, to think of 'cultures' as plural.[13] It would be hard to overestimate the

[9] On recapitulation see Stephen Jay Gould, *Ontogeny and Phylogeny* (Cambridge, MA: Harvard University Press, 1977); Stocking, *Victorian Anthropology*, pp. 228–9; Stocking, *Race, Culture, and Evolution*; Nick Hopwood, 'Pictures of Evolution and Charges of Fraud: Ernst Haeckel's Embryological Reconstructions', *Isis* 97 (2006): pp. 260–301; Robert J. Richards, *The Meaning of Evolution: the Morphological Construction and Ideological Reconstruction of Darwin's Theory* (Chicago: University of Chicago Press, 1992); Gavin R. de Beer, *Embryos and Ancestors* (Oxford: Clarendon Press, 1958); W. Garstang, 'The Theory of Recapitulation: A Critical Restatement of the Biogenetic Law', *Journal of the Linnaean Society of London, Zoology* 35 (1922): pp. 81–102.

[10] Kuklick, *The Savage Within*, p. 100.

[11] For a much-cited and influential treatment of this theme in anthropology see Johannes Fabian, *Time and the Other: How Anthropology Makes its Object* (New York: Columbia University Press, 2002, 1983); see also Roger Sanjek, 'The Ethnographic Present', *Man* n. s. 26 (1991): pp. 609–28.

[12] Stephanie Moser, *Ancestral Images: The Iconography of Human Origins* (Ithaca, NY: Cornell University Press, 1998); Dell Hymes (ed.), *Reinventing Anthropology* (New York: Random House, 1972), pp. 20–4; Stocking, *Victorian Anthropology*, pp. 10–19; Adam Kuper, *The Reinvention of Primitive Society: Transformations of a Myth* (London: Routledge, 2005), pp. 20–36.

[13] Marshall Sahlins, *How 'Natives' Think: About Captain Cook, for Example* (Chicago: University of Chicago Press, 1995), p. 10–15; see also Stocking, 'Matthew Arnold, E. B. Tylor, and the Uses of Invention', in *Race, Culture, and Evolution*, pp. 69–90; Kuper, *The Reinvention of Primitive* Society, p. 30; Christopher Fox, Roy Porter, and Robert Wokler (eds), *Inventing Human Science: Eighteenth-Century Drivers* (Berkeley, CA: University of California Press, 1995).

importance of Tylor's new definition for 'culture', at the very beginning of his 1871 book *Primitive Culture*: 'Culture or Civilization, taken in its wide ethnologic sense, is that complex whole which includes knowledge, belief, art, morals, law, custom, and any other capabilities and habits acquired by man as a member of society.'[14]

EDWARD BURNETT TYLOR AND PRIMITIVE CULTURE

Edward Burnett Tylor (1832–1917) is generally cited as a primary founder of sociocultural anthropology. In fact, the philologist Max Müller referred to anthropology as 'Mr. Tylor's science.'[15] Tylor's work was grounded in and drew on both ancient and recent traditions in the history of thought about human societies, especially from the Enlightenment, and from Hume and Comte. But at the same time, his position in the history of scientific naturalism makes him a key figure in a pivotal moment, and a close look at his work suggests some of the ways in which the scientific naturalism he advocated marks a fundamental change from those Enlightenment thinkers who postulated stages of civilization.

Tylor described scientific naturalism as part of the 'great movement of our time—the introduction of scientific evidence into problems over which theologians and moralists have long claimed exclusive jurisdiction.'[16] Drawing heavily on the idea of fixed laws of nature, and especially on Charles Lyell's principle of uniformitarianism—the idea that the forces shaping the geology of past ages were no different to those at work in the modern world—Tylor praised modern scientific principles, and criticized those who refused to extend the principles of naturalism to human science:

> Our modern investigators in the sciences of inorganic nature are foremost to recognize, both within and without their special fields of work, the unity of nature, the fixity of its laws, the definite sequence of cause and effect through which every fact depends on what has gone before it, and acts upon what is to come after it. They grasp firmly the Pythagorean doctrine of pervading order in the universal Kosmos. They affirm, with Aristotle, that nature is not full of incoherent episodes, like a bad tragedy.... But when we come to talk of the higher processes of human feeling and action, of thought and language, knowledge and art, a change appears

[14] Edward Burnett Tylor, *Primitive Culture: Researches into the Development of Mythology, Philosophy, Religion, Language, Art and Custom*, vol. 1 (London: John Murray, 1871), p. 1. On Tylor's lasting influence see George W. Stocking, Jr., *Delimiting Anthropology: Occasional Inquiries and Reflections* (Madison, WI: University of Wisconsin Press, 2001), p. 105–15.

[15] Stocking, *Victorian Anthropology*, p. 195.

[16] J. W. Burrow, *Evolution and Society: A Study in Victorian Social Theory* (Cambridge: Cambridge University Press, 1970), p. 254.

in the prevalent tone of opinion.... To many educated minds there seems something presumptuous and repulsive in the view that the history of mankind is part and parcel of the history of nature, that our thoughts, wills, and actions accord with laws as definite as those which govern the motion of waves, the combination of acids and bases, and the growth of plants and animals.[17]

He dismissed the question of free will, 'which it need hardly be said is incompatible with scientific argument'.[18] Such concerns were 'obstacles to the investigation of laws of human nature' which 'arise from considerations of metaphysics and theology'.[19] Modern science, however, based on the evidence of experience, not conjecture, was making great progress in discovering and understanding the laws of nature: 'The tendency of modern inquiry is more and more toward the conclusion that if law is anywhere, it is everywhere.'[20]

Tylor argued that theological commitment to the notion of Adam's Fall from the Garden of Eden had misled many people to believe that primitive societies had degenerated from earlier 'golden ages'. Early societies must have been golden ages, on this theory, and so the condition of societies and cultures that seemed to Europeans 'backward' or 'primitive' must have represented a degeneration from the earlier condition. This idea, for Tylor, was a serious misconception: it was inconsistent with the evidence of progress of human society, and it was at odds with the essential scientific principle of uniformitarianism. Crucially, he maintained, it was an argument based on theology, not on science. He insisted in several places that 'the grounds on which this theory has been held have been generally theological rather than ethnological'. And: 'At the present time it is not unusual for the origin of civilization to be treated as a matter of dogmatic theology.'[21] When Tylor wrote about religion, it was to describe his theory of the historical genesis and development of religion; when he pointed to a theory that derived from religion rather than from reason, he wrote of 'theology'. For Tylor, theology was generally dogmatic—an unscientific incursion into the domain of reason, a 'survival' from a religiously inflected mode of thinking.

Tyler coined the term 'survivals' in this context to explain 'processes, customs, opinions, and so forth, which have been carried on by force of habit into a new state of society different from that in which they had their original home, and they thus remain as proofs and examples of an older condition of culture out of which a newer has been evolved'.[22] Animistic beliefs were rational for primitive people, in their ignorance of science; but 'survivals' of those beliefs

[17] Tylor, *Primitive Culture*, vol. 1, pp. 1–2. [18] Ibid., p. 3.
[19] Tylor, *Primitive Culture*, p. 2. [20] Tylor, *Primitive Culture*, p. 22.
[21] Tylor, *Primitive Culture*, p. 32.
[22] Tylor, *Primitive Culture*, p. 15. See also Edward B. Tylor, 'The Philosophy of Religion among the Lower Races of Mankind', *Journal of the Ethnological Society of London (1869–1870)* 2 (1870): pp. 369–81.

persisted in more advanced societies, where they were no longer rational.[23] Tylor's discussion of survivals, however, included clear implications for his attitude toward religion. The understanding of survivals was paramount, both because they constituted evidence upon which to base interpretations of the evolutionary stages of other cultures and because 'most of what we call superstition is included within survival, and in this way lies open to the attack of its deadliest enemy, a reasonable explanation'.[24] Significantly, the predominant subject of *Primitive Culture* was the evolution of religion from 'the lowest races' to the highest Victorian civilization.[25]

In identifying the Tasmanians as the equivalent of Stone Age people Tylor wrote as if the Tasmanians represented a timeless culture understood to be much like that of the earliest human societies, in spite of the fact that it was a society in extreme distress. The Tasmanians familiar to the British anthropologists were a people whose culture had been profoundly disrupted by a particularly brutal form of colonialism. Tylor, who in many ways was meticulous about methodology, could overlook this reality because the model that informed his interpretation seemed to him and his peers so eminently sensible.

Tylor was careful to assure readers that his information had come from reliable eyewitness accounts. He described Tasmanian technology and especially weapons, comparing them with weapons known from Palaeolithic archaeological sites, the Tasmanians coming out somewhat below such early fossil humans in advancement. He used linguistic evidence and accounts of rituals to make inferences about Tasmanian religious ideas, making a point of asserting that these could not have come from contact with Europeans. He found evidence of a 'rude animism, based on the same fundamental principles as the religions of the lower races elsewhere in the world',[26] and concluded in general that 'it seems more likely to consider that in their remote corner of the globe they may have gone on little changed from earlier ages, so as to have remained to our day living representatives of the Stone Age'.[27] It appeared to make sense to conclude that such isolated populations 'may, it seems, remain comparatively unchanged in their level of culture, even from remote prehistoric ages, just as Mollusca of species first appearing far back in the earlier formations may continue to live and thrive in modern seas'.[28] Analogies with non-human biology appear regularly in the explanations of the evolutionary anthropologists.

[23] Tylor, *Primitive Culture*, p. 15.

[24] Ibid., p. 17.

[25] Ibid., p. 24.

[26] Edward Burnett Tylor, 'On the Tasmanians as Representatives of Palaeolithic Man', *Journal of the Anthropological Institute of Great Britain and Ireland* 23 (1894): pp. 151–2.

[27] Ibid., pp. 148–9.

[28] Ibid., p. 150. See also Edward B. Tylor, 'On a Method of Investigating the Development of Institutions: Applied to Laws of Marriage and Descent', *Journal of the Anthropological Institute of Great Britain and Ireland* 18 (1889): pp. 245–72.

Ultimately, his work, especially *Primitive Culture*, was very much concerned with religion. When at the end of the second volume of that work he characterized anthropology as 'a reformer's science', he meant that an understanding of the primitive roots of religious ideas and rituals, which had remained as 'survivals' among the civilized, would liberate people from superstitions clouding the rationality proper to modern scientific societies.[29]

STAGES OF EVOLUTION AMONG THE AMERICANS

The American lawyer turned anthropologist Lewis Henry Morgan (1818–81) took up the theme of the stages of evolution and codified them in more detail than did Tylor. He associated the stages of evolution of cultures with progressive improvements in technology, and every technological stage necessarily connected to a constellation of other things as well, especially language and kinship systems. This reflected a natural order. The technological progression necessarily followed a fixed sequence. The goal of anthropology was to study human history, including morals, kinship systems, and religion as biological natural phenomena. The movement along the trajectory from savage to civilization was not inevitable—the development of some savage societies seemed to have been arrested at a primitive stage—but the direction of the trajectory was inevitable. Societies moved through a series of stages 'ordained by nature'. Essentially, Morgan's model naturalized 'human nature', with an emphasis on the development of kinship systems, marriage, and values often associated with religion and morality.[30] For Morgan, kinship systems were associated with particular stages of development, and the development of private property was a crucial step in the series: without it there could be no progression to civilization.

The ultimate stage, the acquisition of civilization, required the evolution of private property, unknown to primitive societies. (This was a facet of Morgan's theory that earned him a place in the analysis of Marx and Engels, and of later Marxist anthropologists of the twentieth century.) Morgan was just as committed as Tylor to refuting the degenerationist theory he imputed to theology—'it was never a scientific proposition supported by facts'.[31] He was equally a committed

[29] Tylor, *Primitive Culture*, vol. 2, p. 539.

[30] For an astute juxtaposition of Morgan and Franz Boas on the question of nature and culture, see Marshall Sahlins, *Culture and Practical Reason* (Chicago: University of Chicago Press, 1976), pp. 55–73. See also Carl Resek, *Lewis Henry Morgan, American Scholar* (Chicago: University of Chicago Press, 1960); Brad D. Hume, 'Evolutionisms: Lewis Henry Morgan, Time, and the Question of Sociocultural Evolutionary Theory', *Histories of Anthropology Annual*, vol. 7 (2011): pp. 91–126.

[31] Lewis Henry Morgan, *Ancient Society* (London: Macmillan, 1877), p. 523.

monogenist, ultimately adopting a Lamarckian model of cultural as well as physical evolution. In his view, although the 'unity of mankind' pertained, those primitive people lingering in earlier stages of civilization were actually stuck: the progress of cultural evolution became heritable, and the stages fixed by heredity. The stages of society had 'been developed from a few primary germs of thought. Modern institutions plant their roots in the period of barbarism, into which their germs were transmitted from the previous period of savagery. They have had a lineal descent through the ages, with the streams of the blood, as well as a logical development.'[32]

Morgan's influence spread in American anthropology, especially through the agency of the Smithsonian Institution and the Bureau of American Ethnology, led successively by John Wesley Powell (1834–1902) and William McGee (1853–1912), and his ideas came to dominate late nineteenth-century American anthropological practice. Both Powell and McGee published papers including detailed lists of the cultural and technological features that characterized the stages of evolution, and in their hands the hereditary component of the model hardened.[33] This was distinctly and inflexibly a unilinear model of the stages of civilization. Powell, as head of the Bureau of American Ethnology, was among the most powerful American anthropologists of the late nineteenth century; McGee enjoyed some success as a popularizer, culminating in his role organizing the 1904 Louisiana Purchase Exposition in St. Louis, the most extensive ever exhibit of 'native peoples' from the Americas and around the world. McGee intended these exhibits as lessons in anthropology; not surprisingly, the largest of the 'ethnic villages' housed people from the recently colonized Philippines.[34] In retrospect, however, one of the most momentous events was probably the address by Franz Boas (1858–1942) to the Congress at the end of the Fair, signaling a new era in American anthropology.[35]

In it, Boas rejected the unilinear evolutionary model of stages of civilization, and called for a new way of thinking about anthropology as a science: 'we find in anthropology two distinct methods of research and aims of investigation: the one, the historical method, which endeavors to reconstruct the actual

[32] Morgan, *Ancient Society*, pp. 3–4. For more on Morgan, see Philip J. Deloria, *Playing Indian* (New Haven, CT: Yale University Press, 1998).

[33] J. W. Powell, 'From Barbarism to Civilization', *American Anthropologist* 1 (1888): pp. 97–123; W. J. McGee, 'The Science of Humanity', *American Anthropologist* 10 (1897): pp. 241–72; Curtis M. Hinsley, *Savages and Scientists: The Smithsonian Institution and the Development of American Anthropology, 1846–1910* (Washington, DC: Smithsonian Institution Press, 1981).

[34] Nancy J. Parezo and Don D. Fowler, *Anthropology Goes to the Fair: The 1904 Louisiana Purchase Exposition* (Lincoln, NE: University of Nebraska Press, 2007); see also W. J. McGee, 'Anthropology at the Louisiana Purchase Exposition', *Science* n. s. 22, no. 573 (December 1905): pp. 811–26; James Gilbert, *Whose Fair? Experience, Memory and the History of the Great St. Louis Exposition* (Chicago: University of Chicago Press, 2009); Robert W. Rydell, *All the World's a Fair: Visions of Empire at American International Expositions, 1876–1916* (Chicago: University of Chicago Press, 1984).

[35] Parezo and Fowler, *Anthropology Goes to the Fair*, p. 399.

history of mankind; the other, the generalizing method, which attempts to establish the laws of its development'.[36] Referring to the comparative evolutionary stages of the evolution model in anthropology, Boas argued that evolutionary anthropologists had been guilty of 'a subjective valuation of the various phases of development, the present often serving as a standard of comparison. The oft-observed change from simple forms to more complex forms, from uniformity to diversity, was interpreted as a change from the less valuable to the more valuable and thus the historical view in many cases assumed an ill-concealed teleological tinge... a subjective element, emotional in its sources, which leads us to ascribe the highest value to that which is near and dear to us.'[37] It was a comprehensive and stunning attack on the school of evolutionary anthropology and the notion of unilinear and deterministic stages of evolution, and on the idea of grand theories postulating universal laws explaining human nature and human societies. 'The grand system of the evolution of culture, that is valid for all humanity, is losing much of its plausibility. In place of a simple line of evolution there appear a multiplicity of converging and diverging lines which it is difficult to bring under one system.'[38]

Daringly, Boas predicted that such grand theorizing, along with the model of 'unilinear' evolution, was coming to an end, in favour of a new kind of anthropology: 'later than the older sciences, it has outgrown the systematizing period and is just now entering upon the empirical revision of its theories.'[39] He ended with a stirring assertion of the value of anthropology, which was 'its power to make us understand the roots from which our civilization has sprung, that it impresses us with the relative value of all forms of culture, and thus serves as a check to an exaggerated valuation of the standpoint of our own period, which we are only too liable to consider the ultimate goal of human evolution, thus depriving us of the benefits to be gained from the teachings of other cultures'.[40] In addition to rejecting what he perceived as the racist and teleological implications of the evolutionary anthropology model, Boas also insisted that the definition of science emerging in anthropology must be broadened. Anthropology and other historical sciences, Boas insisted, differed methodologically and fundamentally from the nomothetic, law-giving physical sciences. In several subsequent papers, Boas advocated an empirical method for historical sciences that emphasized description and was not predicated on the law-oriented physical sciences. Interpretations of individual cultures should not be tailored to a priori theorizing.[41]

[36] Franz Boas, 'The History of Anthropology', *Science* n. s. 20, no. 512 (1904): pp. 513–24: p. 514.
[37] Ibid., p. 515. [38] Ibid., p. 522.
[39] Ibid., p. 522. [40] Ibid., p. 524.
[41] Stocking, 'The Basic Assumptions of Boasian Anthropology', in *Delimiting Anthropology* (Madison, WI: University of Wisconsin Press, 2001) pp. 24–48: p. 38. For more on the distinction between nomothetic and historical sciences see Jon H. Roberts and James Turner, *The Sacred and the Secular University* (Princeton, NJ: Princeton University Press, 2000), pp. 96–106.

Boas was conditioned by his own upbringing and German intellectual traditions and debates among his mentors to be sceptical of determinism and the observation of cultures through the lenses of grand theories. He described a fundamental insight from his fieldwork in the paper 'On Alternating Sounds'. In transcribing Eskimo vocabularies, Boas discovered that the anomalies and inconsistencies that anthropologists had thought they detected in native speech patterns were actually artefacts of the language of the ethnologist doing the transcribing: people hear in their own languages. These experiences taught him that anthropologists brought their own perceptual frameworks to their encounters with the people they studied. Fieldwork taught anthropologists about their own cultures as well as those of the people studied.[42] This realization influenced the way Boas thought of his own work, and of the epistemology of perception, and, as George Stocking notes, 'it foreshadows a great deal of modern anthropological thought on "culture" '.[43]

By the early twentieth century Boas and his students had become dominant in American cultural anthropology, and the stages-of-evolution model of the evolutionary anthropologists of the late nineteenth century seemed to have been superseded. The Boasians certainly remained dominant in American anthropology for much of the twentieth century, apparently eclipsing the nineteenth-century evolutionary model. Some people have cited the revelations of fieldwork to explain this.[44] Others have cited general changes in the climate of opinion, and a growing ambivalence about the faith in progress. Robert Lowie, one of Boas's more prolific students, wrote in a reminiscence about the early years of the century about the ferment of new ideas in philosophy, mathematics (including non-Euclidean geometry), and social sciences.[45] And yet in 2005, the anthropologist Adam Kuper complained that notwithstanding the dominance of Boasian anthropology, older traditions and vocabularies persisted. Lamenting the persistence of the nineteenth-century evolutionary model, he observed that 'the history of the theory of primitive society is the history of an illusion. It is our phlogiston, our aether.'[46] 'None of this would be particularly remarkable

[42] Franz Boas, 'On Alternating Sounds', *American Anthropologist* 2, no. 1 (January 1889): pp. 47–54; Regna Darnell, *Invisible Genealogies: A History of Americanist Anthropology* (Lincoln, NE: University of Nebraska Press, 2001), p. 40. See also Darnell, *And Along Came Boas: Continuity and Revolution in Americanist Anthropology* (Amsterdam: John Benjamin's Publishing Company, 2000).

[43] Stocking, 'From Physics to Ethnology', in *Race, Culture, and Evolution*, pp. 133–60: p. 159; see also Simon Schaffer, *From Physics to Anthropology and Back Again*, Prickly Pear Pamphlet no. 3 (Cambridge: Prickly Pear Press, 1994).

[44] Brad D. Hume, 'Evolutionisms: Lewis Henry Morgan, Time, and the Question of Sociocultural Evolutionary Theory', *Histories of Anthropology Annual* 7 (2011): pp. 91–126.

[45] Robert Lowie, 'Reminiscences of Anthropological Currents in America Half a Century Ago', *American Anthropologist* n. s. 58 (1956): pp. 995–1016; Robert Lowie, *The History of Ethnological Theory* (New York: Rinehart & Company, 1937); Edward A. Purcell, Jr., *The Crisis of Democratic Theory: Scientific Naturalism and the Problem of Value* (Lexington, KY: University Press of Kentucky, 1973).

[46] Kuper, *The Reinvention of Primitive Society*, p. 10.

if the notion of primitive society was substantially accurate', Kuper wrote. 'But it is not. The whole conception is fundamentally unsound. There is not even a sensible way in which one can specify what a "primitive society" is.'[47]

CONFLICTING LEGACIES

A particularly graphic manifestation of the persistence of older views appeared in a letter that the palaeontologist Henry Fairfield Osborn wrote to a correspondent in 1926 describing two schools of anthropology, one to which Osborn himself belonged and the other represented by Franz Boas. He was not referring, as one might expect, to the bifurcation of the discipline into physical and cultural anthropology. As a matter of fact, the work by Boas that he cited was the famous anthropometric work showing changes in the morphology of immigrants to the United States after periods of residency, suggesting, according to Boas, that the differences between recent immigrants and people born in the United States were not hereditary but at least partly environmental. The implications of this work troubled Osborn. In his letter he wrote:

> There are two schools of anthropologists, one composed entirely of men of the Nordic race like myself, who maintain that there are profound psychical as well as anatomical differences between the races which cannot be altered by education or environment, although they may be modified; the other school, composed chiefly of men of oriental origin like Franz Boas or of strong oriental affinities, tends to brush aside these psychical differences, even claiming that anatomical differences are modified by environment, as shown in Boas' writings. Naturally the attitude of a pure anthropologist such as myself is to be entirely neutral, unprejudiced, unbiased, to truthfully present only the facts ...[48]

By 'oriental', Osborn meant Jewish, and his remarks about Boas are as ironic as his characterization of his own version of anthropology. Boas and his students constituted a dominant school of anthropology by 1926, and Osborn's own ideas about the study of human races would not be called 'neutral' or 'unbiased' by anyone today. He was active in the eugenics movement. His writings on human evolution, and certainly the diagrams he used to illustrate them and to convey

[47] Kuper, *The Reinvention of Primitive Society*, p. 5.

[48] Osborn Correspondence, Osborn Papers, American Museum of Natural History, Box 16, folder 6, Osborn letter to Delancey Nicoll, 22 May 1926. For more on Osborn and the American Museum of Natural History, see Ronald Rainger, *An Agenda for Antiquity Henry Fairfield Osborn and Vertebrate Paleontology at the American Museum of Natural History, 1890–1935* (Tuscaloosa, AL: University of Alabama Press, 1991), and especially on the question of human evolution, Brian Regal, *Henry Fairfield Osborn: Race and the Search for the Origins of Man* (Burlington, VT: Ashgate, 2002). See also Clark, *God—or Gorilla*, and Marianne Sommer, *History Within: The Science, Culture, and Politics of Bones, Organisms and Molecules* (Chicago: University of Chicago Press, 2016).

evolutionary history in exhibits at the museum he headed, the American Museum of Natural History, preserved a way of thinking about the human past that belonged to the Victorian school of evolutionary anthropology, which Boas and his students had decisively rejected. Images redolent of that philosophy, and often including explicitly racial messages, appeared regularly in museum exhibits and publications. Indeed, William King Gregory, who designed the 1929 evolutionary diagram including Trucanini was Osborn's student and colleague at the museum.

Osborn was active among a coterie of scientists who in the face of challenges to evolution from Fundamentalists in the 1920s argued that—far from being a threat to religion—evolution could, if properly understood, confirm one's religious faith.[49] Osborn was an influential man, and his comments suggest that even if Boas and his students had become dominant in cultural anthropology, there were undercurrents.

ANTHROPOLOGY, NATURALISM, AND THE SCIENCE QUESTION

Boas's repeated assertions that anthropology and the other human sciences, as historical sciences, while properly scientific, differed from law-giving nomothetic sciences continued (and continues) to provoke debate among anthropologists, including among his own students. In his reminiscence from the vantage point of mid-century, Boas's student Robert Lowie recalled discussions in his graduate student days about whether anthropology was a science and, if so, what kind of science and whether it should try to be one. This might not be surprising, given Boas's published statements asserting the differences in the types of science appropriate to different disciplines. But some of his other students criticized Boas for what they saw as his reluctance to formulate grand theories of anthropology. In his 1920 review in the *American Anthropologist* of Lowie's *Primitive Society*, Alfred Kroeber, also a former Boas student, gave voice to a conundrum arising from the Boasian model. Praising Lowie's book for its rigorous empiricism, Kroeber called it exemplary of 'the ethnographic method'. What exactly was that method? 'That is, it is descriptive instead of primarily interpretive. It is historical in the sense that it insists on first depicting things as they are and then inferring generalizations secondarily if at all, instead of

[49] Ronald L. Numbers mentions this group, in *Science and Christianity in Pulpit and Pew* (Oxford: Oxford University Press, 2007), p. 134. See also Edward B. Davis, 'Science and Religious Fundamentalism in the 1920s: Religious Pamphlets by leading scientists of the Scopes era provide insight into public debates about science and religion', *American Scientist* 93, no. 3 (May–June 2005): 253–60; Clark, *God—or Gorilla*, pp. 54–5.

plunging at once into a search for principles.'[50] Clearly, this was a description of the principles Boas had so firmly established in his statements on method in anthropology. Principled empiricism was laudable but, Kroeber remarked, 'As long as we continue offering the world only reconstructions of specific detail, and consistently show a negativistic attitude toward broader conclusions, the world will find very little of profit in ethnology. People do want to know why.'[51] In later years, others would echo this complaint about Boas, most conspicuously Marvin Harris, who, in his magisterial tome *The Rise of Anthropological Theory*, took Boas to task for being insufficiently nomothetic.[52]

Given its beginnings, the subsequent history of anthropology has sometimes been troubled and necessarily self-reflective about what it meant to define the study of human societies as a science. By the turn of the twentieth century anthropologists had begun to question earlier evolutionary models, certainly on the grounds of their implications for questions about race and empire, but also as problems of scientific methodology. During the late twentieth century anthropologists went through a series of debates, or episodes of self-reflection, sometimes called 'crises', about what it might mean to treat human cultures as objects of study that should be understood naturalistically, and from what perspective—who should be subjects or objects of study, and who should do the studying. The history of the beginnings of the discipline and its definitions of the objects of study—initially human beings defined as people who were 'other' relative to the civilized Victorian anthropologist—has perhaps made it inevitable that the subsequent history of anthropology would include continuing to ask whether human cultures and histories could be treated scientifically, about what exactly the subject of study ought to be, and whether anthropology was a science at all, or whether the human and social sciences were fundamentally different from the physical sciences.

Anthropologists have sometimes referred to their science as an 'uncomfortable discipline',[53] partly because of the legacy of having invented its object of study in this particular Victorian context. If anthropology were defined simply as the study of humankind, its nature as a science would inspire reflection; because it coalesced as a discipline in the context of debates over race, empire, slavery, and religion, and because it began as the study of humans who were 'others' relative to the people who shaped it, these reflections could indeed be uncomfortable.[54] But this discomfort—especially in the case of sociocultural anthropology—also

[50] A. L. Kroeber, 'Review of *Primitive Society* by Robert Lowie', *American Anthropologist* n. s. 22 (1920): p. 377.

[51] A. L. Kroeber, 'Review of *Primitive Society* by Robert Lowie', *American Anthropologist* n. s. 22 (1920): p. 380.

[52] Marvin Harris, *The Rise of Anthropological Theory: A History of Theories of Culture* (New York: Crowell, 1968; Walnut Creek, CA: Altamira Press, 2001), pp. 277–289.

[53] Raymond Firth, quoted in Kuklick, *The Savage Within*, p. 6.

[54] George W. H. Stocking, Jr., 'Delimiting Anthropology: Historical Reflections on the Boundaries of a Boundless Discipline', *Social Research* 62 (1995): p. 941.

has to do with the complications of defining the study of human societies and human cultures—and by implication human nature—as a form of natural history, and as naturalistic science.

By the 1970s, anthropologists both in the United States and in the United Kingdom had begun to raise questions about the state of the discipline, especially about its history, about the relationship of sociocultural anthropology to science, and about the nature of science itself. If the study of human cultures was a science, what kind of science was it? Did scientific objectivity preclude social activism? Was science too reductive a vehicle for the study of something as complex, and as historical, as anthropology? A 1972 book edited by Dell Hymes, called *Reinventing Anthropology*, began with Hymes asking: 'If anthropology did not exist, would it have to be invented? If it were reinvented, would it be the anthropology we have now?'Hymes answered his own questions: 'To both questions, the answer, I think, is no.'[55]

One of the essays in *Reinventing Anthropology*, by William Willis, was called 'Skeletons in the Anthropological Closet', and in it Willis lamented that 'the goal of Boasian anthropology to see a culture as its members see it was indeed an impossible dream'.[56] Like a number of the authors in the book, Willis argued that the posture of scientific neutrality was irresponsible in anthropology. The emergence of sociobiology and the debates it stimulated added a sense of urgency to the discussions among anthropologists about attributing culture to biology—about naturalizing the concept of human nature. In his critique of sociobiology, the anthropologist Marshall Sahlins contrasted theories of biological determinism with theories of culture, concluding that 'within the void left by biology lies the whole of anthropology'.[57]

There have been obituaries for the discipline, especially from postmodernists; accusations that 'culture theory' was invalid, and arguments over the dichotomy between physical and cultural anthropology, and there were complaints about the complaints.[58] The anthropologist Regna Darnell wrote that 'there is a

[55] Dell Hymes, 'The Use of Anthropology: Critical, Political, Personal', in *Reinventing Anthropology*, ed. Hymes, pp. 3–79.

[56] William Willis, Jr., 'Skeletons in the Anthropological Closet', in *Reinventing Anthropology*, ed. Hymes, pp. 121–52: p. 127; see also George Stocking, 'Delimiting Anthropology', on this book and on the general ferment in the discipline during these years.

[57] Marshall Sahlins, *The Use and Abuse of Biology: An Anthropological Critique of Sociobiology* (Ann Arbor, MI: University of Michigan Press, 1976), p. 16.

[58] Stocking, 'Delimiting Anthropology'; James Clifford and George E. Marcus (eds), *Writing Culture: the Poetics and Politics of Ethnography* (Berkeley, CA: University of California Press, 1986); Matt Cartmill, 'Reinventing Anthropology: American Association of Physical Anthropologists Annual Luncheon Address, April 1, 1994', *Yearbook of Physical Anthropology* 37 (1994): pp. 1–9; Talal Asad (ed.), *Anthropology and the Colonial Encounter* (Amherst, NY: Humanity Books, 1973); Daniel A. Segal and Sylvia J. Yanagisako (eds), *Unwrapping the Sacred Bundle: Reflections on the Disciplining of Anthropology* (Durham, NC: Duke University Press, 2005); Henrika Kuklick, 'The British Tradition', in *A New History of Anthropology*, ed. Henrika Kuklick (Oxford: Blackwell, 1988), p. 75; Tim Ingold (ed.), *Key Debates in Anthropology* (London, Routledge 1996).

fundamental dichotomy between anthropologists who want to be scientists and those who are indifferent to the label for what they do.'[59]

The label 'science', however, has not been the only one at issue. Naturalism too has become problematic, at least for some anthropologists—not in the sense that the 'God hypothesis' has returned as methodology, but in an uneasiness about and disagreements over definitions of culture, and of human nature, in naturalistic, deterministic, reductionist, and biological terms.

It makes a difference that anthropology as a discipline had roots in a time when Edward Bennett Tylor could compare 'simple' societies to molluscs, Lewis Henry Morgan could suggest that human nature was not fundamentally different from the nature of beavers, and the British anthropologist Alfred Cort Haddon could argue that primitive societies could be compared, in terms of evolution, to the *Amphioxus*, or in the case of Australians, to kangaroos.[60]

Recent debates about the culture concept have had much to do with what anthropologists and historians of anthropology have often referred to as the 'original sin' of the discipline.[61] One of the things at stake has been the notion of agency. Marshall Sahlins goes further, arguing that anthropology defined itself as a science, and its objects of study as under the influence of convictions about 'human nature' that belong to the Western theological tradition of the fall from the garden, and especially the European Enlightenment response to that tradition. 'It's all been a huge mistake. My modest conclusion is that Western civilization has been largely constructed on a mistaken idea of "human nature". Sorry, beg your pardon; it was all a mistake. It is probably true that this perverse idea of human nature endangers our existence.'[62]

The perverse idea of human nature, Sahlins argues, is a part of the Judeo-Christian tradition: the consequences of Adam's expulsion from the Garden of Eden have resulted in a 'dogma of human imperfection . . . an inherently wicked humanity, banished from the presence of God to a purely natural and antithetical world of thorns and thistles'. In his exploration of the consequence of this idea, Sahlins inverts the traditional direction of anthropological observation. 'Man cannot but sin, as Augustine said. This kind of self-contempt does not appear to be a general preoccupation of humanity. What makes the Western mythology

[59] Regna Darnell, *Invisible Genealogies: A History of Americanist Anthropology* (Lincoln, NE: University of Nebraska Press, 2001), p. 23.

[60] A. Roldan, 'Looking at Anthropology from a Biological Point of View: A. C. Haddon's Metaphors on Anthropology', *History of the Human Sciences* 4, no. 4 (1991): pp. 21–32: p. 23. See also J. Urry, 'From Zoology to Ethnology. A. C. Haddon: Conversion to Anthropology', *Canberra Anthropology* 5, no. 2 (1982): pp. 58–85.

[61] Henrika Kuklick, 'Introduction', in *A New History of Anthropology*, ed. Kuklick, p. 2.

[62] Marshall Sahlins, 'The Western Illusion of Human Nature' Tanner Lecture, 2005, presented at the University of Michigan, November 4, 2005, *Michigan Quarterly Review* 45 (2006); also published as Sahlins, *The Western Illusion of Human Nature* (Chicago: Prickly Paradigm Press, 2008).

seem even more singular is the cosmological consequences of Adam's crime'.[63] Western society is anomalous among cultures in this way in particular, Sahlins maintains, and turning the anthropological tables on it is in order.

After all the discussions of what kind of science anthropology might be, fundamental questions of values, ethics, and religion remain. Arguments about methodological and metaphysical naturalism in cultural anthropology have become debates not about religion strictly speaking, but about determinism and 'human nature', and about defining the 'subject' of the discipline. In the context of debates over sociobiology, Marshall Sahlins wrote that 'the history of anthropology is a sustained sequitur to the contradiction of its existence as a Western science of other cultures. The contradiction is an original condition: a science of man sponsored by a society which, in a way no different from others, exclusively defined itself as humanity and its own order as culture. Still, I believe that in the anthropological event this society did learn something from others—about itself.'[64] Ever since Boas's insight about 'alternating sounds' anthropologists have been seeking new and better ways to listen.

'Haunted' may be too strong a word. Yet it seems somehow not surprising that so many anthropologists refer to the 'original sin' of their discipline: Trucanini and original sin. Anthropologists and historians have taken apart old notions of 'the primitive' or the timeless 'Other'. They have shed light on what such ideas reveal about their inventors. As Boas wrote long ago, learning to pay attention to the 'subjects' of anthropology can teach anthropologists a good deal about their own cultures. The theory of primitive society, however, unlike the theory of phlogiston, hasn't after all stayed put in the past.

[63] Marshall Sahlins, 'The Sadness of Sweetness: The Native Anthropology of Western Cosmology', *Current Anthropology* 37 (1996): pp. 395–6. I thank Peter Harrison for reminding me of this article. On the implications of the Fall from the Garden of Eden for the history of science, see also Peter Harrison, *The Fall of Man and the Foundations of Science* (Cambridge: Cambridge University Press, 2007).

[64] Sahlins, *Culture and Practical Reason*, p. 554.

13

The Theology of Victorian Scientific Naturalists

Bernard Lightman

Just three years before his death, the biologist Thomas Henry Huxley, a leading agnostic, drew a parallel between a key Christian belief and the scientific theory he had defended so fiercely since the publication of Darwin's *Origin of Species* in 1859. 'If the doctrine of a Providence is to be taken as expression…of the total exclusion of chance from a place even in the most insignificant corner of Nature', Huxley declared, 'if it means the strong conviction that the cosmic process is rational; and the faith that, throughout all duration, unbroken order has reigned in the universe—I not only accept it, but I am disposed to think it the most important of all truths.' But Huxley was not finished. If the doctrine of Providence was held to imply that in some 'remote past aeon' the cosmic process was started by 'some entity possessed of intelligence and foresight' superior in degree to our own, and if it was held that every event was foreknown, 'scientific thought…has nothing to say against that hypothesis'. Such a hypothesis was 'in fact an anthropomorphic rendering of the doctrine of evolution'.[1] Huxley's point, that there was a significant affinity between the Christian concept of Providence and the scientific doctrine of evolution, is telling. It helps to explain why Huxley and other like-minded naturalists of the second half of the nineteenth century had far more in common with their Christian opponents than it might at first appear.

Huxley's contemporaries noted that his enthusiasm for disseminating scientific agnosticism resembled the passion of Christian missionaries preaching to the heathen. Shortly after his death he was described as the 'great apostle of the modern gospel of science', and as one of the 'four Evangelists' of agnosticism.[2]

I am indebted to John Brooke and the editors for their helpful comments.

[1] Thomas Henry Huxley, 'An Apologetic Irenicon', *Fortnightly Review* 58 (1892): p. 567.

[2] Jacob Gould Schurman, *Agnosticism and Religion* (New York: Charles Scribner's Sons, 1896), p. 11; [William Barry], 'Professor Huxley's Creed', *Quarterly Review* 180 (1895): p. 160.

They also noticed that Huxley had a penchant for using religious language in the titles and the texts of his works for a general audience. Edward Clodd, a fellow unbeliever, observed that 'the author of *Lay Sermons*, let it be said, had the making of a preacher in him'.[3] Later, scholars also drew attention to the religious themes in Huxley's writings. The subtitle of one book on Huxley was 'Prophet of Science', while others referred to him as 'prelate' and the proponent of a 'Creed of Science'.[4] Nevertheless, few contemporaries and scholars have seriously claimed that Huxley was profoundly indebted to Christian theology. Huxley's use of religious language, for example, was often explained as an attempt to conceal the more radical dimensions of his unbelief—in order to appease his largely Christian audience—or as a satirical device. However Huxley's point about the parallel between the concept of Providence and the doctrine of evolution would seem to indicate otherwise.

Here I will argue that Huxley, and two of his closest allies within the ranks of the scientific naturalists, the physicist John Tyndall and the philosopher Herbert Spencer, all drew on several concepts closely associated with the Christian theology of their day to articulate, largely in a secularized form, some of their deepest beliefs about nature and the human condition. These beliefs were integral to their science and their vision of scientific progress. They were elements of a 'theology' in line with modern science. Through an examination of Huxley's views on teleology in nature, Spencer's presentation of an evolutionary theodicy, and Tyndall's thoughts on the implications of the first two laws of thermodynamics for the earth's future, the close links between Christianity and naturalism in the second half of the nineteenth century come into clear focus. The Victorian scientific naturalists are particularly important for an understanding of the history of naturalism. They were the first group of scientists and intellectuals to adopt secular naturalism to forge an identity aimed at setting themselves apart from colleagues who remained loyal to Christianity.

SCIENTIFIC NATURALISM IN AN AGE OF PROGRESS

Huxley, Tyndall, and Spencer were all raised in Christian homes and they all slowly lost their faith in the religion of their youth. Spencer discarded the creed of his Wesleyan Methodist parents during the late thirties and forties. In his

[3] Edward Clodd, *Thomas Henry Huxley* (Edinburgh: William Blackwood and Sons, 1902), p. 142.

[4] Houston Peterson, *Huxley: Prophet of Science* (London: Longmans, Green, 1932); Cyril Bibby, *T. H. Huxley: Scientist, Humanist and Educator* (New York: Horizon Press, 1960), p. 45; Franklin L. Baumer, *Religion and the Rise of Scepticism* (New York: Harcourt, Brace, 1960), p. 174.

Autobiography he explained that the 'current creed became more and more alien to the set of convictions formed in me, and slowly dropped away unawares'.[5] Huxley grappled with the chaotic state of his religious emotions while on the *Rattlesnake* voyage during the late forties. 'Morals and religion are one wild whirl to me', Huxley wrote in May 1847, 'of them the less said the better.'[6] In that same year Tyndall expressed strong doubts in his private journal about his Irish family's Protestantism, declaring that he had placed the so-called 'essentials of religion' on 'the same shelf with the swaddling clothes which bound up my infancy'.[7]

Huxley, Tyndall, and Spencer met each other in London and became friends in the 1850s. They became leaders of a significant group of liberal intellectuals, many of them scientists, who rejected traditional Christianity while aggressively pushing for a redefinition of science in the second half of the nineteenth century. Huxley referred to their creed as 'scientific naturalism'. He asserted that scientific naturalism had existed since being formulated by the French philosopher and mathematician René Descartes. Its leading principle was 'the denial of the validity of the evidence adduced in favor of this, or that, extant form of Supernaturalism'.[8] In other words, Huxley argued that proper science excluded any reference to a divine being and scientists should stick to studying observable causes and effects in nature. The scientific naturalists were not just aiming to reform scientific theories and institutions. They were also interested in transforming British culture as a whole. They put forward new interpretations of humanity, nature, and society derived from the theories, methods, and categories of empirical science, especially evolutionary science. They chose to challenge the cultural authority of the Anglican establishment by claiming that scientific naturalists provided the best intellectual leadership for a modernized and industrialized Britain. Besides Huxley, Tyndall, and Spencer, the ranks of the scientific naturalists included the mathematician William Kingdon Clifford, the founder of eugenics Francis Galton, the statistician Karl Pearson, the anthropologists John Lubbock and Edward Tylor, the biologist E. Ray Lankester, the doctor Henry Maudsley, and a group of journalists, editors, and writers such as Leslie Stephen, G. H. Lewes, John Morley, Grant Allen, and Edward Clodd.

Scientific naturalists emphasized that science could provide the basis for a spiritually satisfying alternative to the outmoded Christian creed of the aristocratic, feudal past. Science, Huxley pointed out, 'has found the ideas which can alone still spiritual cravings'. Even religion, Huxley insisted, had been purified

[5] Herbert Spencer, *An Autobiography*, vol. 1 (New York: D. Appleton, 1904), p. 173.

[6] Huxley, *T. H. Huxley's Diary of the Voyage of the H.M.S. Rattlesnake*, ed. Julian Huxley (Garden City, NY: Doubleday, Doran, 1936), p. 278.

[7] Journals of John Tyndall, 26 June 1847, Tyndall Papers, Royal Institution of Great Britain, London, p. 220.

[8] Huxley, *Science and Christian Tradition* (New York: D. Appleton, 1894), pp. 38–9.

as a result of the influence of science, as it had been forced to renounce 'the idols built up of books and traditions and fine-spun ecclesiastical cobwebs'.[9] Scientific naturalists recognized that they could not merely reject the Christian creed they opposed. If they aspired to supplant the clergy as the new cultural authorities, they had to offer a new vision of human life that still provided meaning—a secular creed that offered real hope for the future, since science was the driving force behind all genuine progress. In his 'The Progress of Science 1837–1887' (1887), a review of the scientific achievements since Queen Victoria had taken the throne, Huxley pointed to the 'revolution' that had taken place in the 'political and social aspects of modern civilization'. But social and political progress had been preceded and accompanied, 'and in great measure caused, by a less obvious, but no less marvellous, increase of natural knowledge'. The progress of the physical sciences since the Renaissance was 'due to the fact that men have gradually learned to lay aside the consideration of unverifiable hypotheses; to guide observation and experiment by verifiable hypotheses'.[10] In sum, the adoption of naturalism as the proper method for uncovering truth was responsible for the progress that Victorians currently enjoyed.

The chief obstacle to scientific progress, and therefore to social, political, and religious progress, was Christian dogmatism. When they analysed the history of science, the scientific naturalists were quick to blame the Catholic Church for the many centuries of ignorance between the time of ancient Greece and the Renaissance. In his notorious 'Belfast Address' (1874), a challenge to the authority of the Christian clergy, which Tyndall gave in his role as president of the British Association for the Advancement of Science, he discussed the great scientific triumphs of the ancient Greeks. 'In fact', he declared, 'the science of ancient Greece had already cleared the world of the fantastic images of divinities operating capriciously through natural phenomena.' The Greeks had 'introduced observation with a purpose; instruments were employed to aid the senses; and scientific method was rendered in a great measure complete by the union of Induction and Experiment'. The Greeks had begun to use a naturalistic approach to understanding their world. But, Tyndall affirmed, the 'victorious advance' of the scientific intellect was forced 'to lie fallow for nearly two millenniums' due to the founding of Christianity. The scriptures were made 'the measure of their Sciences', and 'thus reined in, Science was not likely to make much progress'.[11]

Despite their negative attitude towards Christianity as the chief obstacle to scientific progress, the scientific naturalists retained a deep respect for religion. They rejected the notion that science and religion were necessarily in conflict.

[9] Huxley, 'On the Advisableness of Improving Natural Knowledge', *Fortnightly Review* 3 (January 1866): pp. 631–2, 636.

[10] Huxley, *Method and Results* (London: Macmillan, 1894), pp. 43, 65.

[11] John Tyndall, *Fragments of Science*, 8th edn, vol. 2 (London: Longmans, Green, 1892), pp. 145–6.

They believed that science and religion could exist in peaceful harmony as they belonged to two separate spheres. Spencer's avowed purpose in his *First Principles* (1862), the first volume in his ambitious, multi-volume synthesis of all knowledge, was to reconcile science and religion by assigning them separate roles. In the first chapter on 'Religion and Science', Spencer argued that science and religion both 'have bases in the reality of things' and 'how to reconcile them, thus becomes the problem which we should perseveringly try to solve.'[12] For Tyndall, 'no atheistic reasoning can, I hold, dislodge religion from the human heart.'[13] Religion did not live 'by the force and aid of dogma, but because it is ingrained in the nature of man.'[14] Tyndall believed that subjective religious feeling, 'as true as any other part of human consciousness', was safe from scientific attack. But any attempt to objectivize emotions, to thrust poetic conceptions into 'the region of facts and knowledge', is met by science with hostility. Tyndall pointed out that science therefore makes war only on the scenery, not the substance, of religion. 'Let that scenery be taken for what it is worth', Tyndall declared, 'as an effort on the part of man to name what by him is unnamable, to express what by him is inexpressible, to bring in short the mystery of life and its surroundings within the range of his capacities, let it be accepted as a symbol instead of asserted as a fact—a temporary rendering in the terms of knowledge of that which transcends all knowledge—and nine-tenths of the "conflict between science and religion" would cease.'[15] Religion, in its subjective dimension and its articulation through symbol, could be reconciled with the objective facts of science if the boundaries between the two were strictly maintained. Religion belonged to the realm of feeling and was expressed through art and poetry, while science belonged to the realm of intellect and dealt with facts.

Similarly, the main target of Huxley's aggressive rhetoric was actually theology, not religion. This becomes clear once we understand how Huxley viewed the relationship between science, religion, and theology. Huxley insisted, like Tyndall and Spencer, that the spheres of science and religion had to be kept apart from each other. Rightly conceived, Huxley believed, science and religion could never come into conflict because each realm was distinct and without authority outside its proper sphere of interest. In 1859, the year of the publication of Darwin's *Origin of Species*, Huxley was asserting that science and religion were not opposed to each other, rather 'true science and true religion are

[12] Spencer, *First Principles of a New System of Philosophy* (New York: D. Appleton, 1882), pp. 20–1.

[13] Tyndall, *Fragments of Science*, vol. 2, p. 205.

[14] Tyndall, *New Fragments* (New York: D. Appleton, 1898), p. 29.

[15] 'Religion, Carlyle, Political, Etc.', John Tyndall's MS. Note-Books, Royal Institution, MSS T., 2/E8, pp. 15–16. Tyndall offers a similar version, but without the crucial final phrase in *Fragments of Science*, vol. 2, p. 374. The theme of naming the unnamable and expressing the inexpressible is a common one within the Christian fideist tradition. See Bernard Lightman, *The Origins of Agnosticism: Victorian Unbelief and the Limits of Knowledge* (Baltimore, MD: Johns Hopkins University Press, 1987), pp. 53–6.

twin-sisters'.[16] Conflict arose when theology was confused with religion. Huxley distinguished religion from theology, the latter operating in the scientific world of intellect because of its attempt to embody religious feelings in concrete facts. Theology was potentially in conflict with science. In 1894 Huxley wrote, 'Most people mix up "Religion" with Theology and conceive that the essence of religion is the worship of some theological hypostasis or other.'[17] When that happened, there could appear to be a war between science and religion. Huxley declared, 'the antagonism between science and religion about which we hear so much, appears to me to be purely factitious—fabricated, on the one hand, by short-sighted religious people who confound a certain branch of science, theology, with religion; and, on the other, by equally short-sighted scientific people who forget that science takes for its province only that which is susceptible of clear intellectual comprehension.'[18] Huxley and the other scientific naturalists were not the first or the last to distinguish between religious feeling and theological dogma in the nineteenth century. They would have found this distinction in German romantic thinkers of the early nineteenth century, and later liberal Protestants, such as Andrew Dickson White, voiced their support for it.[19]

In their contest for cultural authority with the Anglican clergy, the scientific naturalists promoted the advantages of cultivating the scientific spirit. The growth of scientific knowledge led to social, political, and even religious progress. But the chief obstacle to scientific progress, they insisted, both in the nineteenth century and in the past, was Christianity and its theologians. They were responsible for throttling the naturalistic approach to knowing nature that had begun to develop in ancient Greece. But even though the scientific naturalists were hostile towards the Church, they nevertheless retained a reverence for a religion of feeling. They believed that an emotional response of awe and wonder to nature was appropriate and valuable. Religion was intrinsic to the human condition. Religion based on feeling could never conflict with science, since the two belonged to entirely different spheres. Theology, distinct from religion, belonged in the realm of fact and was more or less a branch of science. The potential for conflict arose when Christian theologians based their theologies on non-scientific grounds. In these cases they confused religion and theology by objectifying their feelings. However, and this point is often overlooked, the scientific naturalists' conception of the relationship between science,

16 [Huxley], 'Science and Religion', *The Builder* 18 (1859): p. 35.

17 Huxley to James Creelman, 11 June 1894, Huxley Papers, Imperial College, vol. 12, p. 343.

18 Huxley, *Science and Hebrew Tradition* (London: Macmillan, 1893), pp. 160–1.

19 Lightman, *Origins of Agnosticism*, p. 132. Andrew Dickson White was anxious to distance himself from John William Draper's emphasis on a conflict between science and religion. In the introduction to *A History of the Warfare of Science with Theology in Christendom* he compared his approach to Draper's, asserting that he saw the struggle as being between science and dogmatic theology, not religion. See Andrew Dickson White, *A History of the Warfare of Science with Theology in Christendom*, vol. 1 (New York: D. Appleton, 1896), p. ix.

theology, and religion left open the possibility of a valid theology based on scientific evidence. They reinterpreted Christian theology in the light of modern science in order to formulate a theology of their own. Combined with their emphasis on religious feeling, the scientific naturalists offered the Victorians a new creed that, they hoped, would be an intellectually, as well as spiritually, satisfying alternative to the Christian one. This was modern science's contribution to the progress of religion.

HUXLEY, NATURAL THEOLOGY, AND A WIDER TELEOLOGY

In 1869 Huxley made a provocative statement that must have puzzled his contemporaries. He insisted, 'there is a wider Teleology, which is not touched by the doctrine of Evolution, but is actually based upon the fundamental proposition of Evolution.'[20] The term 'teleology' was associated in Huxley's time with the study of ends or final causes, especially as related to the evidences of design or purpose in nature.[21] Huxley's endorsement of a 'wider Teleology' would have seemed strange to his readers because the search for design in nature was identified with what was known as natural theology. Natural theology had been a bulwark of Christianity in the first half of the nineteenth century. As formulated by William Paley, an Anglican minister, in his *Natural Theology* (1802), it provided a blueprint for British scientists to understand the natural world. Nature was like a watch. It was composed of pieces designed to fulfil a function. Just as a watch was constructed by human intelligence, nature was created by a divine intelligence, which was omniscient, omnipotent, and benevolent. The designed quality of nature could lead to no other conclusion. The design in an organ of perfection like the human eye could not arise accidentally. It was up to scientists to try to understand what purpose the divine being had in mind when he created each piece of nature. But the very purpose of the naturalism that Huxley and his allies defended seemed to reject the natural theologians' entire approach to science. For the scientific naturalist, it was bad science to use the concept of an unknowable supernatural being in an explanation for the workings of nature. From the point of view of the scientific naturalists, the debate over Darwin's *Origin of Species* was almost entirely about the validity of the form of natural theology in vogue in the first half of the nineteenth century.

[20] Huxley, 'The Genealogy of Animals', in *Darwiniana: Essays* (London: Macmillan, 1894), p. 110.

[21] 'Teleology', in *A New English Dictionary on Historical Principles*, vol. 9, part 2, ed. James A. H. Murray, Henry Bradley, W. A. Craigie, and C. T. Onions (Oxford: Clarendon Press, 1919), p. 149.

So why did Huxley argue in 1869 that a wider teleology was based upon the fundamental proposition of evolution?

Huxley conceived of nature as subject to deterministic laws. His wider teleology was built upon this notion, as was a secularized natural theology. It should be recalled that Huxley's definition of theology as belonging in the realm of intellect, and therefore potentially in conflict with science, did not rule out the possibility of a theology in line with the theories of modern science. He did not argue that all theology should be eliminated, just when it had no scientific evidence to support it. Huxley maintained that Darwin had reinvented teleology. Natural selection was a mechanism for selection among a random set of alternatives—it preserved traits in virtue of their advantageous consequences. It allowed for teleology without a designing divine being. Huxley tried to explain this several times. It was difficult for his audiences to comprehend because evolutionary teleology did not fit any model of teleological explanation existing in the nineteenth century. Before Darwin, teleologists had to choose between goal-directed vital forces or a divinely designed adaptation. But when Darwin proposed a selection-based teleology he provided the doctrine of final causes with a new theoretical base.[22]

Huxley's new teleology was outlined primarily in his essay 'The Genealogy of Animals' (1869) and in his chapter in the *Life and Letters of Charles Darwin* (1887), edited by Darwin's son Francis. The earlier essay was a review of German evolutionist Ernst Haeckel's *The Natural History of Creation* (1868). Huxley rejected Haeckel's assertion that the *Origin of Species* opposed the causal or mechanical view of living nature to the teleological one. Huxley agreed with Haeckel that the theory of evolution was 'the most formidable opponent of all the commoner and coarser forms of Teleology'.[23] Darwin had delivered the 'death-blow' to 'the teleology which supposes that the eye, such as we see it in man or one of the higher *Vertebrata*, was made with the precise structure which it exhibits, for the purpose of enabling the animal which possesses it to see'. Darwin had destroyed Paley's form of teleology, with its emphasis on organs of perfection like the eye. Nevertheless, Huxley insisted, a wider teleology can be established that is based upon the fundamental proposition of evolution. 'That proposition', Huxley wrote, 'is that the whole world, living and not living, in [*sic*] the result of the mutual interaction, according to definite laws, of the forces possessed by the molecules by which the primitive nebulosity of the universe was composed.' A 'sufficient intelligence' in existence when the universe was formed could have predicted the state of living things in Britain in 1869 with as much certainty as anyone living in 1869 could predict what will happen 'to the vapor of the breath in a cold winter's day'.[24] The laws of nature never vary, and

22 James G. Lennox, 'Darwin was a Teleologist', *Biology and Philosophy* 3 (1993): pp. 410, 416–17.

23 Huxley, 'The Genealogy of Animals', pp. 109–10.

24 This and preceding quotations from Huxley, 'The Genealogy of Animals', p. 110.

they determine what will unfold from the beginning of time. To Huxley, the unfolding of nature according to fixed law constituted a teleological system. Since evolution was one of the laws of nature, it was evidence of the determinism that Huxley is defending.

Then Huxley brings in a Paley-like analogy to illustrate his point. When a cuckoo clock is wound up, the phenomena that it exhibits are contained in its mechanism. 'A clever clockmaker could predict all it will do after an examination of its structure', Huxley asserted. If the theory of evolution was correct, 'the molecular structure of the cosmic gas stands in the same relation to the phenomena of the world as the structure of the clock to its phenomena'.[25] Huxley secularizes the analogy, as it is the evolutionary process and the molecular structure of the cosmic gas that produces the phenomena of the world, past, present, and future, not God. But his conclusion is that the 'teleological and the mechanical view of nature are not, necessarily, mutually exclusive'.[26] He even criticizes Haeckel's 'new and convenient' term, 'Dysteleology'. Haeckel intended the term to denote the study of the purposelessness observable in living organisms, such as rudimentary and useless structures. Huxley objected that the facts of dysteleology 'cut two ways'. Either they were useful in the past or they continue somehow to be useful since they haven't disappeared.[27] By focusing on dysteleology, Haeckel gave the mistaken impression that the theory of evolution ruled out the existence of purpose in nature.

When Huxley was editing 'The Genealogy of Animals' for inclusion in his *Collected Essays* in 1893, he added a note. He wrote that he decided to keep in the section of the essay about the reconciliation of teleology with the mechanical view of nature because he had been accused of overlooking this issue. Then he drew the reader's attention to his chapter 'On the Reception of the "Origin of Species"' in the *Life and Letters of Charles Darwin*.[28] Near the end of the chapter he discussed the objections to Darwin's theory expressed in the early days of the evolution controversy based on 'philosophical and theological foundations'. Although these objections had been answered over and over again, they continued to be raised. One 'common objection' was that Darwin's theories 'abolish Teleology, and eviscerate the argument from design'.[29] Huxley pointed out that he had offered some remarks on this subject nearly twenty years earlier, a reference to his article on 'The Genealogy of Animals'. Then Huxley reproduced the pages from 'The Genealogy of Animals' where he discussed the wider teleology of evolution. His subsequent comments are illuminating and provocative. He presented Paley as a proto-evolutionist who anticipated the marriage between

[25] Huxley, 'The Genealogy of Animals', p. 111.

[26] Huxley, 'The Genealogy of Animals', p. 112.

[27] Ibid., p. 113. [28] Ibid., p. 119.

[29] Huxley, 'On the Reception of the "Origin of Species"', in *The Life and Letters of Charles Darwin*, vol. 2, ed. Francis Darwin (London: John Murray 1887), p. 201.

teleology and mechanistic evolution. 'The acute champion of Teleology, Paley,' Huxley declared, 'saw no difficulty in admitting that the "production of things" may be the result of trains of mechanical dispositions fixed beforehand by intelligent appointment and kept in action by a power at the centre that is to say, he proleptically accepted the modern doctrine of Evolution.' Huxley recommended that Paley's successors follow his lead before rushing to see an antagonism between evolution and teleology.[30] Huxley's belief in the compatibility of teleology with evolution, then, was no passing fancy. It was as strong near the end of his life in 1893 as it was when he first wrote 'The Genealogy of Animals' in 1869.

Huxley hinted in several essays what the implications of his teleological emphasis on natural law might mean for the workings of the evolutionary process, particularly in reference to the issue of variation. In 'Mr. Darwin's Critics' (1871) he denied that the theory of natural selection dictated that variation was indefinite or fortuitous. Variation might appear to be accidental, but it was an expression of the operation of molecular forces within the organism. 'And,' Huxley argued, 'as these forces certainly operate according to definite laws, their general result is, doubtless, in accordance with some general law which subsumes them all.' Law to Huxley implied purpose. Huxley had no objection to calling this 'an "evolutionary law" ', though he acknowledged that the law itself remained unknown.[31] Seven years later he developed a similar line of thought in 'Evolution in Biology' (1878). The importance of the theory of natural selection would not be undermined if further research revealed that variability was 'definite' and 'determined in certain directions rather than in others, by conditions inherent in that which varies.'[32]

In 1887 Huxley outlined his vision of what a new natural theology looked like at the end of the century, after evolutionary theory had forced a re-examination of what constituted purpose in nature. The spiritual object of the scientist, he affirmed, was the investigation of a universe that was like a 'sort of kaleidoscope, in which, at every successive moment of time, a new arrangement of parts of exquisite beauty and symmetry would present itself; and each of them would show itself to be the logical consequence of the preceding arrangement, under the conditions which we call the laws of nature.' Gazing upon this sight, a spectator might be filled 'with that *Amor intellectualis Dei*, the beatific vision of the *vita contemplativa*, which some of the greatest thinkers of all ages, Aristotle, Aquinas, Spinoza, have regarded as the only conceivable eternal felicity.' Huxley concluded his hymn to nature with the proclamation that 'order is lord of all.'[33] The new teleology grounding this secularized vision of

30 Huxley, 'Reception,' p. 202.

31 Huxley, 'Mr. Darwin's Critics,' in *Darwiniana*, p. 182.

32 Huxley, 'Evolution in Biology,' in *Darwiniana*, p. 223. I am indebted to Peter Bowler for this reference and the previous one.

33 Huxley, *Science and Christian Tradition*, p. 74.

natural theology was based on the notion of order established by deterministic natural laws. These laws ordained that nature develop in an orderly direction. Evolutionary theory described one of these laws. Modern science therefore provided evidence for this updated vision of design in nature where order itself was the goal. Here was a scientific theology that scientific naturalists could endorse.

THE THEODICY OF HERBERT SPENCER

Just as Huxley proposed the outlines of a secularized version of natural theology to explain how teleology still had a place in evolutionary theory, Spencer looked to a perennial theme in Christian theology to deal with the problem of evil. One of Spencer's main goals in later editions of his *Social Statics*, first published in 1850, was to explain why evil existed in the world. 'All evil results', he declared, 'from the non-adaptation of constitution to conditions.' If a shrub dwindled in poor soil, or became sickly when deprived of light, it was 'because the harmony between its organization and its circumstances has been destroyed'.[34] Evil, to Spencer, could be fully understood only in evolutionary terms. It existed because living beings were not adapted to their environment. The attempt to explain the existence of evil had long been the purview of Christian theologians and philosophers. Since the Christian God was omniscient, omnipotent, and benevolent, theologians and philosophers had to explain how such a being could allow evil and the suffering it entailed. The German philosopher Gottfried Wilhelm von Leibniz coined the term 'theodicy' to describe this justification of the ways of God. In his *Théodicée* (1710) Leibniz argued that this was the best of all possible worlds. In later editions of his *Social Statics*, Spencer reworked the Christian notion of a theodicy by placing it into an evolutionary context. A world governed by the evolutionary process was the best of all possible worlds despite the existence of evil.

Spencer was best known for his attempt to build a comprehensive system of knowledge. In a series of volumes published between 1862 and 1896, which Spencer referred to as the System of Synthetic Philosophy, he demonstrated how the concept of evolution tied together all bodies of knowledge, including astronomy, geology, biology, psychology, and sociology. *First Principles* (1862), the first volume of the system, established the existence of the Unknowable, a shadowy deity that lay behind the phenomenal world governed by the law of

[34] Spencer, *Social Statics, Abridged and Revised; Together with The Man Versus the State* (New York: D. Appleton, 1896), p. 28.

evolution.[35] Spencer's main point in later editions of *Social Statics* was that the evolutionary process automatically moves towards the elimination of all evil. 'Evil', he wrote, 'perpetually tends to disappear'. This tendency is built into the entire evolutionary process. He asserted, 'in virtue of an essential principle of life, this non-adaptation of an organism to its conditions is ever being rectified; and modification of one or both, continues until the adaptation is complete.'[36] For Spencer, the disappearance of evil was identical to the gradual appearance of progress. He stressed, 'progress, therefore, is not an accident, but a necessity.'[37] Since the evolutionary process governed everything that existed, from the inorganic to the organic world, including the development of human thought, progress was inevitable—dictated by a law of nature:

> Instead of civilization being artificial it is a part of nature; all of a piece with the development of an embryo or the unfolding of a flower. The modifications mankind has undergone, and are [*sic*] still undergoing, result from a law underlying the whole organic creation. . . .[38]

Spencer's evolutionary theodicy explained why evil existed—it was a necessary part of a larger progressive process that governed the entire natural world created by the Unknowable.

When Spencer applies his understanding of theodicy to contemporary society, it becomes the justification for his individualistic liberalism. Spencer maintained that because society grew and progressed naturally, it should not be interfered with. In the eighties he vigorously opposed the Liberal party's passing of laws regulating private enterprise and taxing the rich for the benefit of the poor. Spencer's revisions of *Social Statics* in this period reflect his dissatisfaction with the move towards more state intervention. He had always been critical of philanthropists who gave money to the undeserving poor. Though well meaning, these men 'advocate an interference which not only stops the purifying process, but even increases the vitiation'. They were 'blind to the fact that under the natural order of things society is constantly excreting its unhealthy, imbecile, slow, vacillating, faithless members'. Spencer also attacked socialism. In attempting to lessen the suffering of the working class through social reorganization, the socialists merely exacerbated the situation by hampering the evolution of society. 'The process *must* be undergone', Spencer insisted, 'and the sufferings *must* be endured. No power on Earth, no cunningly

[35] In the final chapter of *First Principles*, titled 'Summary and Conclusion', Spencer discusses how the results of the two parts of his work, 'The Unknowable' and 'The Knowable', have come together. An analysis of 'both our religious and our scientific ideas' had shown that 'while knowledge of the cause which produces effects on our consciousness is impossible, the existence of a cause for these effects is a datum of consciousness'. The analysis had led to the ultimate truth that there was a 'Power of which no limit in Time or Space can be conceived'. Spencer, *First Principles*, p. 551.

[36] Spencer, *Social Statics*, p. 28.

[37] Spencer, *Social Statics*, p. 32.

[38] Ibid., p. 32.

devised laws of statesmen, no world-rectifying schemes of the humane, no communist panaceas, no reforms that men ever did broach or ever will broach, can diminish them one jot.'[39] Spencer argued that humanity had to be satisfied with change that was gradual and not violent. Accept the necessary suffering inherent in the evolutionary process, he told his readers, and be reconciled to the ways of the Unknowable. The parallels between Spencer's theodicy and the theodicy put forward by Christian philosophers and theologians are unmistakable. Paley, for example, had urged British workers to be content with their lot in life. God had designed the social, as well as the natural, order. It was impious to complain about 'the necessity to which human affairs are subjected' since it was God who had 'contrived, that, whilst fortunes are only for a few, the rest of mankind may be happy without them'.[40] Both Spencer and Paley try to reconcile their readers to the social order.

The idea that scientific naturalists supported a secular, evolutionary theodicy that rationalized the existing social order is not a new one. At least one scholar, Robert Young, noticed it over forty years ago.[41] But a more detailed examination of Spencer's views, and a comparison of them to Huxley's, reveals that scientific naturalists did not always agree on the nature of their theodicies. In the 1860s Huxley's theodicy resembled Spencer's. In his 'A Liberal Education; and Where to Find It' (1868) Huxley famously used the game of chess as a metaphor for the theodicy that he envisioned at this time in his life. 'The chess-board is the world, the pieces are the phenomena of the universe, the rules of the game are what we call the laws of Nature', Huxley declared. 'The player on the other side is hidden from us. We know that his play is always fair, just, and patient. But also we know…that he never overlooks a mistake, or makes the smallest allowance for ignorance'.[42] Huxley compared this opponent to a 'calm, strong angel who is playing for love', and 'would rather lose than win'.[43] In Huxley's cosmic game of chess a benevolent being works through inflexible natural law.[44] But near the end of his life, in his 'Evolution and Ethics' (1893), he was far

39 This and preceding quotations in Spencer, *Social Statics*, p. 151.

40 William Paley, *The Complete Works of William Paley, D.D.*, vol. 1 (London: J. F. Dove, 1825), p. 429.

41 Robert Young, *Darwin's Metaphor: Nature's Place in Victorian Culture* (Cambridge: Cambridge University Press, 1985), pp. 376, 382. See also James R. Moore, 'Theodicy and Society: The Crisis of the Intelligentsia', in *Victorian Faith in Crisis: Essays on Continuity and Change in Nineteenth-Century Religious Belief*, edited by Richard J. Helmstadter and Bernard Lightman (Basingstoke: Macmillan, 1990), pp. 153–86.

42 Huxley, *Science and Education* (New York: D. Appleton, 1896), p. 82.

43 Huxley, *Science and Education*, p. 83.

44 From the 1840s to the 1870s Huxley was attracted to the theodicy of liberal Unitarianism. At some point in the late 1870s or in the 1880s, he moved away from any recognizable form of Christian theodicy. For an account of Huxley's debts to liberal Unitarianism, and its theodicy, see Bernard Lightman, 'Interpreting Agnosticism as a Nonconformist Sect: T. H. Huxley's "New Reformation"', in *Science and Dissent in England, 1688–1945*, ed. Paul Wood (Aldershot: Ashgate, 2004), pp. 197–214. For Huxley's rejection of theodicies in general, see Huxley, *Evolution and Ethics* (London: Macmillan, 1911), pp. 72–3.

less optimistic about the goodness of natural law. In this essay the evolutionary law of struggle for existence governs the natural world. Progress is only achieved if humanity resists the evolutionary process. Huxley wrote, 'social progress means a checking of the cosmic process at every step and the substitution for it of another, which may be called the ethical process.'[45] Huxley, then, did not agree with Spencer in 1893 that evolution guarantees human progress. It was not clear that Huxley was committed to any form of theodicy near the end of his life. This may have been connected to his change in attitude towards the Unknowable. Although Huxley had initially been comfortable with Spencer's Unknowable in the 1860s, by 1889 he rejected it, saying he did not 'care to speak of anything as "unknowable"'.[46]

Spencer's evolutionary theodicy shared much in common with traditional Christian attempts to explain why an omnipotent, omniscient, and benevolent God allowed the existence of evil. His vision of an evolutionary process, guided by a mysterious deity, treated evil, or the non-adaptation of organisms to their environment, as a necessary part of the nature of things. But his Victorian readers could take comfort in the idea that gradual progress was inevitable. Like Christian theodicies, Spencer's cosmic evolution was designed to reconcile humanity to the natural and the social order. Both orders were actually one and the same, or at least governed by the same natural law. But whereas Spencer encouraged the notion that society should be modelled on nature, at least one scientific naturalist, Huxley, came to another conclusion. The evolutionary process did not lead to social progress. Instead, it prevented the establishment of a truly human society based on ethical principles.

TYNDALL, ENTROPY, AND ESCHATOLOGY

In the concluding chapter of his *Body and Will* (1883), Henry Maudsley, a scientific naturalist who practised medicine, discussed the eventual fate of humanity. Titled 'What Will be the End Thereof?', Maudsley raised the question of whether evolution will go on forever, or if the 'end of life on earth is foredoomed by as certain a fate as the end of the individual life?'.[47] Maudsley asserted that a 'process of degeneracy' was in operation 'alongside a process of evolution' and that eventually it would gain the upper hand.[48] The sun, which supplied earth with its energy, would fail one day, leading to a 'rapidly increasing degeneration of things'. Maudsley went into excruciating detail on what the end would look like as worsening conditions of life made it impossible for

[45] Huxley, *Evolution and Ethics*, p. 81.
[46] Huxley, *Science and Christian Tradition*, p. 311; Lightman, *Origins of Agnosticism*, pp. 136–9.
[47] Henry Maudsley, *Body and Will* (London: Kegan Paul, Trench, 1883), p. 317.
[48] Maudsley, *Body and Will*, p. 318.

humans to continue the struggle for existence. After humans became extinct, a frozen earth would 'be left without energy to produce a living particle of any sort and so death itself is dead'.[49] If science tells us that this is our eventual fate, Maudsley asked, how are we to find meaning in our lives? For Maudsley it was the evolutionary process in nature that inspired idealism, while 'its failure must be the avatar of pessimism'.[50] In his review of Maudsley's *Body and Will*, the philosopher and psychologist G. F. Stout characterized this chapter as an 'eloquent eschatological section' at odds with the rest of the book, which analysed the mind as 'purely the outcome of material processes'.[51] Stout was being satirical when he referred to Maudsley as presenting an eschatology at the end of *Body and Will*. He considered Maudsley to be a materialist. However there was a grain of truth in Stout's observation. Scientific naturalists were concerned about the end of things and presented secular versions of what Christian theologians put under the heading of eschatology.

Eschatology, the branch of Christian theology that focuses on the end of the individual and the world, deals with such issues as human death, the afterlife, the Second Coming, the resurrection of the dead, and the Last Judgement. In the nineteenth century, as Maudsley indicated, evolutionary theory and energy physics dictated very different fates for the human race. For liberal intellectuals who wanted to offer a scientific worldview as a satisfying replacement for Christianity—a worldview that emphasized the concept of progress—the end dictated by energy physics presented a problem. Spencer, for example, envisioned the evolutionary process as eventually leading to the creation of the perfectly adapted human in a utopian society. 'The development of the higher creation is a progress towards a form of being', Spencer declared, 'capable of a happiness undiminished by these drawbacks.' It was in 'the human race that the consummation is to be accomplished' and 'the ideal man is the man in whom all the conditions to that accomplishment are fulfilled.'[52] For Spencer, the progressive evolutionary process would lead to a perfect equilibrium. But, as Tyndall told Spencer, equilibrium, according to energy physics, meant death. Spencer was flabbergasted:

> That which was new to me in your position enunciated last June, and again on Saturday, was that equilibration was death. Regarding, as I had done, equilibration as the ultimate and *highest* state of society, I had assumed it to be not only the ultimate but also the highest state of the universe. And your assertion that when equilibration was reached life must cease, staggered me. Indeed, not seeing my way out of the conclusion, I remember being out of spirits for some days afterwards. I still feel unsettled about the matter, and should like some day to discuss it with you.[53]

[49] Maudsley, *Body and Will*, p. 320. [50] Maudsley, *Body and Will*, p. 325.

[51] G. F. Stout, review of *Body and Will*, by Henry Maudsley, *Mind* 9 (1884): p. 140.

[52] Spencer, *Social Statics*, p. 150.

[53] Spencer to Tyndall, *c.*1858, quoted in David Duncan, *The Life and Letters of Herbert Spencer* (New York: Methuen, 1908), p. 104.

The insights of energy physics seemed to demolish the scientific validity of Spencer's entire evolutionary theodicy and the eschatology that was built upon it.[54] But energy physics was also a problem for Tyndall.

Tyndall was the only physicist among the leading scientific naturalists. Throughout his career he was confronted by the Glasgow professor of natural philosophy, William Thomson (later Lord Kelvin), and his allies, who mounted effective opposition to scientific naturalism. It was this group of scientists who, from the 1850s to the 1870s, virtually created the science of energy. They played a major role in demonstrating that the science of heat was based on two fundamental laws: the law of energy conservation and the entropy law, the first and second laws of thermodynamics. Whereas the conservation law described how energy was transformed from one form into another, the entropy law demonstrated that the circulation of energy had a direction. Though the amount of energy in the universe remained constant according to the first law, the amount of transformable energy continually diminished until equilibrium was reached. The second law suggested that someday in the distant future the sun would burn out. The notion of a final apocalypse did not trouble Thomson or his colleagues. It fitted into their religious beliefs. Bearing the impress of Scottish Presbyterianism, and linked to the industrialists of northern Britain, the North British Physicists group was composed of Scottish natural philosophers James Clerk Maxwell, Peter Guthrie Tait, and the engineers Henry Fleeming Jenkin and William Macquorn Rankine, in addition to Thomson. They promoted a natural philosophy in harmony with Christian belief. Energy flow had a direction that pointed to a universe governed by basic laws organized by a creative and divine being. From their point of view the core doctrine of materialism was reversibility, or the notion of a mechanical system in which there was no difference between running forward or backward. The scientific naturalists, the North British Physicists believed, were materialists who adopted a mechanical system.[55]

Tyndall used the doctrine of the conservation of energy in the service of scientific naturalism. He argued that the fixed quantity of energy in the universe meant that the mechanism of nature remained closed to all external, meaning supernatural, interference. That is how he reconciled his naturalism to the first law. But the second law presented a greater challenge. Like Spencer, Tyndall was uneasy about the cosmological consequences of entropy. As a result, he rarely discussed the second law in his writings, and when he did, he

[54] Darwin was also concerned by the revelations of the physicists. In his *Autobiography* he referred to the thought that humans are 'doomed to complete annihilation after such long-continued slow progress' as being 'intolerable'. See Francis Darwin (ed.), *The Autobiography of Charles Darwin and Selected Letters* (New York: Dover, 1958), p. 65. I am indebted to John Brooke for this reference.

[55] Crosbie Smith, *The Science of Energy: A Cultural History of Energy Physics in Victorian Britain* (Chicago: University of Chicago Press, 1998).

tried to evade its pessimistic conclusions.[56] In his *Heat: A Mode of Motion*, first published in 1863, Tyndall emphasized the stupendous heat generated by the sun and discussed a scientific hypothesis that ruled out the future heat death of the sun. 'The total amount of solar heat received by the earth in a year, if distributed uniformly over the earth's surface', he wrote, 'would be sufficient to liquefy a layer of ice 100 feet thick, and covering the whole earth.'[57] But no decrease in the sun's heat had been detected in historic times. So how did the sun replenish itself? Tyndall brought forward Robert Mayer's hypothesis, first expressed in his *Dynamik des Himmels* (1848), that the sun renewed itself due to the tremendous heat produced by the collision between it and meteors moving at high velocity. 'Here', Tyndall announced, 'then, we have an agency competent to restore his lost energy to the sun, and to maintain a temperature at his surface which transcends all terrestrial combustion.'[58] Although Tyndall acknowledged that Mayer's theory was speculative, he presented it in a positive light.

In the final pages of the book Tyndall depicted the universe as being in a state of permanent cyclical renewal. He reminded his readers of the stupendous energy of the sun and how its 'reservoir of power is infinite'. But the sun is 'a mere drop in the universal sea', and other systems and other suns exist beyond our solar system, each pouring forth energy, 'but still without infringement of the law, which reveals immutability in the midst of change, which recognizes incessant transference or conversion, but neither final gain nor loss.'[59] Tyndall seems to deny the second law in the final phrase of the sentence, highlighting his interpretation of the first law instead. He goes on to depict the energy flowing throughout the universe in poetic terms, transformed from waves, to ripples, and back again into waves, moving through objects and giving them life. 'Asteroids may aggregate to suns', Tyndall rhapsodizes, 'suns may resolve themselves into florae and faunae, and florae and fauna melt in air,—the flux of power is eternally the same. It rolls in music through the ages, while the manifestations of life, as well as the display of phenomena, are but the modulations of its rhythm.'[60] Tyndall's emphasis on a universe of cycles instead of a directional one was not unusual. Cyclical cosmologies were developed in Germany between 1869 and about 1890, mostly by materialists, including Friedrich Engels, Moses Hess, Ludwig Büchner, and Ernst Haeckel.[61]

Near the end of his life Huxley weighed in on the issue of eschatology. He predicted a period of progress followed by a period of decline. Both 'Evolution

[56] In the section on Tyndall on entropy that follows I am indebted to Elizabeth Neswald's 'Saving the World in An Age of Entropy: John Tyndall and the Second Law of Thermodynamics', in *The Age of Scientific Naturalism: John Tyndall and His Contemporaries*, ed. Bernard Lightman and Michael Ready (London: Pickering and Chatto, 2014), pp. 15–32.

[57] John Tyndall, *Heat Considered as a Mode of Motion* (London: Longman, Green, Longman, Roberts and Green, 1863), p. 419.

[58] Tyndall, *Heat*, p. 423. [59] Tyndall, *Heat*, pp. 433–4. [60] Tyndall, *Heat*, p. 434.

[61] Elizabeth Neswald, *Thermodynamik als kultureller Kampfplatz: Eine Faszinationsgeschichte der Entropie 1850–1915* (Frieburg im Breisgau: Rombach 2006).

and Ethics'(1893) and 'Evolution and Ethics: Prolegomena' (1894) end with an endorsement of this gloomy prospect. In 'Evolution and Ethics' Huxley stated that 'the theory of evolution encourages no millennial anticipations. If, for millions of years, our globe has taken the upward road, yet, some time, the summit will be reached and the downward route will be commenced.'[62] In the later essay he wrote that the human race faced a constant struggle to maintain and improve civilization. But progress could only be achieved for so long, 'until the evolution of our globe shall have entered so far upon its downward course that the cosmic process resumes its sway; and, once more, the State of Nature prevails over the surface of our planet.'[63] It was this grim picture of the fate of the earth that inspired H. G. Wells to write his *Time Machine* (1895), in which the human race has degenerated into two species, the effete and doll-like Eloi, and the Morlock, the underground dwellers who feed on the Eloi. When the time traveller moves further into the future, he finds an earth devoid of life due to the slow cooling off of the sun. But Tyndall rejected Huxley's, and Maudsley's, pessimism. He also avoided Thomson's resignation to the heat death of the sun and he did not accept Spencer's belief that evolution guaranteed the coming of paradise on earth where humans were perfectly adapted to their environment. Instead, his secular 'eschatology' was based on the notion that the end would never come. Energy physics, as he interpreted it, predicted a cyclical pattern.

The scientific naturalists were not hostile towards all forms of religion. They lost their faith in institutionalized religion, but not in the possibility of creating a religion that was not in conflict with science. Huxley, Spencer, and Tyndall responded in diverse ways to the task of constructing a new faith complementary to modern science. Nevertheless, they all drew on parallels between modern scientific theory and the theology of Christianity in order to persuade the Victorian public that the creed of scientific naturalism was as intellectually and spiritually fulfilling as their opponents'. Huxley drew analogies between the design of natural theology and the natural order revealed by evolutionary theory; Spencer dealt with the problem of evil in the evolutionary process by presenting a secular theodicy; and Tyndall offered an eschatology informed by his interpretation of energy physics that avoided the pessimism at the heart of the second law of thermodynamics. Although they often secularized theology, the parallels point to the close relationship in the second half of the nineteenth century between British naturalism and the Christianity against which it rebelled. The scientific naturalists did not revolt against Christianity in a vacuum. Their search for a new creed was shaped by the religion of their fathers.[64]

[62] Huxley, *Evolution and Ethics*, p. 85.

[63] Huxley, *Evolution and Ethics*, p. 45. A few pages later in the essay Huxley refers specifically to 'cyclical evolution' (p. 49).

[64] Ronald Numbers has already pointed out that naturalism was largely made in Christendom by pious Christians. So it could be said that the scientific naturalists' attempt to redefine science was also shaped by their fathers' science. See Ronald Numbers, 'Science without God: Natural

Their attempt to build a new theology is related both to their early religious upbringing and to the Christian character of the culture in which they lived throughout their lives. The scientific naturalists were still thinking in Christian terms, even when they proposed a new creed to replace Christianity. Rather than looking at naturalism and Christianity as mortal enemies, it would be more historically accurate to view them as if they were brothers and sisters prone to sibling rivalry. This means that despite their many arguments, they were still members of the same family. Contemporaries may have been impressed by the differences between the scientific naturalists and their Christian opponents. But what stands out more for those of us looking back over a century later is what they shared in common.[65]

Laws and Christian Beliefs', in *When Science and Christianity Meet*, ed. David C. Lindberg and Ronald L. Numbers (Chicago: University of Chicago Press, 2003), p. 284.

[65] Matthew Stanley's book on scientific naturalists and Northern British physicists in the Victorian period provides a comprehensive examination of what was shared in common scientifically by these two groups with opposing religious beliefs. See Matthew Stanley, *Huxley's Church and Maxwell's Demon: From Theistic Science to Naturalistic Science* (Chicago: University of Chicago Press, 2015).

Index

Abano, Pietro d' 48
Abelard, Peter 44–5
Adelard of Bath 40
afterlife 148
Age of Reason (Paine) 127
agnosticism 235–6
agriculture 125
Albert the Great, Saint *see* Magnus, Albertus
alchemy 115–16, 119, 121
Alchymista Christianus (Fabre) 116
Allen, Grant 237
Allgemeine Naturgeschichte (Oken) 186
Allgemeine Naturgeschichte und Theorie des Himmels (Kant) 192
Almagest (Ptolemy) 24, 29
Amphitheatrum Sapientiae Aeternae (Khunrath) 116
Anaxagoras 27
Anaximander 19–21, 26
ancient Greece 7–8, 18, 40
  and gods 19–21, 34–5
  and laws of nature 12
  and religion 31
  and science 80–1, 238
  and the soul 152–3
animism 223–4
anthropic principle 108–9
anthropology 16–17, 216–22, 229–34
  and primitive culture 222–5
  and USA 225–9
anthropomorphism 34–6
anti-Catholicism 157–60
Antipsychothanasia, Or, The Immortality of the Soul Explained and Proved by Scripture and Reason (Wadsworth) 151–2, 154
Aquinas, Thomas 8, 12, 47–8, 50, 52, 81
archaeology 202, 220
Aristotelianism 23, 163
Aristotle 7, 12, 26, 74
  and eternity of the world 49–50
  and Middle Ages 39–40, 42–3
  and the cosmos 28–9, 80–1
  and universities 45–6
arts 38, 43–56
Asclepius 31
associationism 170–1, 173
Astonishing Hypothesis, The (Crick) 124
astronomy 38, 53–4, 73, 79–81, 112
atheism 77, 86, 148–9, 189–90
atomism 29, 32–4, 112, 124, 148–9
Augustine of Hippo 8, 12, 40–1
Augustinian Science 5
Averroes 39–40, 45
Avicenna 40

Bacon, Francis 76, 112, 124
Bain, Alexander 171, 173
Bakewell, Robert 208–11
Bankruptcy of Religion, The (McCabe) 194
Barbour, Ian 110
Bates, Henry Walter 135
Baur, Ferdinand Christian 201
Baxter, Richard 155
Beddoes, Thomas 117
Beiträge zur Einleitung in das Alte Testament (De Wette) 202
Beiträge zur Dynamik des Himmels (Mayer) 251
Bentley, Richard 66–8, 90–1, 119
  and the soul 164
Berthelot, Marcellin 123
Berzelius, J. J. 125
Beyträge zur Naturgeschichte (Blumenbach) 183
Bible, the 15–16, 70, 199–200
  and astronomy 53–4
  and chemistry 117
  and criticism 127–8, 200–8, 211–15
  and geology 182–4, 191–6, 208–11
  and natural philosophy 70–2
  and scholarship 197–8
  and the soul 151–60
biology 27, 131–7, 175, 232
Blake, William 95–6
Blumenbach, Johann Friedrich 183
Boas, Franz 226–31, 234
Body and Will (Maudsley) 248–9
Boethius of Dacia 48
Bologna 43–4
Bonaventure, Saint 50
Boullée, Étienne-Louis 78, 95
Boyle, Robert 68, 71–5, 90, 112
  and chemistry 119–22
  and Christianity 126
  and methodological naturalism 131–2
  and religion 114
Boyle Lectures 90–1, 121
brain, the 172

Brewster, David 145–6, 188
Bridgewater Treatise (Prout) 120, 125
Brief History of Time, A (Hawking) 108
Briefe über Alexander von Humboldt's Kosmos (Cotta) 185
Briggs, Charles A. 213–14
Brightman, Edgar S. 2
Brooke, John 11
Broughton, John 160
Brownson, Orestes 206
Buch, Leopold von 187
Büchner, Ludwig 251
Buckland, William 183
Buckminster, Joseph Stevens 203–4
Budd, Susan 127–8
Buddhism 105
Buffon, Georges-Louis Leclerq, Comte de 183
Burke, Peter 121
Burmeister, Hermann 186
Burnet, Thomas 70–1, 182

Calvin, John 159, 199
Camus, Jean-Pierre 116
canonization 8–9
Capra, Fritjof 105
Carpenter, William B. 173
Cartesianism 63–4, 70–1, 76, 161
Castel, Louis-Bertrand 92
Catholic Church 112, 126, 128, 157–60
  and authority 200
  and science 238
causality 97–8, 103–7
Chadwick, John White 213
Chadwick, Owen 196
Chambers, Robert 138
Châtelet, Émilie Marquise de 93–4
Chemical Essays (Watson) 122
chemistry 112, 114–24
  as catalyst 124–6
  and religion 126–9
Christianity 5, 8–9, 18, 236–8
  and chemistry 126–9
  and evolution 137–8
  and Newton 78–9, 90–1
  and phrenology 172
  and physics 98
  and psychology 175, 177–81
  and science 74
  and the cosmos 80–2
  and the mind 162–3, 165–8
  and the soul 14, 161, 164–5
  see also Bible, the; Catholic Church; God; Protestantism
Christy, Henry 219
Cicero 24–7
civilization 217–8, 221–2, 225–6
Clark, Constance 17
Clarke, Samuel 11, 68–9, 90, 91
Clifford, William Kingdon 237
Clodd, Edward 236–7
Colenso, John William 212–13
Coleridge, Samuel Taylor 112
combustion theory 118
common-sense realism 167–8
Compton, Arthur Holly 104–5
Comte, Auguste 123, 207, 222
Condemnations of 1277 51–3
Conditionalist Faith of Our Fathers, The (Froom) 160
conscience 167
Cooper, Thomas 209–10
cosmos, the 23–6, 28–9
  and Aristotle 28–9, 80–1
  and Christianity 80–2
  and Epicureans 32–4
  and God 82–4, 94–6
  and Humboldt 184–90
  and physics 108–9
  and Spinoza 83–4
  see also astronomy; solar system
Cotes, Roger 64, 66
Cotta, Carl Bernhard von 185
Coward, William 151, 154–60
creationism 2–3, 26–9
Crick, Francis 124
Critica Sacra (Leigh) 154
Critique of Judgement (Kant) 132–3
Crosse, John 188
Cudworth, Ralph 63
culture 222–5, 232–3
Cuvier, Georges 187

Dalton, John 112, 120
Dana, James D. 195
Darnell, Regna 232–3
Darwin, Charles 11, 16, 134, 138–46
  and animal species 220
  and Huxley 141, 144, 243–4
  and religion 113, 115
  and teleology 241–2
  and the mind 173
Darwin, Erasmus 95, 136–7
Das Leben Jesu (Strauss) 203
Davies, Paul 108
Davy, Humphrey 112, 118–20, 122
Dawkins, Richard 133–4, 139, 144–6
'day-age' theory 208–9
De caelo (Aristotle) 45, 47, 49
De generatione et corruptione (Aristotle) 47
de Vries, Paul 2
De Wette, Wilhelm Martin Leberecht 202, 205
Deluc, Jean André 183, 195
Democritus 26–7

Derham, William 75
Descartes, René 10, 12–13, 57, 74, 112, 237
and biology 131–2
and dreams 72–3
and God 60–3
and laws of nature 76
and materialism 161
and Newton 64–5
and the soul 163–4
design 24, 27, 59, 65–6, 70, 99–102, 105–6, 109, 113–14, 117, 121–2, 130–6, 138–42, 159, 183–4, 188–90, 241–5, 247–8, 252
see also final cause, intelligent design, physico-theology, teleology
Dewey, John 176, 178
DG (Diseases of Girls) (Hippocrates) 29–31
Dialogues Concerning Natural Religion (Hume) 134
Diderot, Denis 95
Discourse concerning the Soul of Man, A (anon, 1719) 155, 157–8
Discourse on the Method (Descartes) 61
disease 13, 29, 32, 116
DNA 124, 133
Dobzhansky, Theodosius 144
'double truth' 48
Draper, John William 112–13
dreams 72–3
dysteleology 243

earth 70–1; *see also* geology
eclecticism 23
Eddington, A. S. 105–6
education 106, 109; *see also* universities
Egypt 152, 202
Eichhorn, Johann Gottfried 192, 194, 201, 205
Einstein, Albert 107–8
electricity 118–19
Eléments de la Philosophie de Newton (Voltaire) 93
Empedocles 27
energy 99–101, 249–51
Engels, Friedrich 251
Enlightenment 77–9, 92, 96, 122, 221–2; *see also* Scottish Enlightenment
entomology 135
entropy 250–3
Epicureans 23–4, 29, 32–4, 36, 75
Époques de la nature, Les (Buffon) 183
eschatology 248–9, 251–2
Essai de Cosmologie (Maupertuis) 94–5
Essay Concerning Human Understanding (Locke) 92
Essays and Reviews 16, 193, 211–12
eternity of the world 42, 49–50
Ethics (Spinoza) 83, 87
ethnology 219, 226
eugenics 229, 237
Evagrius of Pontus 74
evangelists 126–8
evil 245–6, 248
evolution 15, 17, 27, 112, 144–7
and chemistry 123–4
and Darwin, Charles 138–46
and Darwin, Erasmus 136–7
and Huxley 242–4, 247–8, 251–2
and Maudsley 248–9
and religion 137–8
and Spencer 245–8
and the mind 173
see also anthropology
Exposition du Système du monde (Laplace) 77

Fabre, Pierre-Jean 116
faculties *see* universities
Fall, the 165–6, 233–4
Faraday, Michael 100, 107, 119
Fatio de Duillier, Nicolas 88
Ferrier, David 177
final cause 63–4, 131–2, 134–5, 138–9, 141, 188, 241–2
see also design, teleology
Finality of the Christian Religion, The (Foster) 214
First Principles (Spencer) 239, 245–6
Fisher, Ronald A. 144
Forbes, James David 188
forms 7, 12
fossils 182–3, 220
Foster, George Burman 214
France, positivism in 123; *see also* French Revolution; Paris
Frankenstein (Shelley) 95
Frankland, Edward 128–9
Franklin, Benjamin 119
free will 104–5
French Revolution 95, 98, 117–18
Freud, Sigmund 180
Froom, Le Roy Edwin 160
Fundamental Fysiks Group 105
Funkenstein, Amos 56–7

Galen 24, 26–7
Galileo 56–7, 61, 84–5
Gall, Franz Joseph 171–2
Galton, Francis 237
Galvani, Luigi 95
Gannett, Ezra Stiles 207–8
'Genealogy of Animals, The' (Huxley) 242–4
'General Scholium' (Newton) 56, 59, 64, 88–91, 97

Genesis *see* Bible, the
Genesis and Geology (Gillispie) 195
geology 15, 70–1, 136, 182–4
and God 191–6
and the Bible 208–11
Geology and Mineralogy Considered with Reference to Natural Theology (Buckland) 183
German Romantics 95, 240
Germany, scholarship and theology in 197, 200–8
Geschichte der Schöpfung (Burmeister) 186
Gillispie, Charles 195
Gladden, Washington 213
Glanvill, Joseph 75
God 5, 7–9, 17
and absolute power 52
and Darwin 139–40
and evolution 138
and geology 191–6
and human beings 15, 162–3
and Humboldt 188–90
and laws of nature 10–13
and Leibniz 68–9
and naturalism 59–66, 72–6
and Newton 58–60, 69, 86–9
and physico-theology 70
and physics 99, 107–8
and solar system 67–8
and Spinoza 83–4
and the cosmos 80–4, 94–6
gods 7, 19–22, 27–9, 34–6
and Epicureans 33–4
and Roman empire 24–7
Goodwin, Charles W. 193
Gould, Stephen Jay 144–6
gravity 67–8, 71, 84–6, 88–9, 91–2
and Newtonianism 89–90, 93–5
and physics 97
Gray, Asa 139, 144
Great Britain
and the Bible 200, 211–12
psychology in 174–7
reviews and criticism in 187–9
Green, Thomas Hill 176
Green, William Henry 213
Gregory, David 67
Gregory, Frederick 126
Gregory, William King 216–17, 230
Grew, Nehemiah 68, 71
Griesbach, Johann Jakob 201, 203

Haddon, Alfred Cort 233
Haeckel, Ernst 182, 242–3, 251
Hall, A. Rupert 130
Hall, G. Stanley 177–8
Halley, Edmund 71
Hammon, George 150, 154, 156–8
Hammond, Henry 154
Harmonics (Ptolemy) 29
Harris, Marvin 231
Harrison, Peter 10
Hartley, David 170
Haught, John 2
Hawking, Stephen 107–8
Heat: A Mode of Motion (Tyndall) 251
Hebrews *see* Judaism
Hegel, Georg Wilhelm Friedrich 201
Hegelianism 189
Helmont, Jan van 73
Hengstenberg, Ernst 201–2, 205
Heraclitus 21–2
Herder, Johann Gottfried von 201
Herschel, John 11, 98–9, 193
Hesiod 19, 21–2, 34–5
Hess, Moses 251
Higgs boson 108
Hinduism 105
Hippocrates 29–32
history 15–18
History of the Conflict between Religion and Science (Draper) 112
Hitchcock, Edward 195
Hobbes, Thomas 57, 75, 112
and the soul 150, 153–4, 156–7
Hoff, J. H. van't 129
Holbach, Paul-Henri-Dietrich d' 149
Homer 21–2, 34–5
Hoyle, Fred 109
human beings
naturalistic theories of 13–14
study of *see* anthropology
Humboldt, Alexander von 184–91
Hume, David 134, 166, 222
Hussey, Thomas John 188–9
Huxley, Thomas Henry 123–4, 128, 133–4
and agnosticism 235–6
and Darwin 141, 144, 243–4
and eschatology 251–2
and evolution 218
and psychology 175
and science 237–40
and theodicy 247–8
and theology 241–5
Huygens, Christiaan 84–5
Hymes, Dell 232

Ibn Rushd *see* Averroes
Ibn Tufayl 39
immigration 229
immortality 148
infidels 46

insects, study of, and the argument to design 135
intelligent design 2–5, 109, 133
Introduction to Entomology: or Elements of the Natural History of Insects (Kirby and Spence) 135
Introduction to Geology (Bakewell) 208–9, 211
Introduction to the Old Testament (De Wette) 205
Islam 37, 43; *see also* Qur'an

James, William 174, 176–9
Jeans, James 105–6
Jenkin, Henry Fleeming 250
Jesus Christ 80–2
John of Gmunden 47
John of Rupescissa 115
Jones, John E. 2
Jouanna, Jacques 29–30
Joule, James 100
Jowett, Benjamin 212
Judaism 37, 43, 161, 200; *see also* Torah
Jussieu, Antoine-Laurent de 187

Kant, Immanuel 70, 94, 132–3, 202
  and Humboldt 189–92
Kekulé, Auguste 129
Kelvin, Lord *see* Thomson, William
Kepler, Johannes 73, 84
Khunrath, Heinrich 116
Kirby, William 135
Kolbe, Hermann 128–9
Kosmos (Humboldt) 184–91
Krebs, Georg 186
Kroeber, Alfred 230–1
Kuhn, Thomas 134
Kuper, Adam 228–9

La Mettrie, Julien Offray de 14, 149, 160
Ladd, George Trumbull 175–6, 178
Lamarck, Jean-Baptiste 138–9
Lambert, Johann Heinrich 187
Lankester, E. Ray 237
Laplace, Pierre-Simon 77–8, 96, 131, 187
Lavoisier, Antoine 117–18, 120, 187
Law, Edmund 151, 153, 156
laws of nature 10–13, 113
  and Descartes 61–3, 76
  and Huxley 242–5
  and Newton 64–7
  and physics 98–9
Laycock, Thomas 173
Layton, Henry 150, 154–9
Lederman, Leon 108
Lehoux, Daryn 7, 61–2
Leibniz, Gottfried Wilhelm 57, 68–9, 76, 245
  and Newton 88–9, 91
Leigh, Edward 154
Levi ben Gerson 53–4
Lewes, G. H. 237
Lewis, Tayler 214
liberalism 246
Liebig, Justus von 125
life sciences *see* biology
Lightman, Bernard 10
Lithographica Helvetica (Scheuchzer) 182
Littrow, Joseph Johann von 187
living things 4–5, 11, 123–4
Locke, John 92, 112, 170, 201
Lodge, Oliver 102
logic, as basis for scientific enquiry 40
Lombard, Peter 8, 12, 52
Lowie, Robert 228, 230
Lubbock, Sir John 218, 237
Lucretius 32–4
Luther, Martin 115
Lyell, Charles 144, 193, 222

McCabe, Joseph 194
McGee, William 226
machine, creation likened to 130–1
Magnus, Albertus 9, 38, 47–8, 55
Maimonides 39–40
Malthus, Thomas Robert 142–3
Manilius, Marcus 24
marvels 54–5; *see also* miracles, natural phenomena
materialism 14, 157–61
Materiality or Mortality of the Soul, The (anon) 154–5, 159
mathematics
  as basis for scientific enquiry 40
  early development of 38–9
  Newton's use of 84–5, 92–3
  subject matter of 7
matter, in relation to mind 163–4
Maudsley, Henry 237, 248–9
Maupertuis, Pierre Louis 93–5
Maxwell, James Clerk 101–4, 109, 250
Mayer, Robert 251
Mécanique celeste, La (Laplace) 77
medicine 13–14, 38, 116–17, 125; *see also* Hippocrates
medieval era *see* Middle Ages
Meditations (Descartes) 60, 74
Mendel, Gregor 139
mental phenomena, study of 162–81
mental philosophy 168–72
metaphysical naturalism 2–6, 16, 131
metaphysics 93
Metaphysics (Aristotle) 29, 45

methodological naturalism 2, 4–5, 16, 131
  and biology 133–4
  and Middle Ages 37–40
  and religion 113–15
Michaelis, Johann David 201, 205
Middle Ages 6–9, 18, 42–8; *see also* chapter 3
Mill, John Stuart 170–1
Millar, Kenneth R. 2
Miller, Samuel 204
Milton, John 151
Mind of God, The (Davies) 108
mind, the 14, 162–72; *see also* psychology
miracles 8–9, 18, 67–8, 71–2, 81
  and biblical criticism 205–6, 212–13
  and Descartes 62
  and physics 104
  see also marvels, natural phenomena
monogenism 219, 226
moon, the 85
Moore, Aubrey 114–15
morality 167, 170, 218
More, Henry 63, 70
Morgan, Lewis Henry 225–6, 233
Morley, John 237
mortalist-materialism 149–59, 160–1
Moses und David keine Geologen (Potter) 192
Muncke, Georg Wilhelm 187
Münsterberg, Hugo 179

Nagel, Thomas 181
Napoleon Bonaparte 77, 96
natural history 65, 135, 182–3, 186–8, 207, 218–20, 230, 232
Natural History (Pliny the Elder) 23
Natural History of Creation, The (Haeckel) 242
natural phenomena 19–20, 97–9, 109, 212–13
natural philosophy 7–8, 14, 20, 26, 38, 58–9, 68, 72, 74, 80, 83–4, 86, 89–93, 117, 149–51, 206–7, 250
  and the Bible 70–2
  and God 56–69, 75–6, 80, 97
  and philosophy 72–6
  in the middle ages 40–7, 50–2, 56, 80, 84
natural theology 15, 59, 99, 120–1, 129, 140, 145, 183, 188–90, 195, 241–5, 252
  see also physico-theology
Natural Theology (Paley) 241
naturalism 1–6, 15–18, 71–2
  and ancient Greece 19–21
  and anthropology 222, 233
  and Aristotle 42–3
  and astronomy 53–4
  and biblical criticism 198, 200–3, 206–7, 212–15
  and Epicureans 32–4
  and God 58–66, 72–6
  and Hippocrates 29–32
  and Humboldt 186–7
  and marvels 54–5
  and Newtonianism 93–4
  and physics 97–8, 109
  and religion 15–16, 40–2
  and Roman empire 24–6
  and supernatural 6–10
  and the cosmos 83–4
  and theology 241
  and universities 45–8
  and USA 208–11
  and Victorians 237–41, 252–3
  see also laws of nature; metaphysical naturalism; methodological naturalism
Natürliche Schöpfungsgeschichte (Haeckel) 182
nervous system 172–3, 177–8
neurophysiology 177
neuroscience 148–9, 160–1
New Age 105
New Theory of the Earth, A (Whiston) 71
Newton, Sir Isaac 13, 56–7, 112
  and chemistry 121
  and Christianity 81, 86–9, 90–1, 113–14
  and Descartes 64–5
  and Enlightenment 77–9
  and geology 70
  and God 58–60, 69, 86–9
  and laws of nature 64–7
  and matter 119
  and physics 97–8
  and solar system 66–8, 71
  and the cosmos 82
  and universal gravitation 84–6, 91–2
Newton (Blake) 95–6
Newtonianism 68–9, 76–8, 89–90, 92–4
Nielsen, Kai 3
Numbers, Ronald L. 13, 18, 111, 128–9

Ockham 52–3
Oersted, Hans Christian 98
Oken, Lorenz 186
On the Causes of Marvels (Oresme) 54–5
On the Nature of the Gods (Cicero) 24–6
Opticks (Newton) 59, 65, 91–2
Oresme, Nicole 54–5
organisms 131–2, 138
Origin of Species (Darwin) 16, 134, 138, 141–2, 145–6
  and Huxley 243
  and teleology 241–2
Origines Sacrae (Stillingfleet) 182
Osborn, Henry Fairfield 229–30
Our Face from Fish to Man (Gregory) 216–17
Overton, Richard 150, 153–4

Owen, Richard 193
oxygen 118

paganism 39, 46
Paine, Tom 127–8
palaeontology 216, 220
Paley, William 141, 241, 243–4, 247
Papineau, David 3, 5–6, 10
Paracelsus 73, 116–17
Paraphrase and Annotations on the New Testament (Hammond) 154
Paraphrase and Notes on the Epistles of Paul (Locke) 201
Paris
  Royal Academy of Sciences 93
  universities in 43–5, 48–50
  see also Condemnations of 1277
Parker, Theodore 197–8, 205–8
Pascal, Blaise 63
Paul of Burgos 53
Paulus, Heinrich 202, 205
Peabody, Elizabeth 207
Pearson, Karl 237
Peckard, Peter 151, 156–7, 159
Pennock, Robert T. 2
Pfeffer, Michelle 14
Pfeiffer, Robert 202
philosophy
  and Aristotle 42–3
  and natural philosophy 10–11, 72–6
  in the classical era 7–8, 20–1, 22–3; *see also* mental philosophy
phrenology 171–2
physico-theology 69–72, 75
  see also design, natural theology
physics
  and anthropic principle 108–9
  and causality 103–7
  and energy 249–50
  laws of 3, 10, 27, 109–10
  and laws of nature 98–9
  and Maxwell 101–2
  and religion 97–8, 102–3, 107–8
Physics (Aristotle) 28
physiology 177–8
Plantinga, Alvin 5
Plato 7, 26–7, 40
Platonism 23
Pliny the Elder 23, 26
Polignac, Melchior de 92
politics, effects of 95, 246–7
Polkinghorne, John 110
Pope, Alexander 77
Potter, David Julius 192
Powell, Baden 192–3
Powell, John Wesley 226
Preliminary Discourse (Herschel) 98–9
Presocratic philosophy 7, 21–2, 26–7
preternaturalism 8
Prichard, James Cowles 219
Priestley, Joseph 117, 119, 122, 126
  and materialism 151, 153, 158–9
Primitive Culture (Tylor) 220, 222, 224–5
primitive society 220–9
Principia mathematica (Mathematical Principles of Natural Philosophy) (Newton) 56, 58–9, 64, 71, 77
  and the nature of God 87–8
  and universal gravitation 84–5
Principles of Philosophy (Descartes) 61
Principles of Psychology, The (James) 176, 178
Pringle, Sir John 117
Prinster, Scott 16
Protestantism 157–60, 199–200, 213–15
Prout, William 120–2, 125
Providence, Huxley on the doctrine of 235–6
psychic phenomena 178–9
psychology 13–14, 162, 168, 173–81
Ptolemy 24, 26, 29, 40
Pyrrhonians 23–4

Quakers 219
quantum physics 104–5
Quatrefages de Bréau, Jean Louis Armand de 187
Quintilianus, Aristides 24
Qur'an 41; *see also* Islam

racism 229–30
Rankine, William Macquorn 250
rationalism 205
Ray, John 121
reasoning, faculty of 167
recapitulation 220–1
Reformation 157–9, 199
Reid, Thomas 166–7
Reinventing Anthropology (Hymes) 232
religion
  and ancient Greece 31
  and anthropology 17, 218–20, 223–5, 233–4
  and biology 132–6
  and chemistry 115–29
  and *Kosmos* 184–5
  and materialism 149–51
  and naturalism 40–2, 252–3
  and Newton 86
  and physics 97–9, 102–8, 110
  and politics 95
  and science 2–4, 111–15, 197–8, 238–41
  see also agnosticism; atheism; Christianity; God; Islam; Judaism; soul, the; theology

religious significance of laws of nature 101–2
Renan, Ernest 188
Riley, William Bell 214
Rio, Martin del 121
Rise of Anthropological Theory, The (Harris) 231
Roberts, Jon 14
Roman empire 22–7
Rupke, Nicolaas 15
Ruse, Michael 10
Ruskin, John 127
Russell, Bertrand 106

Sahlins, Marshall 232–4
Scheuchzer, Johann Jakob 182
schools *see* universities
science
- and ancient Greece 7, 80–1
- and Aristotle 74
- definition of 15, 206–7
- history of 10
- and laws of nature 11
- and Newton 86
- and religion 2–6, 111–15, 197–8
- and the soul 160–1
- and theology 52–3, 242
- and Victorians 237–41
- *see also* anthropology; biology; chemistry; geology; materialism; physics; psychology; social sciences

scientific naturalism *see* naturalism
Scottish Enlightenment 167–8
Scripture, E. W. 176
SD (The Sacred Disease) (Hippocrates) 29–32
secularism 95, 123, 127, 195–6
Sedley, David 20, 26
Selfish Gene, The (Dawkins) 144
Sellars, Roy Wood 1
Seneca 24–6
Sentences, [Book of] (Lombard) 8, 52
Seventh-Day Adventists 160
Shank, J. B. 7, 11
Shank, Michael 9, 58–9
Shelley, Mary 95
Siger of Brabant 48, 50
Silliman, Benjamin 195, 208–11
Simplicius of Cilicia 74
Smith, Adam 143
Smith, Henry Preserved 212–13
Smith, William Robertson 212
social sciences 16–17
Social Statics (Spencer) 245–6
socialism 246
Socrates 80
solar system 66–8, 71, 114; *see also* sun, the
soul, the 13–14, 104, 163–6
- and immortality 148–61

Spain 39
Spencer, Herbert 173, 239, 245–50, 252
Spinoza, Baruch 12–13, 76, 83–4, 87, 90, 107
spiritualism 178–9
Sprat, Thomas 75
Spring, Gardiner 210
Spurzheim, Johann Gaspar 172
Stanley, Matthew 11
Stenger, Victor 148
Stephen, Leslie 237
Stevenson, Robert Louis 221
Stewart, Balfour 102
Stillingfleet, Edward 182
Stoicism 12, 23, 28
Stokes, Sir George 104
Stout, G. F. 249
Strauss, David Friedrich 202–3, 205
strong anthropic principle (SAP) 109
Stuart, Moses 203–4, 210–11
sun, the 249–52
supernatural
- and creation 27
- distinguished from natural 6–10, 17–18, 27
- and Eastern religions 105
- excluded from scientific study 2–5, 15
- and gods 21–2
- and medicine 13
- and natural phenomena 200
- rejected in biblical interpretation 212–13

supernaturalist interpretation of the Bible 201–2, 205–7
'survivals' 223–4

Tait, Peter Guthrie 100, 102, 250
Tallis, Raymond 181
Tao of Physics, The (Capra) 105
Tasmania 216–17, 224
Tatian 198
Taylor, Charles 127
teleology 28, 94, 136, 220, 227, 236, 241–5
- *see also* design, final cause

Telluris Theoria Sacra (Sacred History of the Earth) (Burnet) 70–1, 182
Tempier, Étienne 51
Théodicée (Leibniz) 245
theodicy 245–8
theology 7, 38, 56–7, 235–6
- and Condemnations of 1277 51–3
- contrasted with mere reference to God 58–9
- and Huxley 239–45
- and science 240–1
- and universities 44–7
- *see also* eschatology; physico-theology

thermodynamics, laws of 100, 250

Tholuck, August 205
Thomas, Keith 196
Thomson, William, 1st Baron Kelvin 100–1, 109, 250
Timaeus (Plato) 41
Time Machine, The (Wells) 252
Tittman, Charles Christian 204–5
Toland, John 90
Torah 41
Townsend, Joseph 194–5
Trucanini 216–17, 220, 230
Tübingen School 201, 203
Tylor, Edward Burnett 222–5, 233, 237
Tyndall, John 103–4, 106–7, 109
  and chemistry 112–13, 125
  and Christianity 236–7
  and energy physics 249–52
  and science 238–9
  and the X-Club 128

uniformitarianism 222–3
Unitarianism 204–6
United States of America
  and anthropology 218–20, 225–30
  and the Bible 200, 202–8, 212–14
  and naturalism 197, 208–11
  psychology in 174–6
unity of nature 101–2
universal gravitation *see* gravity, Newton
universities 38, 43–8, 55–6
Unseen Universe, The (Stewart/Tait) 102
urea, synthesis of 125

vegetation, chemistry of 117, 121
Vergil 24
Vitry, Philippe de 54
Vogt, Carl 194
Voltaire 93, 126

Wadsworth, Thomas 151–2, 154–5
Wallace, Alfred Russell 142, 144
Wallace, Anthony 111
Ward, James 176
Ward, Lester Frank 175
Watson, James 124
Watson, John B. 180–1
Watson, Richard 122
Wedgwood, Josiah 143
Weinberg, Steven 108
Wellhausen, Julius 192
Wells, H. G. 252
Welpley, James Davenport 190
Whewell, William 11, 99, 102, 134, 207
Whiston, William 68, 71
White, Andrew Dickson 240
William of Conches 41–2
Willis, William 232
Wisdom of God, Manifested in the Works of Creation (Ray) 135
Wittgenstein, Ludwig 11
Wöhler, Friedrich 125
Woolley, Charles 216
working classes 246–7
World, The (Descartes) 61
Wundt, Wilhelm 174

X-Club, the 103, 106, 128
Xenophanes 22, 35

Young, Robert M. 247